T0392826

A Comprehensive Guide to Managing Municipal Infrastructure Assets

Comprehensive and practical, this book provides an essential resource for educators, researchers, students, and those in public agencies and consultancies who are directly responsible for managing municipal infrastructure such as roads, water, and sewer pipes. This book is thorough in the integration of procedures that establish a cost-effective intervention plan using the latest technologies and management processes. It examines all the aspects of developing an optimal asset management plan for collocated municipal assets. It presents the evolution of asset management from data requirements to investment planning and priority programming of rehabilitation and maintenance. It offers a coordinated approach to effectively manage municipal infrastructure and offers integrated solutions that aid decision-makers in taking informed decisions on (1) when to maintain each asset, (2) which corridors shall be prioritized, and (3) what is the best intervention to undertake for each asset. It also offers a compelling vision of how infrastructure and cities will evolve by 2050, shaped by advancements in digital technology, transportation, governance, sustainability, resilience, and climate change. It provides invaluable insights for practitioners, emphasizing how today's decisions and investments will directly influence the future of urban environments.

Features:

- Presents the most current methodologies and practical applications of managing collocated municipal infrastructure.
- Includes case studies and practical examples for each step, as well as an extensive list of references for each asset class.
- Examines novel approaches for reduced lifecycle costs, enhanced conditions, improved level of service, reduced risk, increased maintenance effectiveness, and reduced service disruptions.
- Explores the future of urban infrastructure in 2050, helping practitioners envision tomorrow's cities and make informed investment decisions in today's infrastructure.

A Comprehensive Guide to Managing Municipal Infrastructure Assets

Soliman Abusamra and Ayman H. El Hakea

CRC Press is an imprint of the
Taylor & Francis Group, an **informa** business

Designed cover image: Shutterstock

First edition published 2025
by CRC Press
2385 NW Executive Center Drive, Suite 320, Boca Raton FL 33431

and by CRC Press
4 Park Square, Milton Park, Abingdon, Oxon, OX14 4RN

CRC Press is an imprint of Taylor & Francis Group, LLC

ISBN: 978-1-032-35578-8 (hbk)
ISBN: 978-1-032-35583-2 (pbk)
ISBN: 978-1-003-32750-9 (ebk)

DOI: 10.1201/9781003327509

Typeset in Times
by codeMantra

Dedicated to all asset managers who have devoted their careers to the provision of public services.

Above all, to my beloved wife, Meriem, and my cherished daughter, Sarah, whose unwavering love and support have been a constant source of inspiration. I love you both deeply.

Soliman Abusamra

Contents

PART 3 Key Performance Indicators (KPIs): Measurement and Modeling

PART 4 Optimization Framework for Municipal Infrastructure

PART 5 Case Studies

PART 6 Future of Cities and Infrastructure in 2050

Preface

Infrastructure is the backbone of our modern society, underpinning every aspect of our daily lives. From the roads we travel on to the water we drink, the electricity that powers our homes, and the communication networks that connect us, infrastructure is essential for the functioning of our communities and economies. This book aims to provide a comprehensive understanding of infrastructure asset management, a field that is crucial for ensuring the sustainability and efficiency of these vital systems.

Infrastructure transcends mere physical structures; it is fundamentally about the people who use these systems and the communities they support. Effective infrastructure asset management is crucial to ensuring that these systems are reliable, safe, and capable of meeting the diverse needs of the population. It involves making informed decisions that balance cost, performance, and risk to achieve the best outcomes for society. As urban areas expand and infrastructure ages, the complexities of maintaining and upgrading these systems intensify. This book provides a detailed roadmap for navigating these challenges, offering practical strategies and tools designed to manage infrastructure assets efficiently and sustainably.

Infrastructure asset management is a dynamic and evolving field. As new technologies emerge and our understanding of infrastructure systems deepens, the strategies and practices for managing these assets must also evolve. This book aims to provide a forward-looking perspective on infrastructure asset management, exploring the potential developments and innovations that will shape the future of this field. One of the key areas of focus in this book is the integration of digital technology into infrastructure asset management. From the use of sensors and data analytics to monitor the condition of infrastructure assets to the application of artificial intelligence and machine learning to optimize maintenance and repair activities, digital technology is transforming the way we manage infrastructure. This book provides insights into how these technologies can be leveraged to improve the efficiency and effectiveness of infrastructure asset management.

Other important themes in this book are sustainability and resilience. As the impacts of climate change become more pronounced, the need for sustainable infrastructure systems has never been greater. Resilient infrastructure systems are those that can withstand and recover from disruptions, whether they are caused by natural disasters, technological failures, or other unforeseen events. This book explores the principles of sustainable infrastructure asset management, providing strategies for reducing the environmental footprint of infrastructure systems and enhancing their resilience to climate-related risks to continue providing essential services in the face of adversity.

In addition to these themes, this book also covers a range of other topics that are relevant to infrastructure asset management. These include the role of public-private partnerships in financing infrastructure projects, the importance of stakeholder engagement in the asset management process, and the challenges and opportunities associated with managing infrastructure in urban and rural settings.

Throughout this book, you will find a wealth of information on the latest trends and best practices in infrastructure asset management. From the use of advanced technologies like artificial intelligence and machine learning to the development of innovative financing models, this book covers a wide range of topics that are relevant to today's infrastructure professionals.

In addition to providing technical knowledge, this book also emphasizes the importance of collaboration and communication in infrastructure asset management. It highlights the need for stakeholders to work together to develop and implement effective asset management plans. By fostering a culture of collaboration, we can ensure that our infrastructure systems are resilient, adaptable, and capable of meeting the needs of future generations.

One of the key themes of this book is the concept of asset management. Asset management is defined as the systematic, coordinated planning and programming of investments or expenditures, design, construction, maintenance, operation, and in-service evaluation of physical facilities to perform their intended function. It involves a range of activities, from allocating staff and collecting data to developing asset management plans and optimizing expenditures. The goal is to maximize benefits, reduce risks, and provide satisfactory levels of service to the community in a sustainable manner.

This book is structured into six parts, each focusing on a different aspect of infrastructure asset management. The first part covers the evolution of infrastructure asset management, tracing its journey from its early years to the present day. The second part explores infrastructure interdependency, coordination, and contractual frameworks, providing insights into how different types of infrastructure assets are interconnected and how their maintenance can be coordinated effectively.

The third part of the book delves into key performance indicators (KPIs) and their measurement and modeling. It discusses various models for deterioration and condition prediction, lifecycle costing, risk assessment, and more. It proposes a mathematical framework to compute the potential improvements of coordinating the planning and execution for the co-located assets as opposed to the conventional silo approach. The fourth part focuses on optimization frameworks for municipal infrastructure, discussing the role of optimization in planning and scheduling maintenance interventions.

The fifth part of the book presents real case studies, showcasing the application of the theories, methodologies, and frameworks discussed in the previous parts. These case studies provide practical insights into how infrastructure asset management can be implemented in real-life scenarios. The final part of the book looks to the future, exploring the potential developments in infrastructure and cities by 2050. It emphasizes the integration of digital technology, sustainability, and resilience in creating a more connected, efficient, and environmentally responsible world.

As you read through the chapters of this book, I hope you will gain a deeper understanding of the complexities and challenges of infrastructure asset management. I also hope that you will be inspired to apply the principles and practices discussed in this book to your own work, contributing to the development of sustainable and resilient infrastructure systems that will benefit our communities for years to come.

In conclusion, this book serves as a comprehensive resource for anyone involved in infrastructure asset management. It provides a thorough understanding of the principles and practices of the field, along with practical tools and strategies for managing infrastructure assets effectively. Whether you are a government official, engineer, consultant, or an interested reader, this book will equip you with the knowledge and insights needed to navigate the complexities of infrastructure asset management and contribute to the development of sustainable and resilient infrastructure systems that will benefit our communities for years to come.

About the Authors

Dr. Soliman Abusamra is a distinguished senior leader with over 15 years of experience spearheading innovative initiatives and executing major infrastructure projects. Currently, he serves as the Director of Commercial and Contracts at Alto - High-speed Rail (HSR), Canada's largest infrastructure undertaking, where he is responsible for overseeing contract management and commercial structuring. His dedication to advancing resilient and sustainable infrastructure reflects his commitment to supporting future generations.

Prior to his role at Alto, he held key positions as a Senior Project Manager and Contract Manager at VIA Rail Canada, managing capital infrastructure and rolling stock projects with a combined value exceeding $1 billion. His global experience includes significant contributions to high-profile transportation projects, such as the Riyadh Metro Project in Saudi Arabia—a design-build rapid transit system featuring 6 lines, 176 kilometers, and 85 stations, with a budget of $22.5 billion—and the REM project in Montreal. His career also includes a role as a Manager in KPMG's Global Infrastructure Advisory Team, where he supported municipalities and clients in developing strategic and tactical asset management plans.

Academically, he earned his Ph.D. in Civil Engineering with a focus on Asset Management from Concordia University. He has authored several books, journal articles, and conference papers on municipal infrastructure asset management. His thought leadership is further demonstrated through speaking engagements at notable conferences, including TED Talks, and other media appearances. He has participated in industry forums such as the IAM and has served as a lecturer at Toronto Metropolitan University. He is also an editorial member and reviewer for prestigious journals like ASCE and CSCE. His active involvement extends to CNAM's board of directors and the Building Transformations' Asset Management and Lifecycle Think Tank.

Ayman H. El Hakea is a Construction Engineering graduate of the American University in Cairo (AUC) in 2007. He has 17 years of experience in various fields of construction. Moreover, he worked with several multi-national companies in Egypt, Morocco, GCC, and the UK, such as Bechtel, Groupe Bouygues, Al-Futtaim Carillion, and Mace. He is currently taking the role of Systems Development Manager at CRC, the construction arm of Dorra Group, the leading construction and real estate conglomerate in Egypt and the Middle East. He completed his MSc degree in Construction Engineering at the AUC in 2015, and he is currently a Construction and Building Engineering PhD candidate at the Arab Academy for Science, Technology and Maritime Transport (AASTMT) in Cairo, Egypt. He published numerous papers in infrastructure asset management, contracts management, building performance simulation, machine learning, lifecycle cost optimization, and computer vision applications in construction. He is a student member of the ASCE (US) and a former graduate member of the ICE (UK). He is also a certified IRCA ISO9001:2015 QMS Lead Auditor.

Acronyms and Abbreviations

A	Area
A	Sub-Catchment Area
AADT	Average Annual Daily Traffic
AC	Asbestos Cement
ADB	Asian Development Bank
AHP	Analytic Hierarchy Process
AI	Artificial Intelligence
ANN	Artificial Neural Networks
ANP	Analytic Network Process
AV	Autonomous Vehicles
BCG	Boston Consulting Group
BOQ	Bill of Quantities
C **or** ***R***	Condition or Reliability
C	Full Coordination Scenario
CAPEX	Capital Expenditures
CAV	Connected and Autonomous Vehicles
CCPA	Canadian Centre for Policy Alternatives
CCS	Corridor Condition State
CHI	Corridor Health Index
CI	Cast Iron
CI	Critical Infrastructure
CIF	Condition Improvement Factor
CIP	Pitted Cast Iron
CIS	Spun Cast Iron
CN	Conventional Scenario
CO_2	Carbon Dioxide
COF	Consequence of Failure
CP	Compromise Programming
CP	Concrete Pressure
CPAF	Cost Plus Award Fee
CPFF	Cost Plus Fixed Fee
CPIF	Cost Plus Incentive Fee
CPLEX	C Programming-Based Solver Using Simplex Method
CRPS	Corridor Resiliency Preparedness State
CS	Corrugated Steel
CT	Clay Tavern
CU	Copper
DHS	Department of Homeland Security
DI	Ductile Iron
DIHY	Ductile Iron Hyprotech
DT	Disruption Time
DW	Decomposed Weight

EU	European Union
EUAC	Equivalent Uniform Annual Costs
EV	Electric Vehicles
F	Demand-Capacity Ratio
F	Fixed Cost
FCM	Federation of Canadian Municipalities
FFP	Firm Fixed Price
FPEPA	Fixed Price with Economic Price Adjustment
FPIF	Fixed Price Incentive Fee
GALV	Galvanized Steel
GAs	Genetic Algorithms
GDP	Gross Domestic Product
GIF	Global Infrastructure Facility
GIS	Geographic Information System
GPA	Grade Point Average
HDPE	High-Density Poly Ethylene
HLA	High-Level Architecture
I	Rainfall Intensity
IaaS	Infrastructure-as-a-Service
IDF	Intensity-Duration-Frequency
IDT	Inter-Disruption Time
IEF	Intervention Efficiency Factor
IFF	Intervention Effectiveness Factor
IPCC	Intergovernmental Panel on Climate Change
IRI	International Roughness Index
ISO	International Standards Organization
IT	Information Technology
ITS	Intelligent Transportation Systems
JD	Joint Duration
JDC	Joint Direct Costs
JIC	Joint Indirect Costs
KPIs	Key Performance Indicators
LCC	Lifecycle Cost
LEED	Leadership in Energy and Environmental Design
LIDAR	Light Detection and Ranging
LIF	Lifecycle Costs Improvement Factor
LOS	Level of Service
MAUT	Multi-Attribute Utility Theory
MCDM	Multi-Criteria Decision-Making
MGI	McKinsey Global Institute
MWh	Megawatt-Hour
NCR	Network Coordination Ratio
NHI	Network Health Index
NPV	Net Present Value
NPW	Net Present Worth2
n_s	Number of Systems

OPEX	Operational Expenditures
OT	Operating Time
P/I	Penalties and Incentives
PaaS	Platform-as-a-Service
PAS	Publicly Available Specification
PBC	Performance-Based Contracts
PC	Partially Coordinated scenario
PD	Parallel Duration
POF	Probability of Failure
PPP	Public-Private Partnerships
PVC	Poly Vinyl Chloride
R&D	Research and Development
RC	Reinforced Concrete
RCM	Regional Climate Models
R_i	Reliability of system i
RI	Risk Index
RIF	Risk Improvement Factor
ROI	Return on Investment
RPIF	Resilience Preparedness Improvement Factor
RTH	Reliability Threshold
SD	Standalone Duration
SDC	Standalone Direct Costs
SIC	Standalone Indirect Costs
SMART	Specific, Measurable, Achievable, Realistic, and Timely to schedule
SoS	System-of-Systems
STDF	Spatio-Temporal Disruption Factor
STIF	Spatio-Temporal Improvement Factor
STL	Steel
T_0	Start of the Planning Horizon
TC	Total Cost
TDT	Total Disruption Time
T_I	Time of Intervention
T_{I+1}	Time after Intervention
UC	Unit Cost
UN	United Nations
UNFPA	United Nations Population Fund
V	Variable Cost
V2X	Vehicle-to-Everything
VC	Vitrified Clay
WHO	World Health Organization
α	Scale Parameter
β	Shape Parameter
γ	Location Parameter
Δ**RI**	Improvement in Reliability

Part 1

Evolution of Infrastructure Asset Management

Part I

1 Introduction

INTRODUCTION

Infrastructure encompasses the essential physical and organizational structures, assets, and facilities that enable economies, households, and businesses to operate effectively. Broadly defined, an asset is anything that contributes value to an organization and its stakeholders. In the context of municipalities, infrastructure includes a wide array of assets such as the roads we travel, water supply networks and their associated piping systems that ensure reliable access to safe drinking water, wastewater, and stormwater management systems that protect our streets from flooding, electric power grids that generate, transmit, and distribute electricity, telecommunication networks that connect us, educational institutions that shape future generations, and healthcare facilities that address our medical needs. These elements, along with various other critical buildings and facilities, constitute the foundation of our daily operations and well-being. In addition to these physical assets, an organization's infrastructure may include intangible assets such as personnel, processes, knowledge, IT systems, and operational workflows. However, this book specifically focuses on the management of physical municipal infrastructure assets.

With a shared understanding of what constitutes infrastructure, this book aims to delve into its fundamental aspects, exploring its significance for nations and economies. It will guide you through the essential concepts of infrastructure management, starting with foundational principles and advancing toward a forward-looking perspective on future management practices. This book is structured into six distinct sections, each building upon the previous, to provide a comprehensive overview of effective infrastructure asset management, as outlined in Table 1.1.

TABLE 1.1
Book Structure

Part	Title	Description
Part 1	Evolution of Infrastructure Asset Management	This part introduces the fundamental concepts of infrastructure and elucidates the critical role of asset management in municipal settings. It traces the development of asset management from its inception to its current state, highlighting key advancements and changes over time. Additionally, it provides a comprehensive review of contemporary contractual practices related to municipal infrastructure, offering insights into the procurement mechanisms employed by governments and municipalities

(Continued)

DOI: 10.1201/9781003327509-2

TABLE 1.1 (*Continued*)
Book Structure

Part	Title	Description
Part 2	Infrastructure Interdependency, Coordination, and Contractual Frameworks	This part addresses the often-perplexing issue of frequent maintenance on the same road. It defines infrastructure interdependency and explores its various forms. Furthermore, it presents a detailed framework for planning and coordinating maintenance efforts, specifically designed for assets that share common corridors. This framework aims to enhance understanding and efficiency in managing interconnected infrastructure
Part 3	Key Performance Indicators (KPIs): Measurement and Modeling	In this part, the focus is on the importance of KPIs in providing a comprehensive overview of infrastructure performance. It details the methodologies for measuring and modeling coordination benefits, including models for deterioration and condition prediction, lifecycle costing, and risk assessment. This part is essential for understanding how to quantify and evaluate the effectiveness of infrastructure management practices
Part 4	Optimization Framework for Municipal Infrastructure	Effective planning and scheduling of maintenance interventions for municipal assets require sophisticated optimization techniques. This part explores the role of optimization in these processes, reviewing various types, methods, and algorithms. It also offers a methodology for developing an optimization model, accommodating multiple co-located assets, to improve planning and execution
Part 5	Case Studies	Theory gains practical significance through application, and this part provides a critical examination of real-world case studies. It demonstrates how the theories, methodologies, and frameworks discussed in previous sections are applied in practice, illustrating their effectiveness and offering valuable insights into their real-life implications
Part 6	Future of Cities and Infrastructure in 2050	Looking ahead, this part explores the anticipated developments in cities and infrastructure by 2050. It emphasizes the integration of digital technology, sustainability, and resilience in shaping the future. Topics include smart cities, renewable energy, digital technology, asset management, transportation innovations, and sustainable design. This part highlights the challenges and opportunities in creating a more connected, efficient, and environmentally responsible world

WHAT IS INFRASTRUCTURE ASSET MANAGEMENT?

Municipalities face the complex challenge of making optimal decisions with limited resources, balancing the needs of current users while considering future generations. Are you operating under a constrained budget, constantly weighing investment decisions against the demands of infrastructure renewal, growth, emerging needs, escalating service expectations, risk management, and unforeseen failures? You are not alone in this dilemma. Effective asset management can provide valuable assistance in navigating these challenges and making informed decisions.

The term "asset management" is probably familiar to many, and you might probably know that there are already numerous tools that could assist you in managing public infrastructure. However, the question is always, "How should we kick off with such limited budgets?". There are lots of activities ranging from allocating staff, collecting data, developing asset management plans, putting together integrated policies and procedures, trading off investments, optimizing expenditures, implementing and monitoring asset management plans, and reacting to sudden failures to a handful of other activities. So, what is "asset management"? The term "asset management" has been defined in many ways by a wide spectrum of governments and non-governmental bodies across the globe. For instance, Haas et al. defined infrastructure asset management as:

> "The systematic, coordinated planning and programming of investments or expenditures, design, construction, maintenance, operation, and in-service evaluation of physical facilities to perform their intended function"
>
> *Haas et al. (1994); Hudson et al. (1997)*

The Federation of Canadian Municipalities (FCM) defined asset management as:

> "An integrated approach, involving all organization departments to effectively manage existing and new assets to deliver services to customers. The intent is to maximize benefits, reduce risks and provide satisfactory levels of service to the community in a sustainable manner – providing an optimum balance. Good asset management practices are fundamental to achieving sustainable communities"
>
> *CNAM (2019)*

While there are countless definitions that one can find for asset management, all of them are based on a set of fundamentals:

- **Value**: In principle, assets exist to provide value to the organization and/or stakeholders. For instance, roads provide users with the means to safely move from one point to another.
- **Integration and Coordination**: First, asset management is an interdisciplinary field that involves governments, municipalities, engineers, construction companies, lawyers, consultants, and public end users all at the same time. This mandates proper coordination to provide value to stakeholders. Second, most of our assets are interdependent. This interdependency could be spatial, physical, functional, or all three together. For instance, roads spatially overlap with water and sewer pipes, which makes it crucial for municipalities to coordinate their planning with the aim to minimize service disruptions. Further details and examples of infrastructure interdependency are provided in Chapter 10.
- **Alignment**: Asset management provides the foundation for organizations to translate end-user service expectations into technical and financial decisions, plans, and activities. It all ties back to the value the asset provides to its stakeholders.

- **Warranty**: Like the warranty we all have for products, buildings, and services, asset management warrants organizations and stakeholders that the asset will always fulfill its intended function.
- **Management**: Asset management oversees all the activities that support achieving pre-defined plans and delivering expected service levels. This includes, among other functions, internal workplace and resource management, financial management of allocated budgets, external management for contractors and builders, and stakeholder reporting.

In simple words, asset management covers all activities that warrant a minimally acceptable infrastructure level of service to stakeholders. These activities range from the initial information acquisition, intended to calculate the demand for a specific type of infrastructure, to the maintenance and rehabilitation, needed to maintain the level of service. These activities begin with infrastructure preliminary design all the way long to construction, monitoring, and performance evaluation. Asset management is not only about managing existing assets to deliver an intended service but also about taking critical investment decisions given limited government resources. Its end goal is to both meet the need for building new infrastructure and keep the existing infrastructure within an acceptable service level. Deferred investments for existing infrastructure systems in many countries led to an extreme decline in the level of service, which has posed a serious need for costly replacements and even caused in some cases sudden catastrophic failures.

ROLE OF INFRASTRUCTURE

Why does the infrastructure fundamentally exist? What is its *raison d'être*? It simply exists to serve our communities, whether through physical assets such as roads to connect, water networks to distribute and deliver potable water, hospitals to treat patients, parks and recreation facilities to provide entertainment, or electricity power grids and telecommunication networks to enlighten and connect. Moreover, infrastructure also serves communities through intangible assets such as land, software, data, and intellectual property. Some communities even include "natural assets" within their asset management practices. For instance, land, drainage channels, rivers, and lakes are included in asset management's scope, along with other tangible infrastructure assets. Managing all those assets to deliver their intended services is what is collectively termed as "asset management". It provides a structured way to track performance, cost, risk, and level of service, in the most efficient and effective manner. Simply put, it is an opportunity for municipalities to perform more with fewer resources. Like car maintenance, failing to timely carry out routine inspections and maintenance might cause certain systems to fail earlier than expected, resulting in greater costs. A sample checklist to manage your assets includes, but is not limited to, the management of various aspects, as shown in Figure 1.1 (CNAM 2019).

Public infrastructure is a vital component in our daily life for a healthy and vibrant community, which encourages economic growth. Developing infrastructure enhances the economy's productivity, consequently increasing the company's competitiveness and boosting the regional economy. Not only does infrastructure enhance the efficiency of production, transportation, and communication, but it also

FIGURE 1.1 Manage your asset checklist.

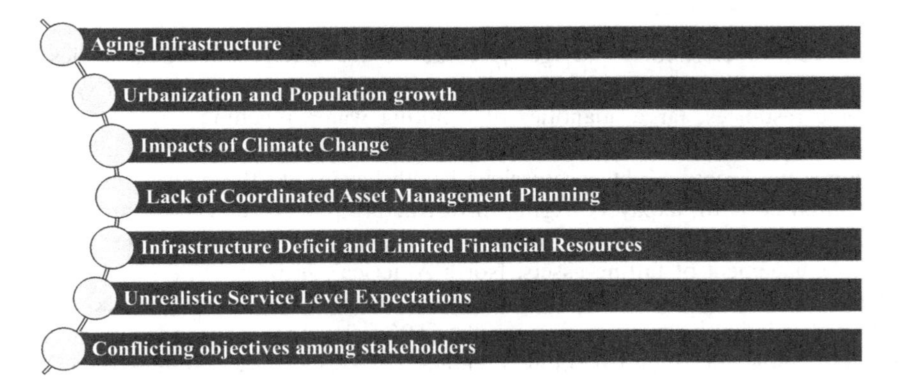

FIGURE 1.2 Challenges in managing municipal assets.

plays a pivotal role in providing economic incentives to public and private sector participants. The accessibility and quality of infrastructure in a certain region greatly help shape investment decisions by domestic firms and determine a region's attractiveness to foreign investors. For instance, Finance Canada's report recently showed that a $1 billion investment in infrastructure can create 16,700 jobs and boost the gross domestic product (GDP) by $1.6 billion (CCPA 2009; MGI 2017).

COMMON CHALLENGES IN MANAGING MUNICIPAL ASSETS

Even though infrastructure can lay the foundation for countries to develop and economies to grow, municipalities face a lot of challenges that complicate the management of such assets. Those challenges are summarized in Figure 1.2.

FIGURE 1.3 Aging infrastructure.

Aging Infrastructure

Infrastructure deteriorates and ages just like we do. Infrastructure deficiencies are abundantly evident in roads, water supply networks, and sewage disposal systems. In some instances, large quantities of drinking water, reaching 50%, are lost in some communities because of severely deteriorating water mains. Any infiltration of aggressive agents might pose serious health hazards to the public, resulting in substantial costs to taxpayers. Aging infrastructure places tremendous pressure on governments and municipalities through steeply growing deficits associated with the repair/replacement of failing assets. North America's infrastructure requires close attention and timely intervention to ensure service continuity. For instance, according to Canada's infrastructure report, more than 50% of Canadian roads are in fair condition or below, thus requiring more attention. Furthermore, according to America's infrastructure report card issued in 2021, America's infrastructure is graded "C-" with roads, wastewater, stormwater, and drinking water receiving grades of D, D, D+, and C-, respectively (ASCE 2021). Failure to take timely action will result in catastrophic failures, including, but not limited to, failing to provide adequate service to users, economic loss, and even fatalities in extreme cases, as shown in Figure 1.3.

Urbanization and Population Growth

Other challenges facing asset managers are urbanization and demographic growth. According to the United Nations Population Fund (UNFPA), the world is witnessing the largest wave of urban growth. In 2008, more than 50% of the world's population resided in urban centers and the figures are expected to exponentially swell throughout the upcoming years (UNFPA 2007). Moreover, the increased density of residential developments substantially increases the number of impervious surfaces, from which rainfall runs off quickly into pipe systems (Mailhot and Duchense 2010). According to Environment Canada, the urbanization of natural drainage basins can increase water runoff by more than 400% (Water Security Agency 2014). Urban growth and population increase continue to put excessive strain on existing infrastructure to meet demand levels while maintaining high service levels (Abusamra 2019).

Another aspect is the urban form and its impact on the cost of infrastructure provisions. There are two categories of urbanization, namely urbanization and sub-urbanization. Urbanization or compact growth denotes a population shift from rural areas to urban centers. Sub-urbanization or urban sprawl is a population shift from central urban areas to suburbs, resulting in the formation of (sub)urban sprawls. As a consequence of the movement of households and businesses out of the city centers, low-density peripheral urban areas grow, encouraging auto-dependency, larger numbers of long-distance trips, and extensive additions to transportation and other infrastructure systems. Studies demonstrated that urban sprawl increases the cost of infrastructure provision, as it requires more lane kilometers that require higher capital investments and maintenance costs, estimated at 30% higher than compact growth patterns. Furthermore, other infrastructure provision costs result from impacts emanating from urban forms such as water, sewage, and electricity services. Studies showed that costs associated with centralized water and sewerage services are directly proportional to the spatial form of development. In simple words, lower-density developments necessitate longer pipelines. As such, it would be more cost-effective to re-urbanize or work in brown-field developments. These trends will continue to put severe pressure on governments to build new infrastructure and accommodate the sprawl. Integrated transportation-land use planning is crucial to maximize the use of infrastructure and optimize the allocation of resources. Thus, there is a need to develop urban evolution models to optimize the use of existing infrastructure and meet the needs of the rising population.

Another consideration is the impact of urbanization on the environment. The transportation sector is a major source of greenhouse gas emissions. Thus, the impacts of travel behavior and the subsequent construction of new infrastructure to accommodate the urban sprawl have detrimental impacts on the environment. It is important for governments to consider that the best long-term solution is not always building new infrastructure. For instance, extending highways to accommodate higher traffic capacities will result in more vehicles and thus increase greenhouse gas emissions. In that case, a more structured solution would consider three steps: (1) adoption of intelligent transportation systems; (2) establishment of smart growth management strategies to reduce congestion; and (3) shifting of consumer behavior and expectations.

Impacts of Climate Change

Over the past few years, natural disasters have become a frequent phenomenon worldwide. Events such as floods, high winds, snowstorms, and earthquakes can have disastrous impacts on infrastructure and its ability to provide its intended function. For instance, precipitation trends from recent decades revealed an increased frequency of extreme rain events in many regions, where climate change is suspected to be the direct cause (Madsen et al. 2009). Furthermore, the Intergovernmental Panel on Climate

Change (IPCC) estimated a temperature increase ranging between 1°C and 3.5°C by the year 2100, due to the increase in greenhouse gas emissions (Houghton et al. 1996; Warrick et al. 1995). They claimed that there is a 90% chance of augmented heavy rainfall event frequency in the 21st century and a probable increase in higher-latitude storms by 40%, because of continuing global warming. Warmer temperatures will, most likely, strengthen the hydrological cycle, resulting in an increase in precipitation intensity, decreased storm return periods, and larger storm intensity. The combination of those challenges could exacerbate the severity of events (IPCC 2007).

Lack of Coordinated Asset Management Planning

Although there is a dire need for better management of existing municipal infrastructure, only a few municipalities in Canada, for example, possess coordinated asset management planning for their infrastructure systems (InfraGuide 2006). While many municipalities have implemented management systems for their horizontal infrastructure, most lack asset management plans for their buried infrastructure systems. Typically, buried infrastructure systems have longer service lives as opposed to their above-ground counterparts. Furthermore, evaluating and assessing the condition of below-ground systems requires complicated technologies.

Coordination and integration is one of the fundamentals of proper asset management. In order to establish holistic asset management planning, InfraGuide (2003) outlined an integrated approach for the assessment and evaluation of municipal road, water, and sewerage networks. This approach consists of five steps: (1) data inventories; (2) investigations; (3) condition assessment; (4) performance evaluation; and (5) renewal planning. Hence after, InfraGuide (2006) outlined the need for coordinated renewal planning of municipal road, sewerage, and water systems, at the network level. It emphasized the same five-step procedure mentioned in InfraGuide (2003) for assessing and evaluating municipal infrastructure. Furthermore, it mentioned that the proposed asset management planning framework should include clear policy objectives and established priorities. Elaborating on these perspectives reveals more integration aspects such as the top-down decision-making approaches, where goals, objectives, and policies are the main decision-making variables, and the bottom-up management approaches, where the technical condition of different assets and the daily intervention aspects are the main decision-making variables. Furthermore, the integration of decision-making across multiple levels (e.g., municipal, provincial, and federal) has not been thoroughly investigated yet.

Infrastructure Deficit and Limited Financial Resources

Before we discuss this challenge, it is important to have a clear understanding of infrastructure lifecycle stages, as shown in Figure 1.4. Most of the challenges discussed earlier emanate from what the authors call the "build-and-forget" approach. Infrastructure systems are designed and built without explicit consideration of their performance over the entire lifecycle span and without consideration of any specific provisions for maintenance and operation. Unfortunately, civil infrastructure systems lag behind in establishing standard maintenance programs similar to other

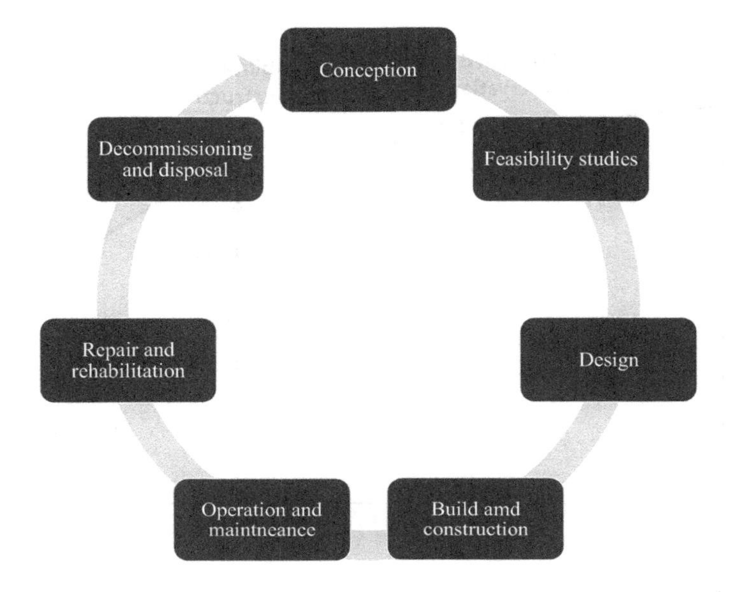

FIGURE 1.4 Infrastructure lifecycle stages.

industries such as automobiles, aircraft, railways, and nuclear power. Management of infrastructure systems shall simply deal with every facet of the asset lifecycle, starting from conception and feasibility studies, all the way long to decommissioning and disposal. Each step within the asset's lifecycle shall be moderated by financial, socioeconomic, and environmental concerns and shall be guided by the principles of sustainable development.

Infrastructure deficit is an accumulation of deferred repair and maintenance over years, as infrastructure systems are left to deteriorate. If the deterioration is not halted or at least slowed down considerably, this deficit continues to grow exponentially, and the deterioration could reach a stage that forces instant replacement of some infrastructure at higher costs. To better explain the infrastructure deficit, consider the case of a pavement section. The cost of delayed repair and maintenance is six to ten times larger than the cost of undertaking scheduled preservations early on, during the pavement's lifecycle, as shown in Figure 1.5.

Consider the case of Canada's infrastructure deficit and its evolution. Most of the Canadian infrastructure was built in the sixties after the Second World War, driven by the significant increase in Canada's population. Around the mid-seventies, the Canadian population started diminishing, and consequently, the need for further infrastructure expansion has considerably decreased. This led to lower expenditures on infrastructure construction, maintenance, and rehabilitation. The lower investments directed toward infrastructure were further compounded by the growing trend of moving into new and large homes in less dense areas. Nevertheless, the urban sprawl creates an increasing demand for new infrastructure and increases the backlog of maintenance and replacement works. Currently, many of Canada's infrastructure facilities have already exceeded their expected service life. Several Canadian water

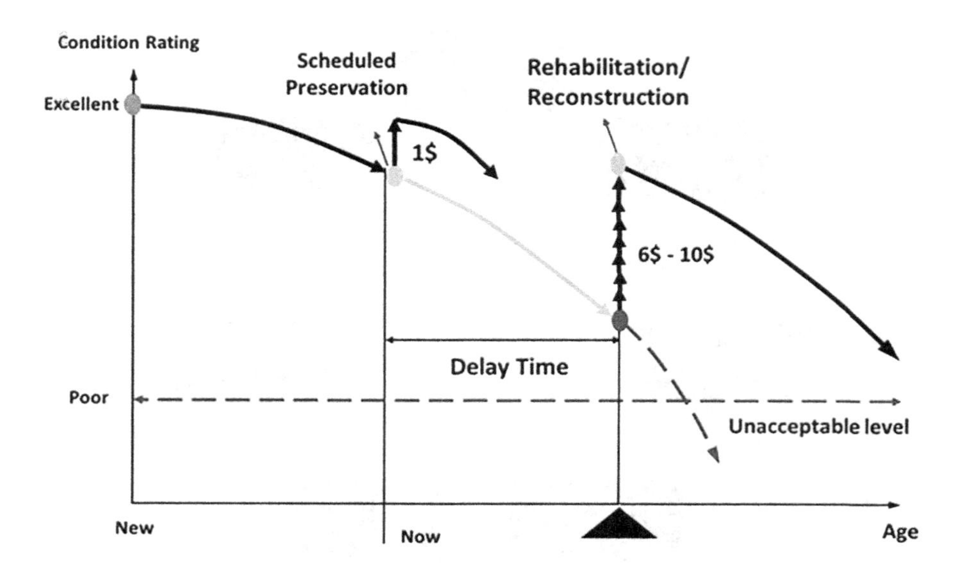

FIGURE 1.5 Cost of delayed repairs – pavements example.

supply and sewerage pipes in older cities are over a hundred years old. This escalating deterioration of infrastructure is attributed to the deferred maintenance that has caused failures and the closing of many services, including roads, bridges, water supply networks, and sewage disposal lines. These service disruptions are becoming common occurrences nowadays and were further aggravated by the aggressive environmental conditions and the lack of environmental considerations in design codes and maintenance and rehabilitation guidelines. To put things in perspective, the overall expenditures on Canada's infrastructure were around $11 billion in the eighties with most of them used for construction and only 20% allocated to rehabilitation of deteriorated infrastructure. This led to the current unacceptable deterioration state of Canada's infrastructure. At that time, the price tag (infrastructure deficit) to upgrade the municipal infrastructure and bring it up to an acceptable condition stood at $12 billion. Since the eighties, the deficit kept rising until reaching $171.8 billion in 2012 for existing infrastructure, steadily growing by $2 billion annually. Moreover, a $115 billion investment is needed for constructing new infrastructure to satisfy the growing population (Mirza 2007; Mirza 2009; Statistics Canada 2017). Studies estimate Canada's infrastructure deficit to lie between $110 and $270 billion (BCG 2017). Over the past few years, the government has been doubling its investment rate for building new infrastructure and repairing existing ones. However, the current investment rates are not sufficient to upgrade all the deterioration of assets and bridge the infrastructure gap created over the past three decades.

Another issue that many municipalities might be facing is the municipal funding model. The municipal funding model is the mechanism by which municipalities receive tax funding to build new infrastructure and repair and replace existing ones. For instance, Canadian municipalities only receive eight cents of every tax dollar collected across Canada and are yet accountable and responsible for the stewardship of 57% of infrastructure assets in Canada. The intent is not to criticize the current fiscal distribution system as it is out of the scope of this book, but rather to highlight the challenge faced by Canadian municipalities due to the lack of financial resources required to maintain the current infrastructure.

Unrealistic Level of Service Expectations

As technology advances, service level expectations for transportation, electricity, drinking water, telecommunication, etc., have become extremely high. However, end users fail to fully comprehend the relationship between where their municipal taxes go as opposed to the services they are getting in return. Thus, it is paramount for municipalities and asset owners to improve the justification of their budget allocation process and to explain its effect on their infrastructure service levels. Furthermore, municipalities could showcase what-if scenarios to end users, to provide them with a better understanding of the short- and long-term impacts of their current funding scenario.

Conflicting Objectives among Stakeholders

Asset management is an interdisciplinary field that involves governments, municipalities, engineers, construction companies, lawyers, consultants, and public end users all at the same time. All those stakeholders share the common goal of ensuring that infrastructure systems deliver their intended function. However, each stakeholder operates under different constraints. For instance, governments have their own political agenda that, in some cases, focuses on increasing investments to build new infrastructure, thus limiting other investments to operate and maintain existing ones. On a different level, municipalities struggle to operate under very tight budget constraints, to manage their aging infrastructure assets, which in turn require billions of dollars to rehabilitate and replace, while being expected to deliver a high service level (Khan et al. 2015). Engineers and construction companies design and build infrastructure systems within very limited project budgets, which, in some cases, force them to design and build those infrastructure systems to meet minimal code requirements. Hence, the result is losing the opportunity for other important considerations, including, but not limited to, adopting environmentally friendly materials, sustainability, maintainability, and lower lifecycle costs. Moreover, lawyers draft rules and policies that align with the government's vision. In addition, public end users possess high service level expectations while lacking appreciation of the tight municipal budgets. Another related aspect is the need created by changing the demographic. For instance, an aging population dictates the type of infrastructure being replaced, while the special needs of such a population will guide the design considerations of new infrastructure. For example, the distance between transit stops, types of transit investments, healthcare facilities, and types of equipment all add more complexity as municipalities do not only have to replace and add new infrastructure, but they are also obliged to design new infrastructure complying with the various demographic needs.

This is just a summary of hundreds if not thousands of considerations and objectives that municipalities must account for while allocating their limited budget. It is simply a combinatorial problem, and there will not be a "*one-size-fits-all*" solution. Municipalities cannot maintain all their infrastructure systems under those conflicting objectives: (1) stringent budget; (2) minimal to no disruption to public services; (3) very short duration; (4) optimal designs and long-lasting sustainable materials; and (5) excellent condition state with minimal to no risk. Thus, it is imperative for municipalities to adopt technological solutions for trading off different scenarios and selecting an optimal scenario that best fits the stakeholders.

IT IS TIME TO ACT

The state of infrastructure around the world has declined steadily over the past three decades. Several of these infrastructure facilities are thus inadequate to meet the present population's needs and certainly the projected future growth and development. Infrastructure everywhere is in dire need of repair and upgrade. The risk of sudden failures and service disruptions drastically increases, forcing municipalities to take immediate corrective action to recover such services. Once the infrastructure asset is built, deterioration sets in and continues to grow exponentially, escalating repair costs and making repair and rehabilitation more difficult and expensive. Furthermore, infrastructure projects carry out a multitude of challenges and risks throughout their lifecycle, due to demand fluctuations, uncertainties, natural disasters, necessity, criticality, etc. This requires decisive intervention to not only be taken at the beginning of the lifecycle but also regularly revised to guarantee the delivery of acceptable service levels, while meeting tight budgets and upholding minimum acceptable conditions. Increasing infrastructure deficits and failing infrastructure systems means increasing the risk of sudden failures and imposing huge financial burdens on governments. Increasing demands on existing infrastructure force asset managers to consider resiliency while taking repair and replacement decisions. Thus, it is time for us to act.

If you own new infrastructure, it is time to adopt a proactive approach for maintaining the assets, and ensuring an asset management plan is in place throughout the asset's lifecycle. If you own an aging infrastructure, it is time to curb the infrastructure gap and bridge the deficit by optimizing your budget allocation to react to failing assets, build new infrastructure, and implement proper asset management practices to preserve the other assets. One of the main objectives of this book is to shed light on solutions that could support municipalities throughout their asset management journey.

REFERENCES

Abusamra, S. (2019). "Coordination and Multi-Objective Optimization Framework for Managing Municipal Infrastructure under Performance-Based Contracts." Doctoral of Philosophy Thesis, Concordia University, Montreal.

ASCE (American Society of Civil Engineers). (2021). "ASCE Infrastructure Report Card." American Society of Civil Engineering. https://www.infrastructurereportcard.org/. June 16, 2022.

BCG (Boston Consulting Group). (2017). "15 Things to Know About Canadian Infrastructure." https://www.caninfra.ca/insights/. Dec 18, 2023.

CCPA (Canadian Center for Policy Alternatives). (2009). "Leadership for Tough Times: Alternative Federal Budget Fiscal Stimulus Plan." https://www.policyalternatives.ca/sites/default/files/uploads/publications/National_Office_Pubs/2009/Leadership_For_Tough_Times_AFB_Fiscal_Stimulus_Plan.pdf. Feb 20, 2023.

CNAM. (2019). "Asset Management 101: The What, Why, and How for Your Community." Canadian Network of Asset Managers (CNAM). https://cnam.ca/new-to-am/am101-introduction/

Haas, R., Hudson, W., and Zaniewski, J. (1994). *Modern Pavement Management*. Krieger Publishing Company, Malabar, FL.

Hudson, W. R., Haas, R., and Uddin, W. (1997). *Infrastructure Management*. McGraw-Hill Publishing, New York.

Houghton, J.T., Meira Filho, L.G., Callander, B. A., Harris, N., Kattenberg, A., and Maskell, K. (1996). *Climate Change 1995: The Science of Climate Change. Contribution of Working Group 1 to the Second Assessment Report of the Intergovernmental Panel on Climate Change*. Cambridge University Press, Cambridge, United Kingdom.

InfraGuide. (2003). *An Integrated Approach to Assessment and Evaluation of Municipal Roads, Sewer and Water Networks*. FCM (Federation of Canadian Municipalities) and (National Research Council), Ottawa.

InfraGuide. (2006). *Managing Risks*. FCM (Federation of Canadian Municipalities) and (National Research Council), Ottawa.

IPCC (Intergovernmental Panel on Climate Change). (2007). "Climate Change 2007: Synthesis Report. Contribution of Working Groups I, II and III to the Fourth Assessment Report of the Intergovernmental Panel on Climate Change [Core Writing Team, Pachauri, R. K and Reisinger, A. (eds.)]." IPCC, Geneva, 104 pp.

Khan, Z., Moselhi, O., and Zayed, T. (2015). "Identifying rehabilitation options for optimum improvement in municipal asset condition." *Journal of Infrastructure Systems*, ASCE, 21(2), 04014037.

Madsen, H., Arnbjerg-Nielsen, K., and Mikkelsen, P. S. (2009). "Update of regional intensity–duration–frequency curves in Denmark: Tendency towards increased storm intensities." *Atmospheric Research*, 92(3), 343–349.

Mailhot, A. and Duchense, S. (2010). "Design criteria of urban drainage infrastructures under climate change." *Journal of Water Resources, Planning and Management*, 136(2), 201–208.

Mirza, S. (2007). "Danger Ahead: The Coming Collapse of Canada's Municipal Infrastructure." FCM (Federation of Canadian Municipalities). https://www.fcm.ca/Documents/reports/Danger_Ahead_The_coming_collapse_of_Canadas_municipal_infrastructure_EN.pdf. Sep 30, 2022.

Mirza, S. (2009). "Canadian Infrastructure Crisis Still Critical." https://ascelibrary.org/doi/abs/10.1061/(ASCE)CO.1943-7862.0001402. Jan 18, 2020.

MGI (McKinsey Global Institute). (2017). "Bridging Global Infrastructure Gaps: Has the World Made Progress?" https://www.mckinsey.com/industries/capital-projects-and-infrastructure/our-insights/bridging-infrastructure-gaps-has-the-world-made-progress. Mar 22, 2023.

Statistics Canada. (2017). "Canadian Demographics at a Glance: Population Growth in Canada." https://www.statcan.gc.ca/pub/91-003-x/2007001/4129907-eng.htm. Jan 18, 2017.

UNFPA (United Nations Fund for Population Activities). (2007). *State of the World Population Report: Unleashing the Potential of Urban Growth*. United Nations Population Fund, New York.

Warrick, R.A., Provost, C. Le., Meier, M. F., Oerlemans, J., and Woddworth, P.L. (1995). "Changes in Sea Level". In: *Climate Change: The Science of Climate Change* (pp. 363–405). Cambridge University Press, Cambridge, United Kingdom.

Water Security Agency. (2014). "Stormwater Guidelines: EPB 322." https://pdf4pro.com/view/stormwater-guidelines-saskh20-ca-5704c6.html. Mar 31, 2019.

2 An Overview of Asset Management

Evolution, Strategies, and Principles

HISTORICAL BACKGROUND

Physical asset management is the practice of managing the physical assets of an organization, such as buildings, equipment, and machinery. The history of physical asset management can be traced back to the early days of human civilization when people began to build infrastructure to support their communities. One of the earliest examples of physical asset management is the construction of pyramids in ancient Egypt. The pyramids were built using a combination of human labor and simple machines, such as levers and pulleys, to lift and move heavy blocks of stones. The workers who built the pyramids were skilled craftsmen who were responsible for maintaining the tools and equipment that were used in the construction process.

As societies grew and became more complex, the need for infrastructure and physical assets increased. The Roman Empire, for instance, built a vast network of roads, aqueducts, and buildings that supported its military and civilian population. The Romans were also known for their advanced engineering techniques, including the use of concrete and arches, which allowed them to build structures that were both functional and aesthetically pleasing. In the Industrial Revolution of the 18th and 19th centuries, physical asset management became increasingly important, as factories and other industrial buildings were constructed to support the growing manufacturing industry. The development of railroads and other transportation infrastructure also played a critical role in the growth of industry, as it allowed raw materials and finished goods to be transported more efficiently.

Today, asset management is a vital part of many industries, including transportation, energy, and manufacturing. It involves the proper maintenance and repair of physical assets, as well as the planning and coordination of their use and replacement. Asset management strategies continue to evolve as organizations seek ways to improve efficiency and reduce costs. Modern asset management strategies often involve the use of advanced technology, such as sensors and data analytics, to monitor the condition of assets and predict when maintenance or repair work is needed. These strategies may also involve the use of specialized software solutions to track and manage asset maintenance and repair. As technology has advanced, the tools and techniques used in asset management have also evolved, allowing organizations to manage their assets more effectively and improve efficiency.

DOI: 10.1201/9781003327509-3

INFRASTRUCTURE GAPS

There are numerous infrastructure gaps that can exist under a variety of contexts, such as gaps in transportation infrastructure, energy infrastructure, or water infrastructure. Some common infrastructure gaps that were identified include but are not limited to (Table 2.1):

TABLE 2.1
Infrastructure Gaps

Infrastructure Gap	Description	Example(s)
Demand management gap	The need to better manage the demand for infrastructure services through behavioral change and new technologies	**Electricity**: Motion-sensor detectors to switch the lights on/off based on people's movement detection **Water**: Users' behavioral change and proximity sensors installed in bathroom faucets to allow water to flow through a valve when hands are held near the faucet **Transportation**: User behavioral change to use public transportation to reduce emission, traffic congestion, and protect the environment
Investment gap	The need for additional funding sources because of ever-increasing social expenditures and aging infrastructure	Increasing infrastructure deficits because of the delayed repair and maintenance to aging infrastructure
Efficiency gap	The need for managing infrastructure in a more cost-efficient fashion	**Transportation**: Preventive maintenance for roads saves ten times more as opposed to rehabilitation
Resilience gap	The need for managing infrastructure in a more informed and intelligent way with respect to risks	**Water and Sewerage**: Build water and sewer systems that are capable of meeting increasing demand, resulting from population growth and climate change
Lack of connectivity	Gaps in transportation infrastructure	**Transportation**: Lack of roads or public transportation, which can make it difficult for people to access jobs, education, and other resources
Aging infrastructure	Many of the infrastructure systems are old and need repair or replacement This can make them less reliable, more expensive to operate, and more vulnerable to natural disasters	Increased vulnerability to natural disasters is a common issue among all infrastructure systems with varying consequences levels depending on various factors
Inadequate capacity	Some infrastructure systems may not have enough capacity to meet the needs of a growing population or economy	**Water**: Water systems may not have enough capacity to meet the needs of a growing community

(Continued)

TABLE 2.1 (*Continued*)
Infrastructure Gaps

Infrastructure Gap	Description	Example(s)
Inequalities	Infrastructure gaps can disproportionately affect low-income communities and communities of color, who may have less access to transportation, clean water, and other essential resources	**Transportation**: Less to no frequencies for public transportation systems
Cybersecurity	With the increasing dependence on digitalization and automation in various critical infrastructure, the risk of cyberattacks is getting higher and higher	**Transportation**: Cyberattacks on the traffic system or autonomous vehicles or trains causing fatal accidents **Water**: Cyberattacks on the processing systems within the water station causing detrimental health impacts to millions of users
Environmental Sustainability	Many countries' infrastructure still lacks the capacity to incorporate greener and more sustainable technologies, making it harder to reduce emissions and combat climate change	**Transportation**: Electric/hybrid vehicles and trains **Roads**: Solar panel roads and sidewalks that could generate electricity as opposed to asphalt/concrete pavements

The needs and gaps vary from one geographical region to another, from country to country, and depends on a multitude of factors, such as population size, natural resources, economic status, and so on. Figure 2.1 summarizes the challenges organizations are facing nowadays throughout the asset lifecycle. Those challenges include, but are not limited to (1) increasingly dynamic end user's expectations that forces organizations to adapt their assets in accordance with such expectations, (2) evolving technology providing insights on how assets are used and managed; (3) stricter regulations governing risk-based asset management and mandating organization to have asset management plans for their asset portfolios; and (4) aging assets and their need for replacement and increase in capacity to cope with population growth, etc. To better understand those challenges, we detailed the factors that have contributed to two of those gaps. The first one is the demand–supply gap illustrated in Figure 2.2. The demand is impacted by urban density, current population, and urban expansion plans. The capacity, however, is impacted by population growth, available budget, resources, and technology. The service level, in its turn, is affected by the aging infrastructure and its ability to provide adequate service, the operational requirements for the infrastructure system to provide the service, the external stressors such as environmental and climatic conditions, and the degree of evolution of asset management practices. To balance the equation, the capacity of infrastructure systems

FIGURE 2.1 Asset lifecycle.

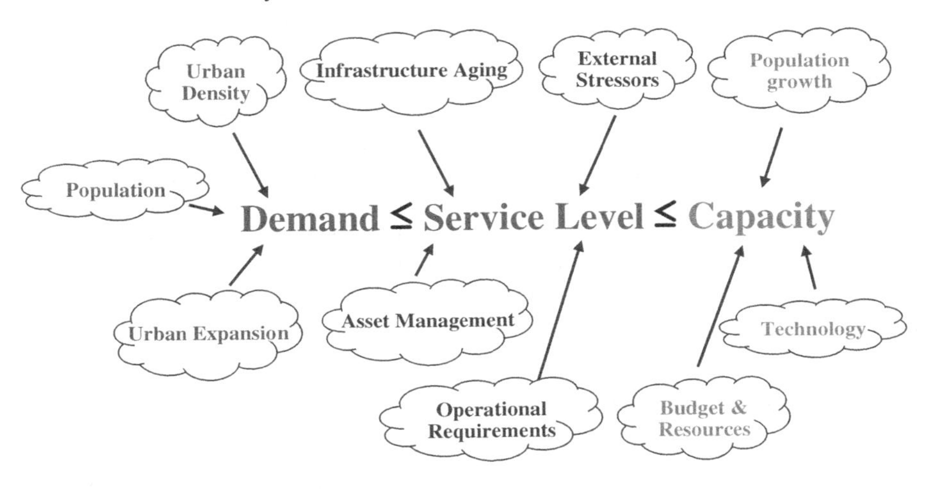

FIGURE 2.2 Demand management and inadequate capacity gaps.

should match –if not exceed – the current demand and should also accommodate future expansions while maintaining adequate service levels.

The other example presented in Figure 2.3 describes the impact of cost escalation and uncertainties on infrastructure systems while showing the current budget deficit. As a quick refresher, a unit cost of a product within a package equals to the total cost of the package divided by the quantity of products within the package. Similarly with infrastructure, the unit cost of providing a certain service is equal to the total cost of providing the service divided by the population receiving the service. The

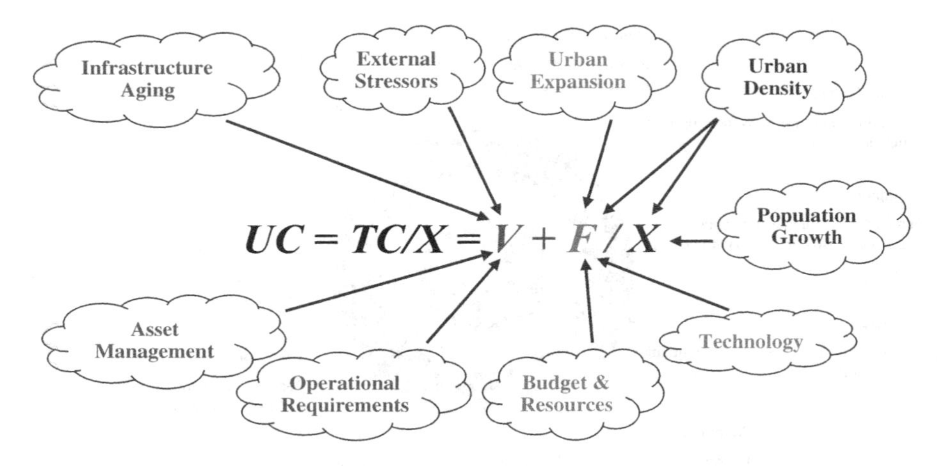

FIGURE 2.3 Impact of cost escalation and uncertainties on infrastructure systems.

total cost is hence the summation of (1) fixed costs to provide the service such as the underlying technology, budget, and resources providing the service (e.g., municipal resources, contractors, labor, equipment, etc…), and urban expansion plans to expand the infrastructure services and accommodate the growing population; and (2) variable costs that could not be directly controlled, such as infrastructure aging that varies depending on countless factors, such as operational requirements for infrastructure systems that keeps changing with technology upgrading, end user's expectations, external stressors (e.g., environmental and climatic conditions), and evolution of asset management practices. Those costs are then divided by the dynamic population. Currently, the unit cost is drastically increasing as the fixed and variable cost rate of increase is much larger than the population. In addition, taxes collected from the end users are not increasing at the same rate as the unit cost.

ASSET MANAGEMENT EVOLUTION

In 1980s, the focus of asset management was mostly centered on Operation and Maintenance (O&M). During this decade, we have realized the development of maintenance management and scheduling systems. Most of the asset replacement decisions were solely based on asset age and came as a reaction to service failures. In 1990s, asset management practitioners shifted the focus to minimizing maintenance costs, and thus in 1999, the *Australian Infrastructure Maintenance Manual* was published. The focus then started to shift a bit from being entirely maintenance-oriented to being preoccupied with fully integrated work and asset management systems. Between 2000 and 2005, performance stood at the forefront of asset management's concern. It was at that stage when some asset management metrics were developed to assess capital expenditure. Furthermore, in 2004, the first standard for asset management PASS 55 (Publicly Available Specification 55) was developed by the British Standards Institute (BSI). It provides guidelines for the management of physical assets in the public sector. PASS 55 was based on the principles of total quality management (TQM) and focused on continuous improvement, customer satisfaction, and integration of asset management activities with the overall goals of an organization. It was designed to be a flexible and adaptable standard that could be applied to a wide range of assets, including buildings, infrastructure, and equipment.

Between 2005 and 2010, the focus shifted this time toward risk-based asset management. Modeling started to evolve, allowing asset management to integrate between asset deterioration and capacity estimation in both replacement and growth planning decisions. Furthermore, BSI published the second edition of its assessment methodology for PASS 55. Between 2010 and 2015, the industry started to focus on multi-objective optimization, integrating cost, performance, and risk in decision-making. In 2014, PASS 55 was superseded by ISO 55000 and the second edition was published in 2024, a series of international standards for asset management that were developed by the International Organization for Standardization (ISO). ISO 55000 represents the culmination of several years of work by experts in the field of asset management and is based on a holistic, systems-based approach to asset management. It covers the entire asset management lifecycle and emphasizes the importance of aligning asset management activities with the strategic goals of an organization. The ISO 55000 series is based on a holistic, systems-based approach to asset management that emphasizes the importance of aligning asset management activities with the strategic goals of an organization. It covers the entire asset management lifecycle from planning and acquisition to operation, maintenance, and disposal (BSIgroup 2008, 2015; ISO 2014, 2024a).

While PASS 55 and ISO 55000 have some evident similarities, ISO 55000 is a more comprehensive standard that has been adopted by organizations around the world as a framework for managing their physical assets. It has also been recognized as an important tool for improving the efficiency and effectiveness of asset management practices, and it has played a key role in driving the evolution of asset management strategies and techniques. In the coming sections, we present the main aspects of ISO 55000 including the framework for asset management strategies, principles for asset management, the asset management system, and key components to measure the asset management maturity of an organization (Woodhouse 2015).

Between 2015 and the present date, technological advancement in both software and hardware has greatly added value to the asset management industry. Embedded sensors can inform asset managers of the asset condition state in real time and provide predictive analysis on asset performance throughout its lifecycle. Nowadays, digital technologies and artificial intelligence (AI) are having a greater influence on each aspect of our day-to-day life. Over the past few years, we have witnessed an unprecedented boom in AI-based asset management tools. For instance, some AI-based tools were developed to predict pavement distresses using image processing and computer vision algorithms, and thus compute the pavement condition index. Other AI-based tools were developed to predict pipe defects and localize the defect, saving thousands of dollars for municipalities and causing less disruption to end users. Another important and rapidly evolving technology is digital twins. A revolution has been taking place in the fields of imagery data collection, processing, and analysis where data acquisition technologies – such as Scan to Building Information Management (BIM) and point cloud modeling – which enabled accurate generation of digital twins of any arbitrary asset. For instance, asset owners can now scan their sites while leveraging their geospatial data and in a relatively short period of time.

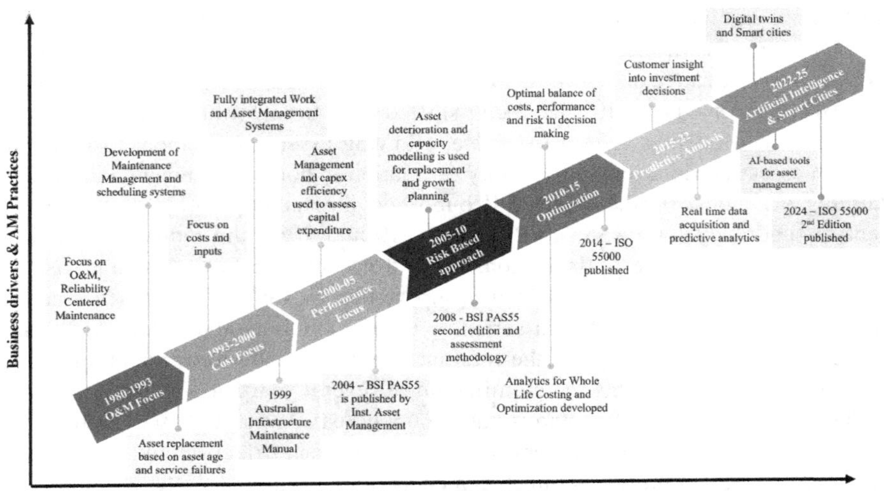

FIGURE 2.4 Evolution of asset management systems and standards.

They can have a digital twin of their sites in an editable format. Afterward, it can be connected to their Enterprise Asset Management (EAM) tool to tag assets and assign work orders in a 3D environment. A summary of asset management systems and standards and their evolution is shown in Figure 2.4.

PRINCIPLES OF ASSET MANAGEMENT

The principles of asset management represent the fundamental concepts and guidelines of effective asset management. These principles are designed to help organizations manage their physical assets in a way that is efficient, effective, and aligned with their overall goals and objectives. There are several principles of asset management that are generally recognized as being important, including:

- **Align Asset Management with Organizational Goals**: Asset management should be integrated with the overall management of an organization and should support the achievement of its strategic goals.
- **Identify and Prioritize Assets**: Organizations should identify and prioritize their assets based on their importance and the risks they pose.
- **Implement an Asset Management System**: Organizations should implement an asset management system that includes processes and procedures for planning, acquiring, operating, maintaining, and disposing of assets.
- **Use Data and Technology to Support Asset Management**: Organizations should use data and technology to support asset management activities, such as monitoring the condition of assets and predicting when maintenance or repair work is needed.

- **Involve Stakeholders**: Organizations should involve stakeholders, such as employees, customers, and regulators, in asset management decision-making to ensure that the needs and concerns of all parties are considered; and
- **Continuous Improvement**: Organizations should continuously review and improve their asset management practices ensuring that they are effective and aligned with the changing needs of the organization.

STRATEGIC, TACTICAL, AND OPERATIONAL LEVELS

Before we discuss the framework for asset management, it is crucial to ensure common understanding of the different levels of planning and decision-making associated with managing physical assets. In asset management, there are three levels of planning and decision-making, namely strategic, tactical, and operational levels. These levels can be described as follows:

- **Strategic Level**: This level involves long-term planning and decision-making (5–25 years) that align asset management activities with the overall goals and objectives of an organization. At the strategic level, decisions are made as to which assets are needed, how they will be used, and how they will be financed.
- **Tactical Level**: This level involves the medium-term planning and decision-making that are needed to implement the developed strategies at the strategic level. At the tactical level, decisions are made as to how to maintain, repair, and replace assets to support the organization's operations. The tactical-level planning horizon ranges from 1 to 5 years depending on several factors, including but not limited to, asset type, service life, etc.
- **Operational Level**: This level involves day-to-day planning and decision-making that is needed to manage the use and maintenance of assets. At the operational level, decisions are made as to when and how to perform maintenance and repair works, and how to ensure that assets are used effectively and efficiently operated and maintained. Operational planning typically ranges between 1 week and 1 year depending on the nature of the asset being managed.

To effectively manage physical assets, it is important to consider all three levels of planning and decision-making. By taking a holistic approach that considers the strategic, tactical, and operational levels, organizations can ensure their assets are used in such a way that supports their overall goals and objectives. Figure 2.5 displays the interconnection between the various levels of planning and decision-making. The operational-level planning uses data, trends, and decision-support systems and undertakes actions and interventions. The tactical level uses asset standards, performance models, risk models, data analytics, and manages projects to meet the desired performance metrics. The strategic level establishes corporate strategy and policy, performs regular strategic assessments, and builds business plans to ensure alignment with organizational goals and objectives. The operational level feeds the

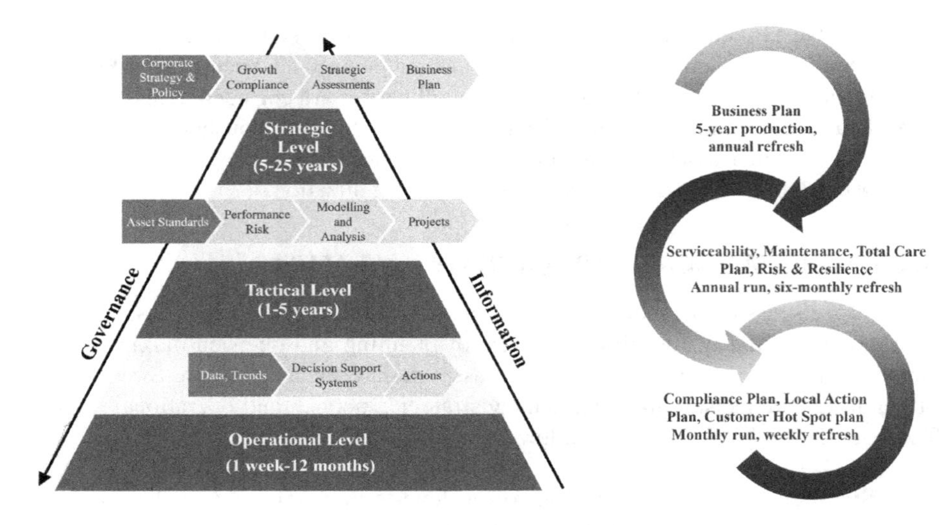

FIGURE 2.5 Interconnection among governance levels.

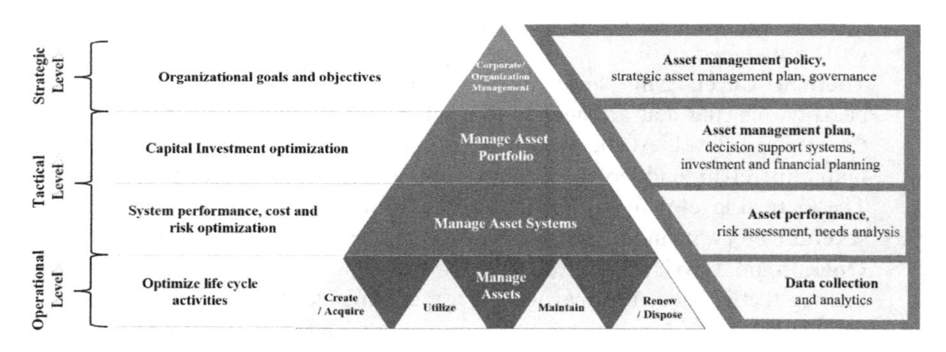

FIGURE 2.6 Output of governance levels.

strategic level with information and performance metrics on their operations. This information flow assists the strategic level to establish and refine long-term priorities. On the other hand, the strategic level provides oversight and governance to both tactical and operational levels to ensure actions and projects are well aligned with the strategic plan and provide tactical and operational-level improvements as needed.

Figure 2.6 outlines the main output from the different levels. The corporate management establishes the organization's goals and objectives. Those goals and objectives then establish the asset management policy and prepare a strategic asset management plan that frames their strategies toward reaching their goals. At the tactical level, asset systems and portfolios are managed. It starts with optimizing performance, cost, and risk for the various asset systems forming the asset portfolio. Thenceforth, based on the outcome of the asset system optimization, optimization of the capital investments for the asset portfolio is carried out by virtue of decision-support systems, investment, and financial planning tools. The outcome is an asset management plan outlining the optimal intervention schedule along with the lifecycle costing plan that best meets the organization's goals and objectives, within the allocated resources. The operational level optimizes the lifecycle activities through management of different assets within the asset systems through continuous data collection and analytics. In parallel, they respond to preventive maintenance work orders, procurement, and delivery of projects for building new infrastructure or rehabilitating existing ones, operation, maintenance activities, etc.

FRAMEWORK FOR ASSET MANAGEMENT

A framework for asset management is a set of guidelines and principles that organizations can use to manage their physical assets effectively. A good asset management framework provides a clear understanding of the roles and responsibilities of different stakeholders, as well as the processes and procedures that should be followed to manage assets effectively. There are key components that are typically included in an asset management framework, including (ISO 2014, 2024a):

- **Asset Management Policy**: This is a high-level document that outlines the organization's overall approach to asset management and establishes the principles and values that will guide asset management activities.
- **Strategic Asset Management Plan (SAMP)**: It is a document that outlines an organization's goals and objectives for managing its physical assets, as well as the strategies and actions that will be taken to achieve those goals. The aim of the SAMP is to ensure that the organization's assets are being used effectively and efficiently, and that they can deliver the desired level of performance. By having a strategic management plan, organizations make better informed decisions, aligning the assets with the goals and objectives of the organization, and optimizing the use of resources for the best outcomes. The plan typically includes information on the current state of the organization's assets as well as the long-term vision for how those assets will be managed in the future. The key elements of a SAMP often include:
    - An inventory of the organization's assets, including information on the condition, location, and value of each asset.
    - An assessment of current and future risks associated with each asset, such as probability of failure, consequences of failure, and repair and replacement costs.
    - A set of objectives for managing the assets, such as maintaining a high level of reliability, reducing costs, and improving the overall performance of assets.
    - Strategies and actions required to meet the objectives, including details on budget, resources, and timelines.

- **Asset Management Plan**: This is a detailed document that outlines the specific strategies and actions required to manage the organization's assets. It typically includes information about the types of assets that will be managed, the risks to which they are subjected, and the strategies that will be used to manage those risks.

- **Asset Management System**: This is a set of processes and procedures that are used to manage the organization's assets. It typically includes information about how assets will be acquired, maintained, and disposed of, as well as the roles and responsibilities of different stakeholders in the asset management process.
- **Performance Measurement and Reporting**: This is a system for monitoring the performance of the organization's assets and reporting on the results. It typically includes metrics and indicators used to assess the effectiveness of asset management activities and identify areas of improvement.

Organizations are operating in an increasingly complex environment, and thus, it is important to adopt a robust asset management framework to help manage their assets effectively, reduce costs, and improve efficiency. Figure 2.7 illustrates the key components in the asset management framework and how they are interconnected. The strategic level of implementation typically starts with putting up the asset management policy that outlines the overall approach to asset management followed by the SAMP, which highlights the goals and objectives for managing its physical assets. At tactical and operational levels, the asset register, condition assessment, criticality/risk assessment, and other performance metrics assessment are fed into the asset management system to be processed. The output of this information processing is an asset management plan that includes asset lifecycle planning, annual interventions plans, and work orders. Furthermore, the asset management system can predict

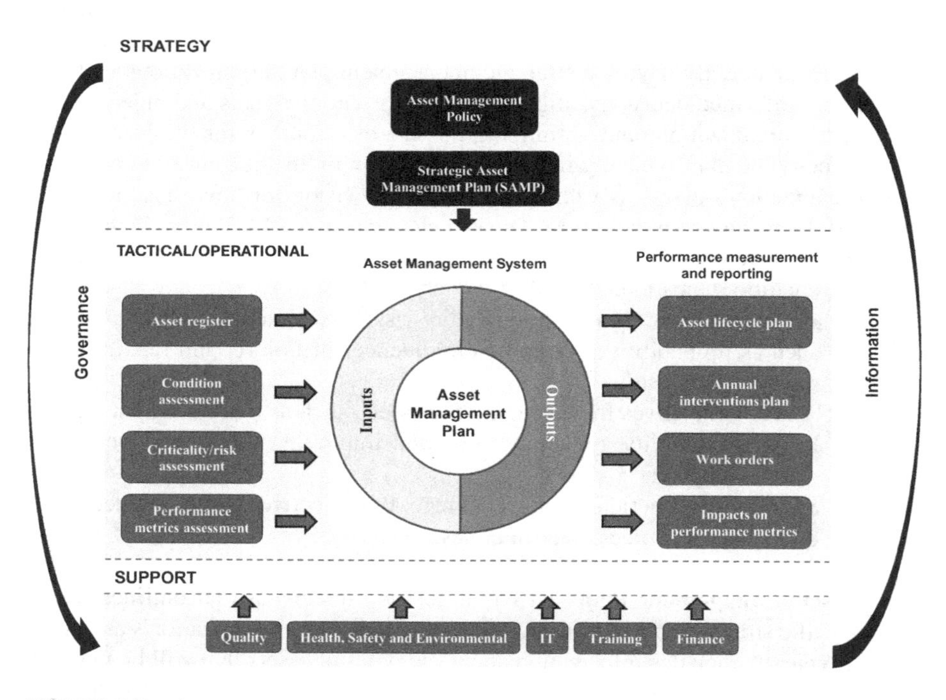

FIGURE 2.7 Asset management framework.

future impacts on performance metrics and subsequently feeds the information back to top management for review and alignment with organizational goals and objectives. There are various supporting groups that are crucial to deliver an asset management plan as detailed in Figure 2.7. The following section discusses the typical key components of an asset management system.

ASSET MANAGEMENT SYSTEM

The asset management system is key in establishing and implementing the asset management framework. Thus, possessing an effective asset management system should be tailored to the specific needs and goals of the organization and should also support the efficient and effective use of assets. There are several key components that are typically included in an asset management system, including:

- **Asset Inventory**: This is a list of all physical assets that an organization owns or manages. The asset inventory should include information on each asset, such as location, condition, value, and the risks that surround it.
- **Asset Data Management**: This is a system for collecting, storing, and analyzing data about an organization's assets. It may include the use of sensors, data analytics, or specialized software to monitor the condition of assets and predict when maintenance or repair work is needed.
- **Asset Maintenance and Repair**: This is the process of maintaining and repairing assets to keep them in good working condition. It may include the use of preventive maintenance programs, repair and maintenance schedules, and specialized equipment and tools.
- **Asset Acquisition and Disposal**: This is the process of acquiring new assets and disposing of old or obsolete assets. It should include policies and procedures for evaluating the needs of an organization and choosing the most appropriate assets to meet those needs.

ASSET MANAGEMENT STRATEGIES

Asset management strategies are the approaches used to manage their physical assets. As shown in Figure 2.4, there are several different types of asset management strategies that organizations can use to manage their physical assets, including:

- **Corrective-Based Asset Management**: This approach focuses on fixing problems as they arise, rather than proactively managing assets to preempt problems and involves waiting until an asset fails or experiences a problem before taking an action to repair or replace it. In corrective-based asset management, the assumption is that assets will perform reliably until they experience a problem, at which point they will be fixed or replaced. This approach can be relatively simple and inexpensive, as it does not require the implementation of preventive maintenance programs or other proactive measures. However, corrective-based asset management can also be less effective than other approaches, as it does not address issues before they turn into complex problems. This can lead to unexpected failures and downtime, which can be costly and disruptive to an organization's operations. As a result, corrective-based asset management is generally less popular than other approaches, such as condition-based or reliability-based asset management.

- **Time-Based Asset Management**: This approach focuses on the age and expected life span of an organization's assets and involves scheduling maintenance and repair work based on the age of the asset, with the goal of extending the asset's life span as much as possible. In time-based asset management, the assumption is that assets will experience a certain amount of wear and tear over time and will eventually need to be replaced. By scheduling maintenance and repair work at regular intervals, organizations can prevent minor issues from becoming major problems and extend the life span of their assets. Time-based asset management is a simple and effective approach that can help organizations to reduce costs and improve efficiency. It is particularly useful for assets that have a predictable life span, such as equipment and vehicles, as it allows organizations to plan and schedule maintenance and repair work in advance. However, it may not be as effective for assets that are subject to unpredictable wear and tear, such as infrastructure and buildings, as it may not address issues that arise unexpectedly.
- **Condition-Based Asset Management**: This approach focuses on the current condition of an asset and uses this information to make decisions about when to repair or replace it. This approach also involves regularly monitoring the condition of assets and using this data to predict when maintenance or repair work is likely to be needed. In condition-based asset management, the goal is to prevent asset failures and downtime by proactively identifying and addressing issues before they become serious problems. This may involve using sensors or other monitoring technologies to collect data about the condition of assets, as well as implementing preventive maintenance programs and repair schedules. By taking a proactive approach to asset management, organizations that use condition-based asset management can reduce costs, improve efficiency, and increase the life span of their assets. It is a particularly useful approach for organizations that rely on critical assets or equipment.
- **Reliability-Based Asset Management**: This approach focuses on the reliability and performance of an organization's assets and involves identifying the key factors that affect the reliability of assets and implementing measures to improve their performance. In reliability-based asset management, the goal is to ensure that assets can function consistently and effectively over time, with a minimum of downtime and maintenance. This may involve implementing preventive maintenance programs, using specialized tools and equipment to diagnose and repair problems, and regularly inspecting and testing assets to identify potential issues. By improving the reliability of their assets, organizations that use reliability-based asset management can reduce costs, improve efficiency, and increase customer satisfaction.
- **Risk-Based Asset Management**: This approach focuses on risks that an organization's assets pose to the organization and their associated response measures and involves identifying key risks associated with assets and implementing measures to mitigate those risks. In risk-based asset management, the

goal is to minimize the potential impact of risks on the organization's operations and financial performance. This may involve regular inspection and testing of assets to identify potential issues, preventive maintenance programs, and replacement of high-risk assets. By taking a proactive, risk-based approach to asset management, organizations can improve the reliability and performance of their assets and reduce the likelihood of costly failures or disruptions.

- **Resilience-Based Asset Management**: Resilience comes from the Latin root "resilio" or "resilire," which means to bounce back to the initial position after receiving shock. As such, this approach focuses on the ability of an organization's assets to withstand and recover from disruptions and other challenges. This includes the ability of assets to adapt to climate change impact. This approach to asset management involves identifying the risks and vulnerabilities that assets may face and then implementing measures to mitigate those risks. In resilience-based asset management, the goal is to ensure that assets can continue to function effectively in the face of disruptions, whether they are caused by natural disasters, equipment failures, or other events. This may involve designing assets to be more resistant to damage or implementing contingency plans to ensure that critical assets can be quickly repaired or replaced in the event of a failure. Resilience-based asset management also involves regular review and updating of asset management plans and processes to ensure that they are aligned with the changing needs and goals of the organization.

A summary of the evolution of asset management strategies could be found in Figure 2.8. It is worth noting that all the asset management strategies provide an effective way to manage physical assets, depending on the needs and goals of an organization. However, it is paramount to understand that selecting the best-fit approach relies on several factors, including asset type, risk appetite, etc. Some organizations may resort to a combination of these approaches to create a comprehensive asset management plan.

ASSET MANAGEMENT MATURITY

Asset management maturity refers to the extent to which an organization has developed and has implemented effective asset management processes and practices. A mature asset management system is one that is well organized, efficient, and aligned with the overall goals and objectives of the organization. There are several factors that contribute to the level of maturity of an asset management system, including:

- The extent to which asset management is integrated with the overall management of the organization.
- The level of understanding and commitment to asset management among employees and management.
- The quality and effectiveness of asset management processes and practices.
- The use of data and technology to support asset management activities.
- The level of transparency and accountability in asset management decision-making.

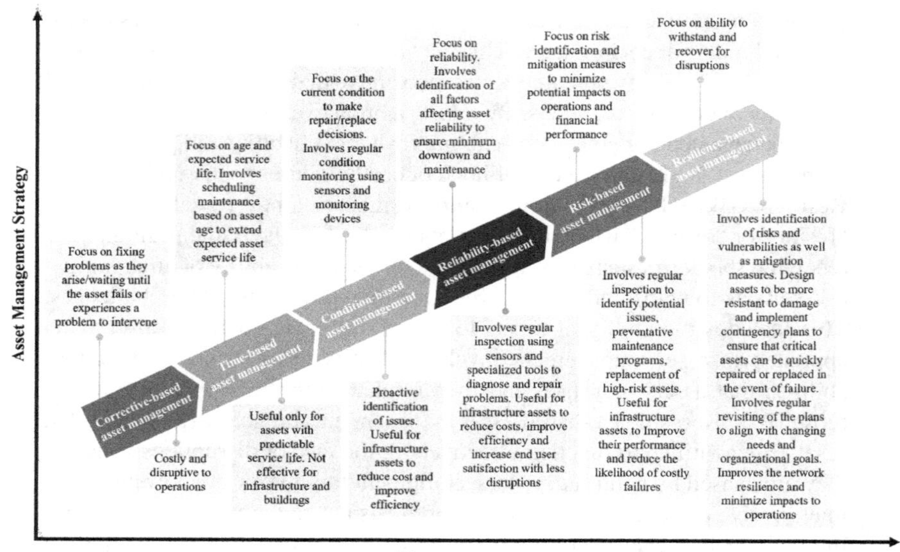

FIGURE 2.8 Evolution of asset management strategies.

An organization's level of asset management maturity can be assessed using a variety of methods, including self-assessment, benchmarking against industry standards or best practices, and external audits. By identifying areas for improvement and implementing targeted changes, organizations can improve the maturity of their asset management systems and increase the efficiency and effectiveness of their asset management practices (ISO 2016, 2024b; GFMAM 2024).

REFERENCES

BSIgroup. (2008). "Publicly Available Specification for the Optimal Management of Physical Assets." https://www.irantpm.ir/wp-content/uploads/2014/01/pass55-2008.pdf. Feb 23, 2021.

BSIgroup. (2015). "Moving from PAS 55 to BS ISO 55001 - The New International Standard for Asset Management - Transition Guide". https://www.4bt.us/wp-content/uploads/2018/01/Moving-from-PAS-55.pdf. June 28, 2020.

GFMAM. (2024). *The Asset Management Landscape*. GFMAM, 3rd Edition. https://gfmam.org/sites/default/files/2024-06/GFMAM_AM_Landscape_v3.0_English_2024.pdf

ISO. (2014). *ISO 55000 Series: Asset management— Overview, Principles and Terminology*. ISO, Geneva.

ISO. (2016). *Asset Management Maturity Scale and Guidance*. ISO, Geneva.

ISO. (2024a). *ISO 55000 Series: Asset Management— Vocabulary, Overview and Principles*. ISO, 2nd Edition, Geneva.

ISO. (2024b). *Pathway to Excellence—Maturity Scale and Guidance*. ISO, Geneva.

Woodhouse, J. S. (2015). "Briefing: Standards in asset management: PAS 55 to ISO 55000." *Infrastructure Asset Management Journal,* ICE Publishing, 3(1), 57–59.

3 Municipal Contractual Practices

INTRODUCTION

Public infrastructure impacts the communities from the water they drink, sewer systems they use, roads they drive on, electricity they use, and many other services. Thus, the quality and reliability of those assets play a pivotal role in determining the quality of life. Municipalities are continuously challenged with maintaining the assets to meet the taxpayers' expectations. Given the fact that assets have different deterioration natures, maintenance cycles, and procedures, and are heavily impacted by external factors such as environmental and climatic conditions, traffic, etc., it is financially inefficient for municipalities to hire permanent resources for maintaining their assets. Thus, municipalities usually undertake procurement to select contractors to complete the works. There are different procurement strategies and types depending on the scope and scale of works, funding levels, level of service, and asset type.

Before we get started with the current practices, let us start with what is a contract and why is it important? A contract is a binding document that guides the relationship between two parties (in that case municipality and contractor) to perform certain service(s) and/or work(s). It also allocates the roles and responsibilities of each party, remuneration for the scope of work, schedule to complete the scope of work, etc. The purpose of this chapter is not to discuss the different types of contracts and the pros/cons of each one, as this is widely available, but to highlight the currently used contractual practices and shed lights on PPP and PBC along with their different models as a potential solution to close the infrastructure financial gap.

CURRENT CONTRACTUAL PRACTICES

Contractual practices vary from one municipality to another and even within the same municipality, different departments can have different practices. There are different types of contracts used for public works as highlighted in Figure 3.1.

IN-HOUSE

In-house contracts refer to the works that are carried out from within the municipality by their own workforce. Municipalities mostly use this type of contract to respond to typical preventive maintenance and emergencies (e.g., potholes repair, pipe breaks).

TENDERING

The other types of contracts always require some sort of tendering to be subject to competition and ensure the best value for taxpayers' money. Tendering is a

DOI: 10.1201/9781003327509-4

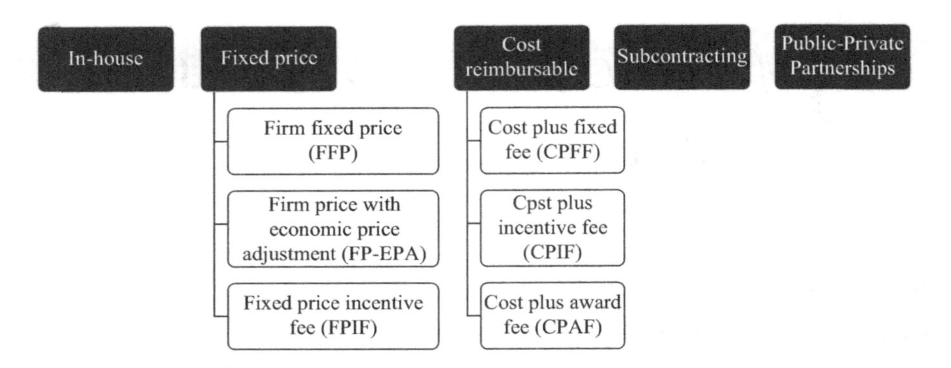

FIGURE 3.1 Types of contracts for public works.

commonly used type of contract and it has four types: (1) single-source tendering or direct award, (2) open tendering, (3) selective tendering, and (4) negotiated tendering (Constructor 2017). Single-source tendering or direct awards exist when a contract is awarded to a contractor without competition, or where there is a material change to an existing contract. In most cases, municipalities are obliged to publish all the direct award contracts for which they are responsible for, with values greater than a pre-set value, which varies from one municipality to another. However, this type of contract is rarely used to ensure transparency for public projects or guarantee fair competition to obtain the best value for the taxpayers' money.

The second type "open tendering" allows employers to advertise their proposed projects and permit all interested contractors to apply for tender documents. Sometimes, employers call for a deposit from applicants, which is returned on receipt of a bona fide tender. However, this method wastes the contractors' resources since many contractors may spend time preparing tenders and lose. Furthermore, knowing that their chances of gaining the contract are small, contractors may not study the contract in detail to work out their minimum price but simply quote a price that will be certain to bring them profit, if they are awarded the contract. Thus, employers may receive only "a lottery of prices," which is not necessarily the lowest price. If they choose the lowest tender, they take the risk of an improperly studied contract to appraise the risks involved, or the tenderer might not have the technical or financial resources to successfully complete the work. Cost consultants may think about the risks that all such low bids could prove unsatisfactory, but they cannot advise the employer what another bid to accept because they lack the certainty of information on the submitted bids.

The third type "selective tendering" allows employers to advertise their projects and invite contractors to apply and be shortlisted to bid for the project. The selected contractors should meet the pre-qualification criteria, set prior to the bidding invitation. The advantage of this type is the guarantee that the selected contractors have adequate experience, are financially sound, and have the resources and skills to carry out the work. Moreover, it motivates the contractors to thoroughly study the tender documents and put forward their keenest prices, given that they have reasonable chances of gaining the contract. However, since contractors have all been pre-qualified, it is difficult to reject the lowest bid, even if it appears dubiously low unless it is due to some obvious mistake. One of the key issues with both open and selective tendering is that contractors' circumstances can change after tender submission. They are prone to either lose other contracts, which might affect their financial stability, or succeed in other tenderers and thus would not have enough skilled resources to deal with all the work they were awarded.

The last type "negotiated tenders" functions through employers inviting contractors of their choice to submit prices for a project. In most cases, this type of tender is used for specialized work or when there is a very tight deadline or emergency work. In that case, the selected contractors have a good chance of being satisfactory given the previous satisfactory work performed with the municipality. When invited to tender, the contractor submits their price, and all the queries are discussed and usually settled without difficulty. Thus, mistakes in pricing can be reduced, as both both the engineer (cost consultant), advising the employer, and the contractor are confident that the job should be completed within the budget, if no unforeseen troubles arise. However, negotiated tenders for public works are rare because the standing rules of the public authorities do not normally permit them.

Fixed-Price Contracts

As we have a solid understanding of tendering, it is time to investigate the different categorizations of contracts, namely fixed-price and cost-reimbursable contracts. Fixed-price contracts involve setting a fixed total price for a defined scope of work. They are normally used when the scope of work is well defined and there are no significant changes to the scope expected during the execution of the contract. There are different forms of fixed-price contracts: firm fixed price (FFP), fixed price economic price adjustment (FPEPA), and fixed price incentive fee (FPIF). The first variation, and the most used type for municipal contracts, is FFP. In FFP contracts, the price for the scope of work is set at the outset and is not subject to change unless the scope of work changes. The second variation of fixed-price contracts is FPEPA. FPEPA is very similar to FFP but with a special provision allowing for the pre-defined final adjustments to the contract price due to changed conditions, such as inflation-related cost increases or decreases for certain commodities (e.g., asphalt, concrete, steel, wood, etc.). Over the past few years and due to the unstable economic situation worldwide arising from COVID-19, the Russia-Ukraine war, etc., this type of contract has been increasingly used for construction works, especially large projects with works that will take years to complete. The last variation of fixed-price contracts is FPIF. In FPIF contracts, the municipality and contractor have some flexibility to allow for deviation from performance, with financial incentives tied to achieving agreed-upon metrics. In most cases, those financial incentives are tied to cost or schedule, as the municipality does not want to minimize service impacts on the end users. Under FPIF contracts, there is usually a guaranteed maximum price set in the contract, and all the costs above that guaranteed maximum price are solely incurred by the contractor, unless it is resulting from a variation order or changes to the original scope of work.

The key advantages of fixed-price contracts are: (1) assurance for both parties over the contracts, (2) increased confidence level that the aspects of the contract including the duration and costs will not change, and (3) greater understanding and control of their own expenses, allowing the project to remain within the budget. From a contractor's standpoint, these types of contracts require contractors to disclose a great amount of detail on how they will fulfill their responsibilities under the contract, thus providing less opportunity for misinterpretation by either party that can expose the contractor to unnecessary litigation and fines. On the other hand, the key disadvantages of fixed-price contracts are: (1) they are mostly awarded based on price, and so if the contractor fails to precisely estimate the price, they will likely incur a loss, and (2) they are highly competitive and the contractor shall submit a bid price that covers work involved and a reasonable profit, while still being lower than the competitors.

Cost-Reimbursable Contracts

Cost-reimbursable contracts involve payments (cost reimbursements) to the contractors for all legitimate actual costs incurred for completed work, plus a fee representing the contractor's profit. There are different forms of cost-reimbursable contracts: cost plus fixed fee (CPFF), cost plus incentive fee (CPIF), and cost plus award fee (CPAF). The main difference among those types is the form by which the fee is calculated. The first variation of cost-reimbursable contracts is CPFF. In CPFF contracts, the contractor is reimbursed for all costs associated with performing the scope of work and receives a fixed-fee payment calculated as a percentage of the initial estimated project costs. In most cases, the fee amounts do not change unless the project scope changes. Another variation of the cost-reimbursable contractors is CPIF. In CPFF contracts, the contractor is reimbursed for all costs associated with performing the scope of work and receives a pre-determined incentive fee based on achieving certain performance objectives as defined in the contract. If the final costs are less or greater than the original estimated costs, both the municipality and contractor share costs based upon a pre-negotiated cost-sharing formula (e.g., 40%/60% split over/under target costs based on the actual performance of the contractor). The last variation of the cost-reimbursable contracts is CPAF. In CPAF contracts, the contractor is reimbursed for all legitimate costs, but most of the fee is earned based on the satisfaction of certain subjective performance criteria that are defined and incorporated into the contract. The determination of the fee is based solely on the subjective determination of the contractor's performance by the municipality.

Cost-reimbursable contracts are beneficial when uncertainties make it difficult for costs to be estimated (e.g., lack of knowledge of the work needed to meet the contract requirements, significant changes in the scope of work are expected during the execution of the contract). There is minimal risk for contractors under cost-reimbursement. Furthermore, unexpected costs/resources the contractors might incur during the performance of the works are paid for by the municipality. Finally, contractors are not forced to cut corners to complete projects with the added stress of ensuring they do not diminish their profit margin. From the municipality's standpoint, the key disadvantage is the inability to know the cost of completing the works and the uncertainty as to what it pays for.

SUBCONTRACTING

Subcontracting is another way of transferring the risk to a subcontractor. It is used in mega-scale projects when the range of required capabilities for a project is too diverse to be carried out by a single general contractor. In such cases, subcontracting parts of the project may assist in keeping costs under control and mitigate the overall project risk.

PUBLIC–PRIVATE PARTNERSHIPS

Public–private partnerships (PPP) or (P3) are *long-term, performance-based approach to procuring public infrastructure that can enhance governments' ability to hold the private sector accountable for public assets over their expected lifespan* (PPP Canada 2017). The main objective of this type of contract is to engage the private sector in delivering and operating huge infrastructure projects to benefit from their expertise, innovation, discipline, and incentives of capital markets. It transfers a huge share of the risks, associated with infrastructure development (e.g., cost overruns, schedule delays, unexpected maintenance, and hidden defects in assets) to the private sector. The idea is to engage the private sector in a bundled contract throughout the asset lifecycle. The contractual payments for operations and/or maintenance are linked to the quality of the original construction. The pros and cons of this type of contract are summarized in Table 3.1.

TABLE 3.1
Pros and Cons of PPP

Pros	Cons
Use of private expertise and capital	Varying priorities between public and private sector
Generation of cost--efficiency gains	Partnership agreements are hard to reach with no one-size-fits-all model
Risk transfer to the private sector	Public sector is exposed to financial risks of private sectors
Competition and creation of incentives for innovation	Transparency issues and discrepancies in the selection of private partners
Establishment of long-term commitments from private sector	
Improved understanding of public needs and regulations	
Creation of jobs	
Governments do not pay for the asset until it is built and operational	
Huge portion of the contract is paid out over a long term, and only if the asset is properly maintained and performs well under the pre-defined set of metrics	
Operation and maintenance costs are known upfront, and thus, taxpayers are not bearing the risk of any costs that might unexpectedly arise during the contract period	

There are three main aspects to consider prior to selecting PPP. Those aspects are (1) ownership (privatization models), (2) operations (partnership model), and (3) revenue costing (business model). In the coming sections, we will be discussing the different privatization, partnership, and business models for PPP.

Privatization Models

Privatization models are the models that distribute the ownership between the public and private sectors and highlight the tasks each party will undertake. There are four privatization models: (1) formal, (2) functional, (3) partial material privatization, and (4) material privatization. In all the privatization models except material privatization, the ownership of the assets is distributed between the public and private sectors. The ownership percentage is agreed upon as a part of the terms and conditions and depends on many other factors, such as business model, scope of work, asset lifecycle, PPP duration, etc. The difference between those models is the tasks of the private sector throughout the asset lifecycle. In the formal privatization model, the private sector is responsible to finance and transfer the ownership after the contract duration. Mortgage is an example of a formal privatization model where the banks lend the end users money to purchase a house for an interest rate over a long period of time. Once the end users pay all the costs and interest of the house, the bank fully transfers the ownership back to the end user. In the functional privatization model, the private sector is responsible for designing, financing, building, operating, maintaining, and transferring at the end of the contract duration. The contract duration typically ranges between 10 and 30 years depending on the asset type. In partial material privatization, the private sector designs, finances, builds, operates, and maintains the asset without transferring it back to the public. This type is usually used when assets will be at the end of their service life at the end of the contract duration or when the public still needs to be involved in certain ways. Finally, in the case of the material privatization model, the private sector fully holds the ownership of the asset and is responsible for designing, financing, building, operating, and maintaining the service. Examples of material privatization models are parking, hospitals, and airports that are built and operated by private companies. The public sector is only involved in defining the service levels to ensure that the end users receive adequate service. A summary of the privatization models is shown in Table 3.2.

TABLE 3.2
Types of Privatization Models

	Ownership		Tasks				
Model	**Public**	**Private**	**Design**	**Finance**	**Build**	**Operate/Maintain**	**Transfer**
Formal	X%	100 – X%		x			x
Functional	X%	100 – X%	x	x	x	x	x
Partial material privatization	X%	100 – X%	x	x	x	x	
Material privatization		100%	x	x	x	x	

Partnership Models

Partnership models describe the type of partnership between the public and private sectors. There are three partnership models: (1) contractual partnership, (2) institutional partnership, and (3) horizontal partnership. The contractual partnership is a form of vertical partnership, where PPP is used as an outsourcing contract. In such a case, functional privatization is the ideal ownership model, as the public sector keeps ownership and the private sector performs their tasks while meeting the predetermined metrics. An example of a contractual partnership is Tampa Bay Water and Veolia Environmental Services. To improve the quality of the surface water treatment facility's raw water supply, Tampa Bay Water awarded a third agreement to Veolia Environmental Services for the operation and maintenance of a 15-billon-gallon raw water reservoir. Tampa Bay Water owns the facility and Veolia is paid for the environmental services provided over the contract duration. The institutional partnership is a mix of vertical and horizontal partnerships, where a special-purpose private company with mixed public–private shares is created. The special-purpose private company has a vertical partnership with the public partner. Within the special-purpose private company, shareholders agree on the distribution of public and private shares. In this partnership model, functional privatization is the optimal ownership model as the public sector shares the ownership with the private sector, while this special-purpose private company designs, builds, finances, operates, and maintains the asset. Examples of services running under that model are most of the metro and passenger train operators such as Metrolinx in Toronto and Sociėtė Transport de Montreal. Those private companies were created by the provincial government to operate the collective transportation services (e.g., buses and metro fleets) in their cities. The horizontal partnership is a form of partnership, where the public partner permanently transfers the ownership to the private partner to design, build, finance, operate, and maintain. In this partnership model, partial material privatization is the sole ownership model, as the private sector does not transfer back the ownership to the public partner and an agreement on the shares is discussed as a part of the contract. An example of this privatization model is Düsseldorf International Airport located in the city of Düsseldorf. The city of Düsseldorf has made a horizontal partnership with Hochtief and Aer Rianta to operate and maintain the airport. The city still owns a 50% stake in Düsseldorf International Airport and the private stakeholders own the other 50%. Figure 3.2 summarizes the various partnership models.

Business Models

Business models are the mechanisms through which the private sector generates revenues. There are two types of business models, namely (1) owner-based income and (2) user-based income. For owner-based income, the Capital Expenditures (CAPEX) and Operational Expenditures (OPEX) are compensated by the public owner. Within the owner-based income business model, there are four forms of payments as follows.

- **Performance-based payments**: Payments are made based on meeting certain performance metrics. This form of payment is mostly used for maintenance services. Most of the maintenance contracts for municipal

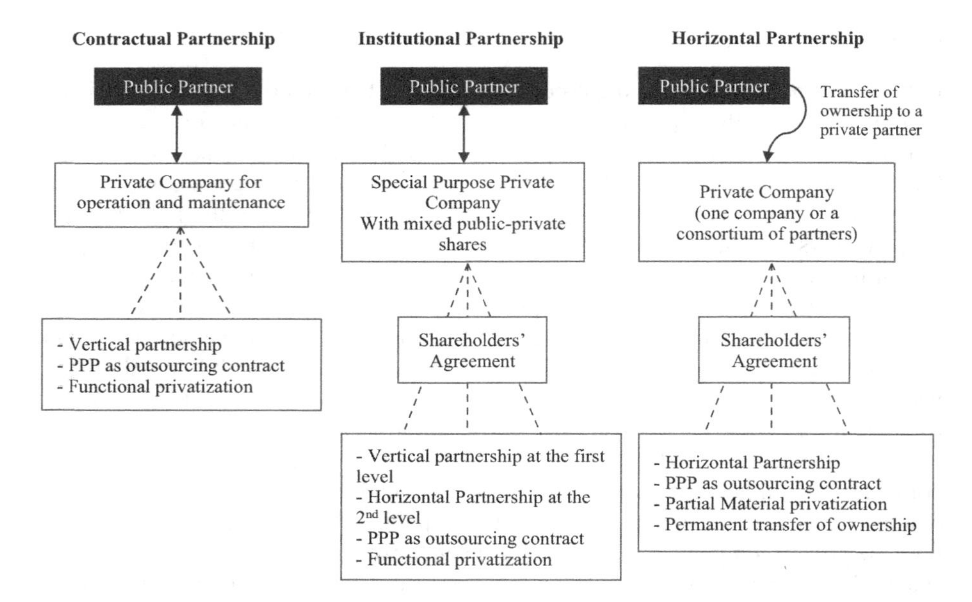

FIGURE 3.2 Privatization models.

infrastructure are based on performance-based payments. Examples of performance metrics could be the pavement condition index for roads, the building condition index for facilities, the number of leaks/breaks per month for water pipes, and so on.

- **Availability-based payments**: Payments are made based on the availability of the service. The most common example of this form of payment is for buses and trainsets, where the availability of the buses or trainset to enter service is the sole indicator for the contractor to receive their payment. Failing to meet the availability target results in deductions to the service payment. Another example is the security service, where the contractors are paid for the availability of their resources on-site.
- **Volume-based payments**: Payments are made based on the volume delivered. The most common example of this form of payment is catering, where the owner pays the contractor based on the delivered volume.
- **Results-based payments**: Payments are made based on the results of the service provided. The most common example of this form of payment is cleaning, where the owner pays the contractor on the basis of the results.

On the other hand, for the user-based income, the CAPEX and OPEX are compensated by the user fees. Within the user-based income business model, there are three forms of payments as follows.

- **Compulsory usage**: Payments are made by the public end users to use the service. The most common example of a compulsory usage form of payment is the toll roads. In this case, the tolls are used to reimburse the contractor for CAPEX (construction costs), OPEX, and profit margin.

- **Quasi-compulsory usage**: Payments are made by the end users to use the service. The most common example of a quasi-compulsory usage form of payment is parking. In this case, the parking fees are used to reimburse the contractor for CAPEX, OPEX, and profit margin.
- **Free choice of usage**: Payments are made by the end users to use the service. The most common examples of free choice of usage form of payment are the metro or passenger trains. In this case, the fee is collected by the operator to cover the capital purchase costs, OPEX to maintain the fleet, as well as the profit margin.

PERFORMANCE-BASED CONTRACTS (PBC)

PBC is a special type of contract that was conceptually designed to increase both the efficiency and effectiveness of infrastructure maintenance. It is like the PPP but limited to the operation and maintenance of the assets, without construction. Thus, it targets the operation and maintenance of the already-built infrastructure. It is *a type of contract that focuses on the outputs, quality, and outcome of the service provision and may tie at least a portion of the contractors' payment as well as any contract extension or renewal to their achievement.* (Martin 2003). In other words, it is a *type of contract under which the maintenance contractor undertakes to plan, program, design, and implement maintenance activities to achieve specified short and long-term condition standards for a fixed price, subject to specified risk allocation* (Frost and Lithgow 1998). It sets forth the final expected performance rather than directing the maintenance contractor with the methods and materials to achieve the expected performance.

It has been extensively applied to the pavement sector in numerous countries such as Canada in 1988, Argentina in 1990, Finland in 1998, and Zambia in 1999. Given its limited application to other infrastructure sectors, the roads will be taken as an example to explain the contractual scheme and setup. Typically, PBC covers an array of activities needed to maintain a road service quality level for users. The main PBC activities are displayed in Figure 3.3. Those activities include (1) initial rehabilitation works, which are carried out prior to signing the contract for bringing up the road to certain pre-defined standards; (2) regular maintenance services, which include all the activities related to the management and evaluation of the road under the contract, as well as the physical activities to maintain the agreed service quality levels; (3) improvement works, which are specified by the employer to add

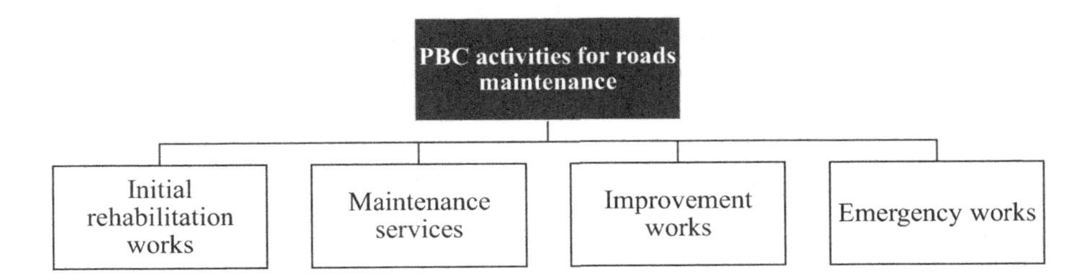

FIGURE 3.3 Key PBC activities – Pavements.

new characteristics to the roads (e.g., adding a lane for new traffic, safety, or any other considerations); and (4) emergency works, which include any activity needed to reinstate the roads after any damages resulting from unforeseen natural phenomena with imponderable consequences, such as extreme freeze and thaw, strong storms, flooding, earthquakes, etc. (Abu-Samra et al. 2017).

The financial tender should be presented under the following five categories: (1) initial rehabilitation works, which is represented through a lump-sum amount. The maintenance contractor should indicate the quantities of measurable outputs that will be executed to achieve the pre-defined contractual performance standards; (2) maintenance services, which is represented in the form of monthly lump-sum payment in case of meeting the pre-defined contractual metrics; (3) improvement works, which is represented in a form of unit prices for each improvement work type. The payment for the improvement works will be calculated based on these unit prices defined by the maintenance contractor in the signed contract documents; (4) emergency works, which are represented through unit prices in a conventional bill of quantities. The payment for the emergency works is decided on a case-by-case basis given the uncertainty of the estimated quantities; and (5) price adjustment, which is a clause defined in the contract to compensate the contractors for any increase in the cost indices. This clause is applicable to all the above-mentioned prices and activities.

The successful implementation of PBC is prerequisite on both well-defined metrics and a balanced penalties and incentives (P/I) system. The metrics aim at accurately evaluating and assessing the service quality level performed by the contractor. Those metrics shall be Specific, Measurable, Achievable, Realistic, and Timely to schedule (SMART). Throughout the last decade, it has been noticed that municipalities are lacking proper control over the contractors through poorly defined metrics, inadequate incentives, and high penalties. The tendency to place high penalties associated with low incentives indirectly forces the contractors to significantly increase their cost contingencies to cover any risks they might encounter throughout the contract duration.

PBC has been globally applied in many nations and it has proven to be cost-effective for the municipalities, as it displayed a fair amount of financial and administrative savings (Abu-Samra 2015). A summary of the benefits could be condensed into the following points.

- Partially transfers the risk of non-compliance with the service quality standards to the contractor.
- Reduces the overall maintenance cost through the economy of scale. In addition, it secured long-term funding for maintenance programs.
- Introduces the concept of performance risk sharing through the P/I system.
- Expands the role of the private sector by introducing a new area of work, where road maintenance was always the role of the public sector. This creates an opportunity for contractors to efficiently plan their work to both meet the agreed metrics and increase their profit margins.
- Increases the efficiency and effectiveness of maintenance operations with the municipalities having the upper hand opportunity through well-defined metrics.
- Provides municipalities with better budget certainty, as the monthly expenses are pre-defined in the contract along with pre-defined rates for urgent/emergency works.

Numerous scholars suggested that PBC is a potential solution to close the infrastructure financial gap for many reasons including but not limited to (1) contractors have a wider knowledge base and expertise as opposed to municipalities and thus are aware of cheaper and better processes to achieve the desired outcomes in a more efficient manner; and (2) competitiveness between the contractors motivates them to submit the least financial offer. The main performance warranties for PBC to reduce the risk on the municipalities could be summarized in the following points.

- Allows the contractor to deliver the project using their own best practices.
- Maximizes the contractors' innovation as the contractors get incentives in case of promoting any innovation throughout the contract duration.
- Risk is fully transferred to the party having much control over the project.
- Cost-effective for municipalities and contractors and increased probability of attaining cost savings while reaching the desired metrics, enhancing the network condition, and transferring the risk to the contractor.
- Minimizes the negative impact of the infrastructure projects on the public, as strict metrics are in place on the contractor to reduce the service disruption duration, resulting in shorter driving times through and around work zones.
- Minimizes the administrative costs needed for bidding, administrating, and managing numerous short-term individual contracts.

As discussed in the previous chapter, risk is an important consideration for municipalities. Thus, it is important to showcase the degree of risk associated with the different types of contracts, as shown in Figure 3.4. In-house maintenance has increased the risk to the municipality as opposed to PBC, which poses minimal risk to the municipality and increased risk to contractors. The figure shows the risk distribution among the different contractual approaches.

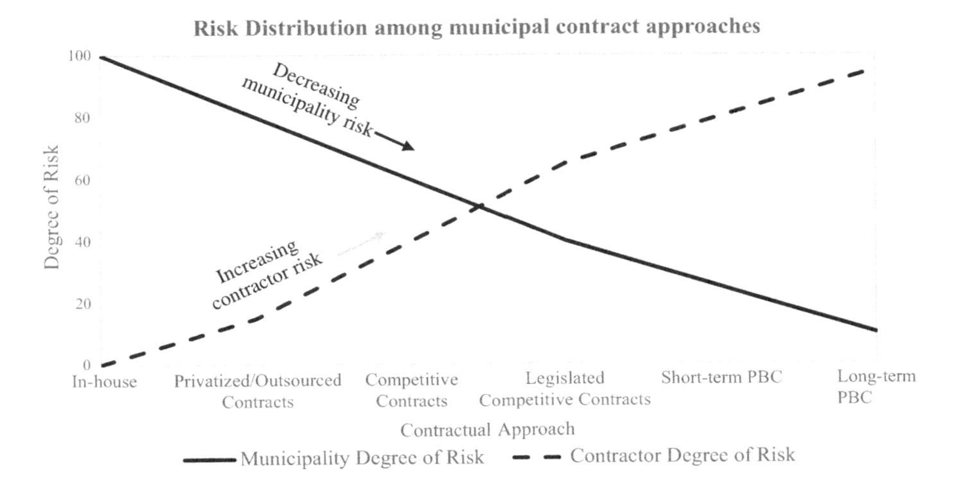

FIGURE 3.4 Risk distribution among different types of contracts.

REFERENCES

Abu-Samra, S. (2015). *Integrated Asset Management System for Highways Infrastructure*. LAP LAMBERT Academic Publishing, Saarbrcken.

Abu-Samra, S., Osman, H., and Hosny, O. (2017). "Optimal maintenance and rehabilitation policies for performance-based road maintenance contracts." *Journal of Performance of Constructed Facilities,* ASCE, 31(1). https://ascelibrary.org/doi/10.1061/%28ASCE%29CF.1943-5509.0000928#core-collateral-references

Constructor. (2017). "3 Types of Tendering Method in Construction." https://theconstructor.org/construction/types-of-tendering-methods-in-construction/6372/. Jan 24, 2023.

Frost, M. and Lithgow, C. M. (May 1998). "Future Trends in Performance Based Contracting-Legal and Technical Perspective." *IRR Conference* (p. 2), IRR, Sydney.

Martin, L. L. (2003). *Making Performance-Based Contracting Perform: What the Federal Government Can Learn from State & Local Governments* (pp. 1–20). The Price WaterHouse Coopers Endowment for the Business of Government, Washington, DC.

PPP Canada (Public Private Partnership). (2017). "About P3." https://www.p3canada.ca/en/about-p3s/. Oct 19, 2020.

Part 2

Infrastructure Interdependency, Coordination, and Contractual Frameworks

4 Infrastructure Interdependency

INTRODUCTION

Continuity and reliability of infrastructure systems are key to enhancing nations' economies and providing citizens with the quality of life they deserve. As per the U.S. President's Commission (PCCIP 1997), an infrastructure system is defined as *a network of independent, mostly privately-owned, man-made systems and processes that function collaboratively and synergistically to produce and distribute a continuous flow of essential goods and services.* Most countries have a defined set of critical infrastructure systems, and any incapacity or destruction of those infrastructure systems would have a debilitating impact on the nation's defense and economic security. This criticality is even further amplified as those systems are congested due to the increasing population and demand growth. Even though the set of critical infrastructure systems might vary from one country to another depending on a multitude of factors, most the countries' critical infrastructure systems include but are not limited to transportation networks, water supply systems, natural gas and oil, electric power systems, telecommunications, banking and finance, government services, and emergency services (Hasan and Foliente 2015).

Coordination of intervention activities has been extensively considered within the broader context of dependency and interdependency relationships among infrastructure systems as discussed earlier. Various computational approaches have been employed to study these interdependencies, including simulation, econometrics, network-based analysis, and system-of-systems modeling. Ouyang (2014) conducted a comprehensive review of interdependency modeling approaches, categorizing them into the following.

- **Simulation-based approaches**: Agent-based simulation, continuous simulation, and system dynamics (Macal and North 2002).
- **Econometric approaches**: Input–output models (Haimes and Jiang 2001; Santos and Haimes 2004).
- **Network analysis approaches**: Advanced geospatial analysis, hydraulic network modeling, and geometric proximity analysis (Islam and Moselhi 2012; Jeong et al. 2006; Duenas-Osorio et al. 2007).
- **System-of-systems approaches**: Integrated systems-of-systems (SOS) modeling, high-level architecture (HLA) simulation, and integrated generalized transportation network and system of systems (Eusgeld et al. 2011; Friesz et al. 2001).

DOI: 10.1201/9781003327509-6

TYPES OF INFRASTRUCTURE INTERDEPENDENCY

Infrastructure systems are dependent and interdependent in many ways. So, what is the difference between dependency and interdependency? Simply, dependency indicates the unidirectional relationship between infrastructure systems while interdependency refers to bidirectional interaction among infrastructure systems. In most cases, scholars use interdependency to describe the relationship among the infrastructure systems whether unidirectional or bidirectional. There are different schools in classifying infrastructure interdependency as detailed in Table 4.1. Furthermore, Figure 4.1 summarizes the types of interdependencies among the infrastructure systems.

TABLE 4.1
Categorization and Types of Infrastructure Interdependencies

Authors	Interdependency Types	Definitions	Examples
Rinalidi et al. (2001)	Physical	The state of one infrastructure system is dependent on the material output(s) of another infrastructure system	E1, E3
	Cyber	The state of one infrastructure system depends on information transmitted through the information infrastructure	E2, E3
	Geographical	A local environmental event can cause state changes in two or more infrastructure systems	E6
	Logical	The state of one infrastructure system depends on the state of another via a mechanism that is not physical, cyber, or geographic	E4, E5, E7, E8, E9, E10
Zimmerman (2010)	Functional	The operation of one infrastructure system is necessary for the operation of another infrastructure system	E1, E2, E3
	Spatial	The proximity between infrastructure systems	E6
Dudenhoeffer et al. (2006)	Physical	There is a direct linkage between infrastructure systems from a supply/consumption/production relationship	E1, E3
	Geospatial	There is co-location of infrastructure components within the same footprint	E6
	Policy	There is the binding of infrastructure components due to the policy of high-level decisions	E4, E5, E7
	Informational	There is a binding or reliance on information flow between infrastructure systems	E2, E3
Wallace et al. (2003)	Input	The infrastructure systems require as input one or more services from another infrastructure system to provide some other service	E1, E2

(*Continued*)

TABLE 4.1 (*Continued*)
Categorization and Types of Infrastructure Interdependencies

Authors	Interdependency Types	Definitions	Examples
	Mutual	At least one of the activities of each infrastructure system is dependent upon each of the other infrastructure systems	E3
	Shared	Some physical components or activities of the infrastructure systems used in providing the services are shared with one or more other infrastructure systems	E7, E10, E8
	Exclusive or (XOR)	Only one of two or more services can be provided by an infrastructure system, where XOR can occur within a single infrastructure system or among two or more systems	E8
	Co-located	Components of two or more systems are situated within a prescribed geographical region	E6
Zhang and Peeta (2014)	Functional	The functioning of one system requires input from another system or can be substituted, to a certain extent, by the other system	E1, E2, E3, E10
	Physical	Infrastructure systems are coupled through shared physical attributes, so that a strong linkage exists when infrastructure systems share flow right of way, leading to joint capacity constraints	E6
	Budgetary	Infrastructure systems involve some level of public financing, especially under centrally controlled economies or during disaster recovery	E4
	Market and Economic	Infrastructure systems interact with each other in the same economic system or serve the same end users, who determine the final demand for each commodity/service subject to budget constraints or are in the shared regulatory environment where government agencies may control and impact the individual systems through policy, legislation, or financial means such as taxation or investment	E5

EXAMPLES OF INFRASTRUCTURE INTERDEPENDENCY

How are those infrastructure systems interconnected and interdependent? A question many of us might be asking. For instance, the supply of electric energy is needed for the water, telecommunication, and traffic systems to maintain their normal operations, and the other way around where electric power systems need water and

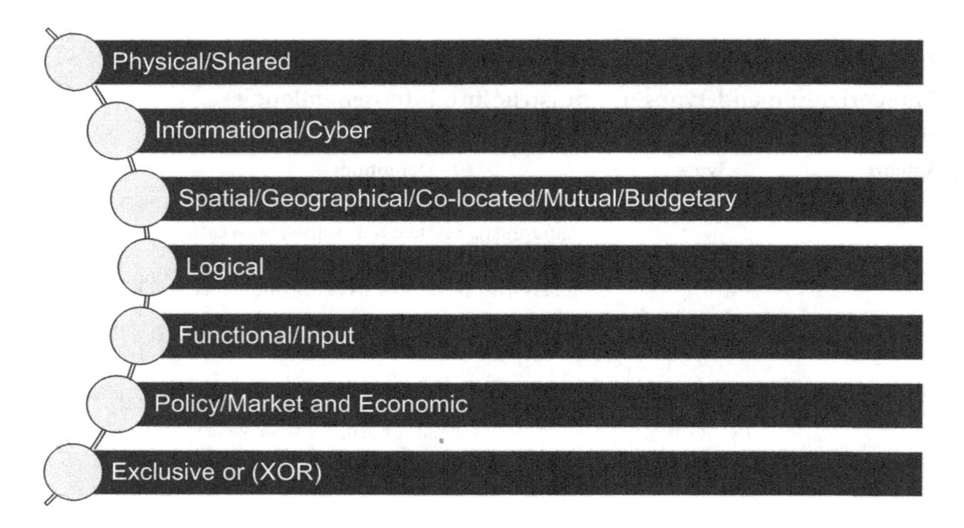

FIGURE 4.1 Types of infrastructure interdependency.

multiple telecommunication services to generate and deliver the electric power to users. To reinforce your understanding of the different types of interdependencies and given the fact that some interdependencies are invisible in normal operations and only emerge and become obvious under disruptive scenarios, some extreme events along with the evidence for each interdependency type will be used for illustration. The extreme events include the 1998 Ice Storm in Canada that caused parts of Ontario, Quebec, and New Brunswick, and the Northeastern United States experience one of the worst ice storms in recent history (Senesac 2022), the 2001 World Trade Center Attack that led to the collapse of the twin towers and damage of numerous other buildings and utilities at the World Trade Center site (Britannica 2021), the 2003 North American Blackout that lasted up to 4 days in various parts of the eastern USA and Canada (Minkel 2008), the 2004 hurricane season in Florida that included a series of hurricanes, such as Charley, Frances and Jeanne, within a short period of approximately 2 months (National Weather Service 2024), the 2007 UK floods that struck much of the country during June and July (Wright 2017), and the 2010 Chile Mw 8.8 Earthquake that caused coastal regions to both uplift and subside, and tsunami waves to hit the low lying Chilean coastline as well as distant shores across the Pacific Ocean (Pallardy and Rafferty 2024). All those events cost billions of dollars in economic losses. Table 4.2 lists some examples of interdependencies during and after the events.

While many scholars have studied infrastructure interdependency during the operational phase, focusing on how system disruptions propagate through related networks, less attention has been given to interdependencies during interventions (e.g., repair, rehabilitation, replacement) in terms of geographical and temporal dimensions. Despite the temporal dimension's direct impact on spatial, physical, and financial aspects, it has not been thoroughly explored.

TABLE 4.2
Examples of Interdependencies During and After Major Events

Example	Description
E1	Outages in power systems would result in the failure of traffic signals, water supply pumping stations, automated teller machines, and electric vehicles, and might even result in the closure of businesses
E2	Disruptions in communication services would impact the awareness and control of electric power and water systems and result in partial failures due to a lack of observability
E3	Electricity loss would lead to the interruption of communication services, which would further impact emergency communications and restoration coordination of power systems
E4	During the restoration process, electric power systems and communication services are usually given repair priority relative to other infrastructure systems and receive more investment for improvement and retrofit
E5	Outages in power systems led to price changes in food and fuels
E6	Water-main breaks flooded co-located utility systems (e.g., in the case of the World Trade Center attack, the water flooded rail tunnels, a commuter station, and the vault containing all the cables for one of the largest telecommunication nodes in the world)
E7	Emergency services distribute emergency resources to restore various types of damaged utility systems (e.g., in the case of the World Trade Center attack, the New York Waterway with 24 boats dispatched some to work as floating ambulances from piers in Lower Manhattan, and others to go to Hoboken, Hunts Point in Queens, and the Brooklyn Army Terminal)
E8	Debris-covered streets would hinder the emergency response personnel, end users, and financial district workers and might accordingly disrupt the financial services
E9	Failure of gas stations to pump fuel would result in drivers scrambling to find other functional gas stations and cause traffic congestion
E10	Closure of some metro stations would increase the traffic load on the bus transportation system and result in longer than usual lineups at bus stops

REFERENCES

Britannica, T. (2021). "September 11 Attacks Summary." *Encyclopedia Britannica.* https://www.britannica.com/summary/September-11-attacks. Oct 20, 2023.

Dudenhoeffer, D., Permann, R., and Manic, M. (2006) "CIMS: A Framework for Infrastructure Interdependency Modeling and Analysis." *Proceedings of the 2006 Winter Simulation Conference,* Institute of Electrical and Electronics Engineers, Piscataway, NJ.

Duenas-Osorio, L., Craig, J. I., Goodno, B. J., and Bostrom, A. (2007). "Interdependent response of networked systems." *Journal of Infrastructure Systems*, ASCE, 13(3), 185–194.

Eusgeld, I., Nan, C., and Dietz, S. (2011). "System-of-systems approach for interdependent critical infrastructures." *Reliability Engineering and System Safety,* Elsevier, 96(6), 679–686.

Friesz, T. L., Peeta, S., and Bernstein, D. H. (2001). "Multi-Layer Infrastructure Networks and Capital Budgeting." Working Paper TF0801A, George Mason University and Purdue University, Fairfax, VA.

Haimes, Y. Y. and Jiang, P. (2001). "Leontief-based model of risk in complex interconnected infrastructures." *Journal of Infrastructure Systems*, ASCE, 7(1), 1–12.

Hasan, S. and Foliente, G. (2015). "Modeling infrastructure system interdependencies and socioeconomic impacts of failure in extreme events: emerging R&D challenges." *Natural Hazards Journal*, Springler Link, 78, 2143–2168.

Islam, T. and Moselhi, O. (2012). "Modeling geospatial interdependence for integrated municipal infrastructure." *Journal of Infrastructure Systems*, ASCE, 18(2), 68–74.

Jeong, H., Qiao, J., Abraham, D., Lawley, M., Richard, J.P., and Yih, Y. (2006). "Minimizing the consequences of intentional attack on water infrastructure." *Computer-Aided Civil and Infrastructure Engineering*, 21(2), 79–92.

Macal, C. and North, M. (2002). "Simulating Energy Markets and Infrastructure Interdependencies with Agent-Based Models." *Proceedings of the Workshop on Social Agents: Ecology, Exchange and Evolution* (pp. 195–214), Chicago, IL.

Minkel, J. R. (2008). "The 2003 Northeast Blackout--Five Years Later." *Scientific American*. https://www.scientificamerican.com/article/2003-blackout-five-years-later/. Oct 24, 2023.

National Weather Service. (2024). "20 Year Anniversary of the 2004 Hurricane Season." National Ocean and Atmospheric Administration. https://www.weather.gov/tbw/hurricanes2004. Jul 24, 2024.

Ouyang, M. (2014). "Review on modeling and simulation of interdependent critical infrastructure systems." *Reliability Engineering and System Safety*, Elsevier, 121, 43–60.

PCCIP (1997). Critical foundations: Protecting America's infrastructures: The report of the president's commission on critical infrastructure protection. *Journal of Global Information Technology Management*, 1(1), 49–50.

Rinalidi, S., Peerenboom, P., and Kelly, T. (2001). "Identifying, understanding and analyzing critical infrastructure interdependencies." *IEEE Control Systems*, 21(6), 11–25.

Santos, J. and Haimes, Y. (2004). "Modeling the demand reduction input–output (I– O) inoperability due to terrorism of interconnected infrastructures." *Risk Analysis*, 24(6), 1437–1451.

Senesac (2022). "The Great Ice Storm of 1998." National Weather Service Heritage. https://vlab.noaa.gov/web/nws-heritage/-/the-great-ice-storm-of-1998. Oct 19, 2023.

Pallardy, R. and Rafferty, J. (2024). "Chile Earthquake of 2010." *Encyclopedia Britannica*. https://www.britannica.com/event/Chile-earthquake-of-2010. Jul 24, 2024.

Wallace, W., Mendonça, D., Lee, E., Mitchell, J., and Chow, J. (2003). "Managing disruptions to critical interdependent infrastructures in the context of the 2001 World Trade Center Attack." Beyond September 11th: An Account of Post-Disaster Research.

Wright (2017). "The Wettest Summer on Record -10 Years on from the 2007 Floods." *BBC News*. https://www.bbc.com/news/uk-england-40548635. Oct 21, 2023.

Zhang, P., and Peeta, S. (2014). "Dynamic and disequilibrium analysis of interdependent infrastructure systems." *Transportation Research Part B: Methodological*, ScienceDirect, 67, 357–381.

Zimmerman, R. (2010). "Social implications of infrastructure network interactions." *Journal of Urban Technology*, Taylor & Francis, 8(3), 97–119.

5 Coordination Framework for Municipal Infrastructure

INTRODUCTION

Despite recent increases in public infrastructure investments, municipal infrastructure is deteriorating faster than it is being renewed. Several factors contribute to this decline, including inadequate funding, population growth, stricter health and environmental standards, subpar quality control during installation, insufficient inspection and maintenance, and inconsistencies in design and operation practices. Additionally, the burden on infrastructure intensifies due to significant growth in certain sectors, accelerating the aging process and increasing the social and financial costs associated with service disruptions, maintenance, repairs, or replacements (Brinkman and Sarma 2022).

The provision of reliable services is paramount for the people's quality of life and nation's economy. As discussed in the previous chapter, infrastructure systems are independent and function collaboratively and synergistically to provide services to end users. In this chapter, we will consider three of the key services that we use daily (transportation, drinking water, and sewer/wastewater). The infrastructure systems providing those services are interdependent in different ways, but the fact that they share the same corridor (spatial location) is key. For instance, if a watermain or sewer main pipe breaks, would it impact the laid-above road section? Absolutely. If the municipality is replacing a water or sewer pipe, would that not disrupt the road? Wouldn't you consider that a waste of taxpayers' money if the road had just been replaced few years ago? Those are just examples of three assets, there are many other interdependencies among various infrastructure systems that should not only be considered while constructing those systems but also throughout their service life.

This chapter starts by outlining the current practices for infrastructure works. Then, it discusses the best practices and benefits of coordinating infrastructure renewal programs. Furthermore, it outlines a systematic program for rehabilitation or reconstruction of an infrastructure system near the end of its physical life while considering the other interdependent infrastructure systems to minimize disruption and maximize value. It also defines the various coordination scenarios among the roads, water, and sewer networks. Finally, it presents a framework for coordinated asset management planning and decision-making and highlights the typical activities required to build a decision-making system.

DOI: 10.1201/9781003327509-7

COORDINATION BENEFITS

Infrastructure is simply chaos, and we are dealing with the government, municipalities, construction companies, lawyers, consultants, and public end users (taxpayers) all at the same time. Thus, adopting new practices shall result in benefits for most of the involved stakeholders, especially taxpayers. In this section, we will highlight the benefits of coordinating the infrastructure renewal programs, especially for interdependent infrastructure systems, as detailed in Table 5.1:

TABLE 5.1
Coordination Benefits

Benefit Category	Benefit Sub-Category	Description
Financial and temporal savings	Reduced repair, rehabilitation, and replacement costs	Reductions in the lifecycle costs arise from the economy of scale, avoidance of duplicated activities, primarily in pavement repair. Those savings could enable more projects to be implemented and thus, reduce the huge infrastructure deficit
	Reduced disruption time and user costs	Infrastructure projects often cause unavoidable physical disruptions, resulting in unanticipated costs for users. These costs encompass various factors, such as lost time, missed business opportunities, increased fuel consumption, and environmental and social impacts (such as traffic disruption, noise, and air pollution). By enhancing coordination, we can potentially reduce the frequency and duration of corridor disruptions, leading to a significant reduction in these adverse effects
Funding benefits	Potential coordination with new development work	Coordination of the long-term infrastructure works with development works could capitalize on possible efficiencies and have the new development work fund some long-term infrastructure priorities. For instance, development works related to land-use change could potentially fund the infrastructure works required to bring up the existing infrastructure systems to the state of good repair and later expand their capacity
	Full cost accounting	Roads have always been a challenging infrastructure system to maintain appropriate funding levels, as its traditional funding source is a highly sensitive tax base. Historically, roads have topped the list in terms of the percentage of roads in fair-condition states and the increasing deficits arising from the delayed repair and rehabilitation of those assets. For instance, the impact of the repair works to underground utilities on the lifecycle of the road infrastructure is not captured in most traditional cost-sharing practices even though those utilities can be on a user-pay basis and have a more secure funding source. Some practices would transfer some long-term funding requirements from the more sensitive road area to the underground utilities as they have dedicated funding sources. This shift is fair as it would increase the balance in funding infrastructure priorities and improve the overall level of service to the community

(*Continued*)

TABLE 5.1 (*Continued*)
Coordination Benefits

Benefit Category	Benefit Sub-Category	Description
	Funding approval process	Several practices involve approval processes, which can significantly increase the flexibility and coordination procedures surrounding these issues. Better funding approval procedures that allow planning for individual projects to occur earlier and more cost-effectively result in significant benefits. They also have the potential to reduce administrative costs associated with the approval process and to increase the opportunities for coordination mid-year, which have direct financial benefits
Municipalities benefits	Enhanced decision-making	The inevitable result of coordination techniques is improved education, and sensitivity of infrastructure providers and project managers in one utility area of the needs and considerations in other areas. This, in turn, led to improved decision-making, even before any specific coordination efforts were undertaken
	Lower risk of failure to infrastructure systems	Poor coordination among infrastructure systems led to unplanned failures and increasing costs of maintenance. Coordinated planning leads to enhanced capital expenditures, lower unplanned failures, and accordingly lower risk of failure to the infrastructure systems
Public benefits	Enhanced perception of infrastructure providers	Inadequate coordination negatively impacts the public image of infrastructure providers. Since public perception often influences a local council's decisions and behavior, any enhancements in coordination efforts yield long-term benefits for all public works service providers
	Improved awareness for the need of lifecycle rehabilitation and replacement strategies	Budgeting procedures are becoming stricter, requiring justification for the needs (e.g., maintenance, rehabilitation, and replacement projects). This led to increased awareness of infrastructure needs and long-term benefits of undertaking lifecycle costing analysis
	Enhanced level of service	Better budget allocation and reductions to maintenance costs will result in an enhanced level of service. This enhancement is a result of municipalities having access to more funds for undertaking more projects to enhance the infrastructure level of service

RISKS ARISING FROM COORDINATION

Most of us working in the municipal world have been exposed to significant public complaints about the lack of effective coordination among various infrastructure systems. Depending on the methods of applying the infrastructure renewal programs' coordination, there is a possibility that several risks and consequences might arise, as described in Table 5.2.

TABLE 5.2
Coordination Risks

Risk Category	Risk Sub-Category	Description
Financial risks	Increased administrative costs	Coordination will add a layer of administrative costs arising from the staff time and direct funding. The increase in the staff workload and costs associated with establishing some of the committees is even higher in larger urban areas, where several committees will be set up on various aspects
	Lost opportunity costs	Failing to follow the best coordination practices can lead to higher costs for individual projects and diminish the resources available to address essential requirements
	Premature replacement	Coordination may result in replacing some individual infrastructure systems before the end of their service life. This may offset some benefits gained through increased coordination, reduced disruption, and reduced maintenance costs. Therefore, undertaking proper analysis is paramount for deciding the degree to which these practices should be followed
Decision-making risks	Skewed priorities and imbalanced funding	In numerous communities, inadequate funding makes it challenging to distribute infrastructure renewal efforts across different program areas effectively. In extreme situations, a significant portion of program funding may be entirely allocated to coordinating works related to development or other specific program areas. As a result, there are limited resources available to address the remaining utility needs. This lack of flexibility can hinder coordination with other utilities while addressing individual utility requirements
	Reduced flexibility	Implementing certain restrictive practices, such as "moratorium" may decrease operational flexibility but could also invite criticism. It is essential to manage expectations carefully to ensure they align with what can realistically be achieved
Stakeholders' risks	Opposition from external utilities	Given that various utilities operate with distinct cost centers and mandates compared to municipalities, resistance can arise. This challenge can escalate into a significant issue, depending on the level of resistance, and accordingly consume substantial time and resources. Therefore, it is crucial to handle coordination practices carefully to prevent any deterioration in relationships

CURRENT PRACTICES IN ASSET MANAGEMENT

It is paramount to study the current practices to better understand the state-of-the-art methodologies and technologies for municipal infrastructure planning, design, construction, management, assessment, maintenance, and rehabilitation that consider local economic, environmental, and social factors (FCM 2013).

Corridor Upgrades

A fully coordinated upgrade involves upgrading all the infrastructure systems on a specific road or in a geographic area at the same time. It has significant benefits with respect to maximizing coordination and minimizing disruption. However, special considerations shall be taken prior to undertaking premature replacements to ensure that the economic life lost to early replacement does not exceed the economic benefits resulting from improved coordination. In situations where an underground utility is approaching the end of its remaining service life, extensive economic analysis should be undertaken to evaluate and justify complete corridor renewal and rehabilitation. Refinement on the corridor approach includes the installation of utilidors, a linear utility chamber constructed to accommodate a variety of utilities (e.g., hydro, telephone, cable, and steam heat), and the upgrading of many blocks on a particular street or an entire neighborhood at the same time.

Most of the municipalities look for full coordinated upgrade opportunities to upgrade all the infrastructure systems on a specific road. However, the trigger to qualify the corridor seems to vary significantly depending on the specifics of the municipality. For instance, some municipalities start with the roads program and once a road is identified, specific reviews are conducted for the other internal programs, such as water, sewer, and drainage, with priority given to upgrading as many elements as possible. Other municipalities start with the water program. In those cases, the full corridor upgrade starts with the specific underground program, and the opportunity is taken to repave the entire roadway when the underground utility is complete. Even though municipalities are taking steps to look for full coordinated upgrade opportunity, the practice varies significantly depending on several factors including but not limited to balanced funding availability, the age and condition of infrastructure systems, varying service lives, etc.

While most of the municipalities believe that complete corridor upgrades are the best practice for their community as it maximizes the coordination benefits and minimizes disruption to their community, some municipalities have articulated their concerns. Those concerns are centered toward the economic life lost due to premature replacement of some infrastructure systems. In many cases, the economic benefits of the fully coordinated upgrade are not sufficient to offset the lost life of some infrastructure systems. To take an informed decision, municipalities shall conduct an economic analysis to trade-off economic life lost due to premature replacement and the cost avoided by the repeated pavement repairs and social disruption to the area. Furthermore, the effects of a full coordinated upgrade on revitalizing the area and encouraging other investments in the area should also be considered, especially in areas with high-growth potential.

Another approach is undertaking partial corridor upgrades where some but not all the infrastructure systems on a specific road are upgraded. In those cases, it is recommended to complete a check on all other utilities and rectify any deficiencies before closing out the corridor. Another approach that provides enhanced economies of scale is to seek approval for upgrading for many blocks of a particular street or an entire neighborhood at the same time, which we call "horizontal integration". This provides construction efficiencies and concentrates the disruption to the community to a very specific time frame.

Coordination Practices

Efficiently coordinating co-located assets and utilities is crucial. Across different municipalities, a diverse range of coordination practices aims to deliver the necessary level of service (value) while optimizing costs. These practices encompass various approaches, including management, financial strategies, economic considerations, engineering, and other relevant factors applied to physical assets.

- *Multi-Year Plans*
  Effective coordination of various programs relies on developing multi-year plans with specific projects. However, this practice varies significantly among municipalities. Some municipalities project plans up to 10 years ahead, while others focus on shorter 1-year horizons. Unfortunately, 1-year horizons lack sufficient lead time for effective long-term coordination and joint opportunities. Most municipalities aim for 3-to-5-year plans. Some external utility owners struggle to produce plans beyond 1 or 2 years due to factors like unpredictable customer demand. Despite this, municipalities often project further ahead, even though strict service levels can hinder the development of effective long-term intervention programs (beyond 5 years) for coordinated planning.

  Once municipalities finalize their multi-year plans, they establish a formal circulation system. Each system's plan is shared with others to ensure that pending underground work is completed before street work begins. Additionally, municipalities communicate extensively with external stakeholders, including users and overlapping utilities, to highlight project specifics. While this practice is more common in smaller areas with fewer projects, it ensures a final check on coordination issues before construction starts. In high-growth municipalities, development-related committees coordinate ongoing works with capital programs. Some municipalities accept cash in lieu of required work from developers to align development-related efforts. Occasionally, an annual budget is allocated for capital works alongside development projects.

  In summary, multi-year capital plans not only facilitate program coordination but also mitigate the impact of political shifts on upcoming priorities.
- *Governance*
  Governance is one of the techniques that municipalities build to adopt coordination. Several municipalities establish committees with representation from asset owners and utility companies supporting the city's program. This approach ensures that there are open lines of communication between the asset owners. For instance, the committee would have representative from the water, roads, sewer, and drainage service areas that would be able to communicate and jointly develop a consolidated and coordinated asset management plan to consider the potential coordination of works while accounting for the assets' varying deterioration patterns, service life, etc. Another form is to establish what we call a "joint utility coordination committee" to focus on the relationship with utilities' companies to the municipal

program. Those would include all the agencies that are responsible to maintain infrastructure that the municipality does not own. In some cases, those committees are combined in forms like "neighborhood improvement committees" and the focus is on coordinating the works within neighborhood to streamline the works and minimize the disruptions.

Restrictive Practices

Another way of reinforcing the coordination is through law, rules, and governing procedures. Several municipalities employ restrictive practices to enhance coordination and, more crucially, reduce the impact on a recently completed project over an extended period. It is worth noting that all the restrictive practices form incentives to minimize disruption to a particular road surface and enhance the coordination of various infrastructure systems.

- *Permit Requirements*
  Most of the cities would require excavators to obtain a permit from the local municipality before they can begin digging in roads. While the cost of the permit is usually minimal, it allows the municipality to regulate road excavation and implement additional policies as needed.
- *No-Cut Rules or Moratorium*
  Numerous municipalities implemented some form of a no-cut rule within their jurisdiction. This rule, also known as a moratorium on excavations, prohibits any digging or excavation activities for a specified period after pavement overlays, except in emergency situations. Typically, the ban on excavations lasts for 3 years, although in certain cases, it may extend to 5 years. The prevalence of the no-cut rule varied based on factors such as municipal culture, urban development levels (with high-development areas having fewer restrictions), and community sensitivity to disruptions. Each municipality had its own approval processes; some required council approval, while others allowed exceptions under specific circumstances. Recent studies revealed that even proactive cities experienced a significant percentage of their moratorium streets being re-excavated within 2 years of resurfacing. To avoid creating false expectations and negative public perceptions, a solid understanding and strict enforcement of the no-cut rule are essential.
- *Pavement Restoration Procedures*
  There are various mechanisms used by various municipalities when it comes to the actual road repair procedures. For instance, some municipalities oblige utility operators to repair the excavation in accordance with municipal specifications. Other municipalities undertake the final pavement restoration at the utility operators' expense. Other municipalities charge a flat fee for pavement repair, which transfers the responsibility for the final repair to the municipality in exchange of a per square meter fee to the utility operator. Even though all those approaches have their pros and cons, and it is difficult to indicate a preferred approach, municipalities tend to pay more

attention to the quality of the final repair than outside excavation agencies, as the municipality will ultimately inherit the road along with any deficiencies in the repair process.

- *Pavement Degradation Fees*
  A fee charged to agencies cutting the pavement in addition to the repair cost. It accounts for the reduction in the pavement infrastructure service life resulting from the excavation process. Reduction in the pavement infrastructure service life is an inherent by-product of utility cuts, and no matter how well a utility cut is repaired, the nature of the excavation and the disturbance to the subbase layer have a significant effect on reducing the overall service life of the pavement infrastructure. It has been widely recognized in the past infrastructure reports issued by America and Canada that road infrastructure is always in poorer condition than the underground utilities and is usually the more difficult area for raising funds due to the lack of a dedicated funding source, which is funding raised for a specific utility and restricted by a policy framework for use on one infrastructure component. Even though few Canadian municipalities are applying the pavement degradation fees, significant interest and support from the many municipalities has been realized as it assists in moving toward full cost accounting, which is a process that relates all the associated costs and effects of a particular program to its funding source, and appropriately charges the agencies responsible for long-term costs. It also encourages coordination among the various co-located infrastructure systems to avoid repeat fees.

There are several ways to quantify the pavement degradation fee. Some municipalities relate the fee to the age of the last pavement overlay. Others have a flat rate for ease of administration as it does not require a large overlay database. It is obvious that relating the fee to the age of the last overlay is a more accurate method of reflecting the true impacts of utility cuts on pavement life. Thus, we strongly recommend adopting the concept of a pavement degradation fee in addition to proper road repair procedures, as it is a worthwhile practice for municipalities to pursue. The choice of a flat or variable rate can be left to the discretion of each municipality, depending on their technological advancement and administrative processes.

Funding and Approval Process

A way of enforcing coordination is through adopting joint funding requests and approvals versus silo asset-based funding. Establishing dedicated funding sources is an ideal way to secure partial funding for assets through a direct link between the users and funding (e.g., the more you use the road and the more fuel/electricity you will consume). Another type of funding is "block funding". It focuses on establishing budgets and funding for an asset portfolio level as opposed to the individual asset class funding to enforce coordination among the service areas and ensure that service areas work jointly to estimate their needs and lifecycle costs.

- *Dedicated Funding Sources*
  Existing budgets often fall short of covering the lifecycle costs of replacing infrastructure components. However, funding challenges vary significantly among urban areas. Roads and drainage systems typically struggle more to secure adequate funding compared to sewer and water infrastructure as they benefit from dedicated funding through utility rates. Roads and drainage projects rely on general tax revenue and face competition from other program areas. Notably, roads receive higher funding levels when there is a specific funding source such as fuel tax. Establishing dedicated funding for various infrastructure services should be a top priority for all providers, as a direct link between users and funding tends to drive better support.
- *Block Funding*
  Block funding is the approval of budgets on a program level for roads, drainage, water/sanitary, etc., as opposed to at an individual project level. This allows for significant flexibility with respect to changing priorities among individual projects. The timing of approvals for different funding programs does not significantly hinder coordination efforts, as key coordination activities occur independently of the approval process. However, early approvals are crucial for effective coordination. The approval process for individual programs and projects also impacts coordination throughout the year. Some cities focus on block funding approvals, with individual projects submitted for informational purposes. These flexible arrangements allow municipalities to adjust individual projects based on late-breaking information during planning. This enhances coordination across program areas. In contrast, other cities specify which projects will be constructed each year, requiring council approval. This practice can limit coordination with external influences. To improve coordination, seeking program-level approvals and providing project details for information is recommended. Achieving this depends on the community culture and council dynamics, which may evolve over time.

Communication and Presentation of Infrastructure Needs

Many organizations regularly present their long-term infrastructure plans to their councils, providing a wealth of detail. As a minimum standard, each municipality should include the replacement value, expected lifespan, calculated lifecycle replacement target, proactive initiatives to meet the target, and the benefits of achieving it in their presentations. Comparing this budget to actual expenditures highlights infrastructure needs. Some organizations even formalize this difference as the "infrastructure deficit," reporting it annually. The level of political support varies widely, with some communities strongly backing infrastructure-related issues while others take little action upon receiving the information. The goal is not merely to secure specific funding levels but to raise awareness among councils and communities about infrastructure challenges. Continued deferral of these issues constitutes a deficit, and long-term support is crucial. Presentations emphasizing positive outcomes such as reduced emergency repairs and disruptions tend to resonate better than those focusing on negatives. Overall, awareness of infrastructure-related issues has significantly increased, with numerous initiatives driving progress.

Technical Aspects

In addition to all the previously mentioned practices, there are some technical aspects that shall be considered to establish a coordinated asset management plan.

- *Social and Environmental Costs*
  Many of the municipalities are aware of the social and environmental costs of their projects, but few of them attempt to quantify those costs. Environmental reviews are mostly undertaken because of a formal mandate from a higher authority. Social costs are mostly acknowledged but not considered in the planning process. One of the approaches to start quantifying the social costs is to consider them in the lifecycle costs including any penalties from impacted businesses or a lane rental charge that represents the traffic disruption to some degree. Over the past few years, several environmental review and assessment frameworks for infrastructure works have been established. This includes the "Envision" framework developed by the Institute for Sustainable Infrastructure.
- *Pre-installation of Lateral Service Connections*
  In areas slated for future land development, pre-installing lateral service connections can save time and costs by reducing the need for pavement cuts. Future development layouts help municipalities install the correct number of connections in the right locations. If layouts are unavailable, estimates with larger-than-standard building service connections can accommodate higher-than-expected demand. Often, the cost of pre-installations can be recouped from future developers. Municipalities experiencing high-growth rates sometimes install interim-sized utilities initially, upgrading them as demand increases. Interim utilities may be placed under temporary roads, which will be expanded later. In such cases, the final locations of utilities should be predetermined to minimize pavement cuts during corridor upgrades.
- *Utilidor and Trenchless Technologies*
  To minimize disruption to surfaces, installation of a utilidor to house a variety of utilities such as fiber optics, telephone, cable, and hot water for central heating could be a satisfactory solution. This type of installation is uncommon in North American urban infrastructure and is usually justified only in downtown cores, where utility space is at a premium or under extreme weather conditions. The benefits of utilidors include but are not limited to: (1) one-time construction of the corridor; (2) long-term access to utilities; (3) ease of maintenance; and (4) minimal disruption to above-ground infrastructure such as roads. Another approach to minimize the pavement cuts and minimize public disruptions is adopting trenchless technology for the pipes' replacement. There are various trenchless construction techniques to

rehabilitate or install underground utilities depending on many factors such as pipe material, depth, length, soil properties, etc.

PLANNING AND DECISION-MAKING

Enhancing lifecycle planning and fund allocation is becoming increasingly crucial for municipalities and utility operators to address growing challenges. Municipalities are financially strained due to a significant increase in underperforming and deteriorating assets, coupled with insufficient funds to cover the rising infrastructure deficit debt (Mirza 2007, 2009; Uddin et al. 2013; FCM 2019). In addition to aging and deteriorating assets, municipalities face other challenges: (1) infrastructure deficit; (2) growing population and urbanization; (3) increasing demands for higher service levels from taxpayers; and (4) smaller share of taxes compared to provincial and federal governments, despite being responsible for the largest share of public assets.

Given these financial challenges, coordinated planning of co-located municipal infrastructure networks, such as roads, water, and sewer, could enhance economic efficiency and effectiveness. However, coordinating interventions for multiple co-located assets presents challenges for decision-makers. The common practices, requirements, and technologies for roads, water pipes, and sewer pipes differ due to their unique characteristics, such as deterioration rates, design service lives, environmental reactions, usage nature, and lifecycle costs. Additionally, there are varying requirements related to quality, safety, environmental, and health regulations.

Integrated planning and funding of municipal projects should focus on optimal coordination of intervention timing, locations, types, and alternatives, as well as fund allocation. Higher coordination among co-located assets will improve municipal asset management efficiency and effectiveness, reducing economic losses. Increased coordination can also help close the infrastructure deficit in the long term. Studies suggest that even a 1% savings on the annual infrastructure budget could save billions over the next decades, both globally and nationally. However, inadequate coordination among municipal projects remains a common source of economic losses. Municipalities need further research on optimized planning and funding to maximize project coordination throughout asset lifecycles, minimizing economic losses, and increasing financial savings.

Recently, it has been observed that there is no clear definition of integrated asset management. The following definitions illustrate various perspectives on integrated asset management over the past decade:

- Danylo and Lemer (1998) viewed the primary role of an infrastructure asset management system as an "integrator," a system that can interact with and analyze outputs from various systems.
- Shen and Spainhour (2001) emphasized that tools and methodologies for infrastructure lifecycle management should integrate environmental, economic, and technical issues into a comprehensive solution.
- InfraGuide (2003a, 2003b, 2003c, 2004, 2006) recommended that all municipalities across Canada adopt an integrated approach to assess and evaluate their roads, sewer, and water systems. Treating these assets as one

integrated system is considered best practice for managing multiple infrastructure networks.
- Adopting integrated multi-disciplinary approaches has become essential for implementing efficient and sustainable asset management (Halfawy 2008, 2010; Halfawy et al. 2008; Shahata and Zayed 2010).
- The Institute of Asset Management, in collaboration with the British Standards Institution, published the first publicly available specification for optimized management of physical assets in 2004. This was updated by 50 organizations from 15 industry sectors in 10 countries (BSIgroup 2008). The Publicly Available Specification (PAS 55) defines asset management as "the systematic and coordinated activities and practices through which an organization optimally and sustainably manages its assets and asset systems, their associated performance, risks, and expenditures over their life cycles to achieve its organizational strategic plan" (BSIgroup 2015; Hastings 2014).
- Following the widespread adoption of PAS 55 in utilities, transport, and manufacturing industries, the International Standards Organization (ISO) used PAS 55 as the foundation for developing the ISO 55000 international standards series. Initially published in 2014, with a second edition released in 2024, ISO 55000 defines asset management as "coordinated activity of an organization to realize value from assets" (ISO 2024; NAMS 2006, 2007, 2014, 2015).

From this standpoint, integrated asset management research has focused on several perspectives, summarized as follows:

- **System Complexity**: Enhancing the management and balance of factors at both micro/project and macro/network levels.
- **Financial Reporting**: Optimizing asset intervention schedules to achieve near-optimum funding decisions and balancing different maintenance alternatives/schedules over the planning horizon (Halfawy 2008).
- **Information Management**: Addressing the "Big Data" issue by improving data integration processes across multiple departments (Michele and Daniela 2011).
- **Integration Factors**: Focusing on specific areas such as Geographic Information System (GIS) integration systems (Halfawy 2010), integrated risk-based decision-making (Shahata and Zayed 2016), and integrated condition rating (Elsawah et al. 2014).
- **Conflicting Perspectives**: Considering both community and municipal perspectives with changing objectives (Khan et al. 2015).

In light of those issues and with an objective of closing the gaps, it is paramount to take informed decisions, which requires the definition of the proper funding model and using the right planning tool. Many municipalities use planning tools such as computerized pavement management systems to prioritize pavement segments, pipe degradation, and capacity models for sewer and water pipes. Few cities are developing pilot projects to develop an integrated infrastructure management upgrade program that takes into consideration all the co-located infrastructure systems into one program. They have also restructured their public works and engineering departments to include an asset management branch. This focuses on integrated planning to holistically resolve the ongoing infrastructure-related issues and is effective to ensure the long-term focus on resolving the root cause of those issues.

To develop a planning and decision-making tool, there is a need to assess the condition of each infrastructure system and establish deterioration patterns that consider the social, environmental, climatic, and operational factors that impact the infrastructure continuously and systematically. There are many tools and technologies to assess the infrastructure condition. It varies depending on the type of infrastructure (e.g., above-ground versus underground infrastructure, and material). The assessment could be made through field observation and maintenance records to condition rating equipment. Thenceforth, the condition rating data are compiled into computer software. In conjunction with the planned deterioration patterns and the capacity upgrading needs defined through field monitoring and capacity models, municipalities use those condition rating data to develop capital programs. Normally, in case the infrastructure capacity requires an upgrade, municipalities consider the projected future demand growth in the range of 30 years to avoid early upgrades arising from failing to meet rising demands.

In summary, even partial integration among infrastructure works facilitates coordination among departments and avoids repetitive work. However, there is a need for a coordination framework that considers both silo, partial, and full coordination scenarios while developing the intervention program for the co-located infrastructure systems.

Coordination Scenarios

Before we start listing the coordination scenarios, it is important to recall the concepts of nCr formula, as this is how the coordination scenarios are derived from. It is simply the selection of things without considering the order of the arrangement. It is used to find all the possible arrangements where selection is done without order consideration. The nCr formula is presented in the equation below:

$$nCr = \frac{n!}{r!(n-r)!} \tag{5.1}$$

where n is the total number of assets; and r is the number of assets to be chosen out of n assets.

In our case, there are three assets. Thus, the total number of assets (n) is always equal to three. Let's now study the different possible combinations. The first case is maintaining only one asset at a time. Through applying the nCr formula, there are 3C1 combinations that equal to 3. This is just a simple application of the formula. The second case is maintaining more than one asset at a time but not all. Given that we have three assets only, the only possible scenario would be two assets at a time. Through applying the nCr formula, there are 3C2 combinations that equal to 3. The third case is maintaining all the assets at a time. Through applying the *nCr* formula, there are 3C3 combinations that equal to 1. As described, there are three main coordination scenarios are (1) conventional approach, (2) partial coordination; and (3) full coordination. All the possible combinations within those coordination scenarios are detailed in Table 5.3. In simple terms, the conventional approach represents the current municipal practices where the interventions' planning and implementation are undertaken for each asset separately, discarding their interdependencies and potential coordination savings. The partial and full coordination represents the coordination of more than one asset while planning and implementing the interventions.

Let's further demonstrate how it works if we add the sidewalks and underground electric transmission lines; the total number of assets (*n*) is always equal to five. By applying the *nCr* formula, decision-makers will have to trade off 31 scenarios for each corridor within the network every year, making the decision-making process complex.

TABLE 5.3
Coordination Scenarios and the Potential Combinations

Coordination Scenarios	Description	Combinations	Coordination Scenario Description
Conventional	The maintenance of each asset is planned and undertaken separately without considering the interdependencies among those systems	**Scenario 1**: Roads	The roads' interventions are planned separately
		Scenario 2: Water pipes	The water pipes' interventions are planned separately
		Scenario 3: Sewer pipes	The sewer pipes' interventions are planned separately
Partial coordination	The maintenance planning and implementation of more than one asset, but not all the potential assets, within the corridor are coordinated	**Scenario 4**: Roads and water pipes	The roads and water pipes' interventions are coordinated whereas the sewer pipes' interventions are planned separately
		Scenario 5: Roads and sewer pipes	The roads and sewer pipes' interventions are coordinated whereas the water pipes' interventions are planned separately
		Scenario 6: Water and sewer pipes	The water and sewer pipes' interventions are coordinated, whereas the roads' interventions are planned separately
Full coordination	The maintenance planning and implementation of all the existing assets within the corridor are coordinated	**Scenario 7**: Roads, water, and sewer pipes	The roads, water, and sewer pipes' interventions are fully coordinated

Multi-Objective Coordination Framework

Upon considering the added value the private sector can bring to close the infrastructure gaps and realizing the conflicting objectives of the various infrastructure stakeholders, a multi-objective coordination and decision-making framework for corridor infrastructure under PBC has been developed. The methodology rests on three core foundations as follows: (1) integrated PBC contractual scheme; (2) multi-dimensional assessment models; and (3) multi-objective optimization for PBC-based asset management. It spins around five main phases, as shown in Figure 5.1. The first phase is investigating the literature review with the objective of gathering information about various aspects such as optimization for asset management; municipal contractual practices, and asset management. The main outcomes of this phase are: (1) defining the municipal corridor infrastructure interdependencies and the current asset management contractual practices and (2) identifying the multi-objective optimization techniques and the coordination dimensions along with their assessment criteria. Hence after, the second phase is centered on the PBC parameters, which aims at analyzing the distinct factors that affect the PBC from asset owners, municipalities and maintenance contractors' perspectives. In this phase, the KPIs will be defined along with their deterioration patterns, inspection frequencies, and degrees of importance for being inputted into the multi-objective optimization model. The third phase is the contractual foundation, where an integrated PBC contractual scheme is established. The contractual scheme functions through a consortium between the municipal departments, a joint venture between the maintenance contractors, and a PBC between the municipal departments' consortium and maintenance contractors' joint venture. To assess the contractual scheme, the fourth phase takes place to combine the outputs of the second and third phases and develop multi-dimensional performance assessment models that rest on eight dimensions: (1) spatial, (2) temporal, (3) financial, (4) physical, (5) risk, (6) resilience preparedness, (7) efficiency, and (8) effectiveness. Those dimensions are the PBC multi-objectives, in other words KPIs. The fifth and final phase is building a PBC-based asset management system that functions through five main models as follows: (1) central database model that contains the data of the corridor infrastructure under study; (2) deterioration model that predicts the future condition state of each asset; (3) financial model that calculates the lifecycle costs (LCC) of the systems, corridors (e.g., group of systems), and network (e.g., group of corridors); (4) multi-dimensional performance assessment models that compute the state of the pre-defined KPIs throughout the planning horizon; and (5) optimization models that function through several optimization engines and act as a decision-support system for both municipalities and maintenance contractors in PBC structuring and maintenance planning phases.

Decision-Making System

What does the decision-making system offer to asset managers? It answers the three W's (Where, When, and What):

- Where should I intervene at each point of time? In simple words, which corridor should I pick at any point of time.

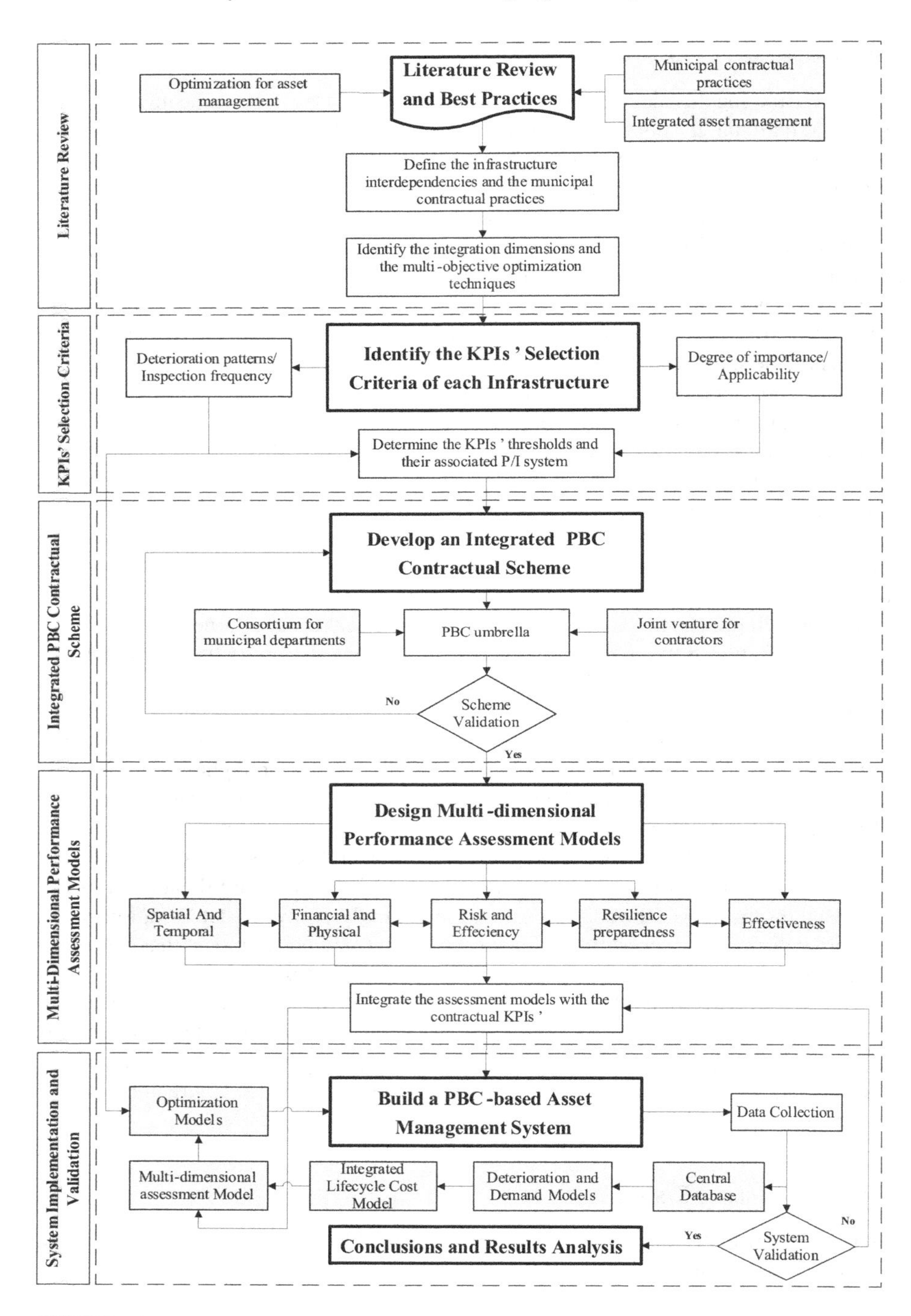

FIGURE 5.1 Multi-objective coordination and decision-making framework (Abusamra 2019).

- Which type of intervention should I choose for this particular asset? It depends on the available budget, the current state, the desired quality of service, etc.
- When should I intervene for each asset?

The answers to those questions would support asset managers to: (1) allocate the necessary budget to undertake those interventions; (2) understand the impacts of those interventions on the quality of service and communicate with the stakeholders; and (3) plan for their interventions (e.g., procurement activities for pipe replacements and road works, resources allocation, and resources loading). Now that we realize the benefits of having a reliable decision-making system, let's go through its' components:

- *Asset Inventory*
  Asset managers need to know what they have to be able to manage it. You cannot simply manage what you do not know and thus, failure to have a complete asset inventory will result in misleading decisions.
- *Inspection*
  Asset state changes frequently with physical, environmental, and climatic conditions. Thus, it is important to inspect the assets on a regular basis, recognize the distresses, and evaluate their deterioration pattern, and current state.
- *Condition Rating*
  Upon inspection, the inspection results are translated into a condition rating through uniform set of rules depending on each asset type. Those ratings are then used to decide the type of intervention required to bring back the asset to an acceptable condition state.
- *Future Prediction*
  Asset managers usually establish long-term plans for their assets and thus, knowing the current state is not enough to make informed decisions. Therefore, it is important to develop prediction models to forecast the asset state across the planning horizon.
- *Modeling and Metrics Evaluation*
  Modeling the impacts of aging, interventions, environmental and climatic conditions along with other events on the asset is a foundation for performing optimization and what-if scenarios. Those models are dependent on the metrics asset managers would like to track. For instance, if asset manager would like to track the asset condition state, they should develop deterioration models. Another example is financial models for tracking the asset lifecycle costs. Other examples include but are not limited to temporal models, spatial models, and risk models.
- *Optimization*
  The complexity arising from the uncertainties and the number of assets each municipality is managing mandates the need to undertake trade-off scenarios and compare the impacts on the asset portfolio. Given the fact that it is humanly impossible to undertake all those computations and come up

with an optimal solution that yields the best use of the limited budget, there is a need to use optimization algorithms that simulate various intervention scenarios based on your set of objectives and constraints and selects an optimal one.

- *Reporting and Visualization*
  Reporting and visualization play crucial roles in decision-making. Reports have transformed from being presented in tables, making analysis time-consuming and challenging, to creative data visualization. Asset managers now consolidate various data types into visually intuitive representations. Heatmaps, color-coded matrices, and interactive dashboards help convey complex concepts like risk exposure, diversification, and performance relative to benchmarks. These visualizations enhance both the decision-making processes and the stakeholders' understanding of asset portfolio performance by ensuring effective communication of asset reporting results.

REFERENCES

Abusamra, S. (2019). "Coordination and Multi-Objective Optimization Framework for Managing Municipal Infrastructure under Performance-Based Contracts." Doctoral of Philosophy Thesis, Concordia University, Montreal.

Brinkman, M. and Sarma, V. (2022). "Infrastructure Investing Will Never Be the Same." *Mckinsey and Company*, London, United Kingdom. https://www.mckinsey.com/industries/private-capital/our-insights/infrastructure-investing-will-never-be-the-same#/. Jun 10, 2023.

BSIgroup. (2008). "Publicly Available Specification for the Optimal Management of Physical Assets." https://www.irantpm.ir/wp-content/uploads/2014/01/pass55-2008.pdf. Feb. 23, 2021.

BSIgroup. (2015). "Moving from PAS 55 to BS ISO 55001 - The New International Standard for Asset Management - Transition Guide." https://www.4bt.us/wp-content/uploads/2018/01/Moving-from-PAS-55.pdf. Jun 28, 2020.

Danylo, N. and Lemer, A. (1998). "Asset Management for the Public Works Manager: Challenges and Strategies, Findings of the APWA Task Force on Asset Management." American Public Works Association (APWA). https://spectrum.library.concordia.ca/id/eprint/985084/. Jan 4, 2021.

Elsawah, H., Guerrero, M., and Moselhi, O. (2014). "Decision-Support Model for Integrated Intervention Plans of Municipal Infrastructure." *International Conference on Sustainable Infrastructure 2014* (pp. 1039–1050), Nov 6–8, ASCE, Long Beach, CA.

FCM (Federation of Canadian Municipalities). (2013). *Coordinating Infrastructure Works—A Best Practice by the National Guide to Sustainable Municipal Infrastructure*. Federation of Canadian Municipalities and National Research Council, Ottawa. https://fcm.ca/sites/default/files/documents/resources/guide/infraguide-coordinating-infrastructure-works-mamp.pdf. Jun 10, 2023.

FCM (Federation of Canadian Municipalities). (2019). "The Canadian Infrastructure Report Card." https://fcm.ca/en/resources/canadian-infrastructure-report-card-2016. May 12, 2022.

Halfawy, M. (2008). "Integration of municipal infrastructure asset management processes: Challenges and solutions." *Journal of Computing in Civil Engineering*, ASCE, 22(3), 216–229.

Halfawy, M. (2010). "Municipal information models and federated software architecture for implementing integrated infrastructure management environments." *Automation in Construction*, Elsevier, 19(4), 433–446.

Halfawy, M., Dridi, L., and Baker, S. (2008). "Integrated decision-support system for optimal renewal planning of sewer networks." *Journal of Computing in Civil Engineering*, ASCE, 22(6), 360–372.

Hastings, N. A. J. (2014). *Physical Asset Management: With an Introduction to ISO55000*, 2nd Edition. Springer International Publishing, Cham.

InfraGuide. (2003a). *An Integrated Approach to Assessment and Evaluation of Municipal Roads, Sewer and Water Networks*. FCM (Federation of Canadian Municipalities) and (National Research Council), Ottawa.

InfraGuide. (2003b). *Coordinating Infrastructure Works*. FCM (Federation of Canadian Municipalities) and (National Research Council), Ottawa.

InfraGuide. (2003c). *Planning and Defining Municipal Infrastructure Needs*. FCM (Federation of Canadian Municipalities) and (National Research Council), Ottawa.

InfraGuide. (2004). *Managing Infrastructure Assets*. FCM (Federation of Canadian Municipalities) and (National Research Council), Ottawa.

InfraGuide. (2006). *Managing Risks*. FCM (Federation of Canadian Municipalities) and (National Research Council), Ottawa.

ISO. (2024). *ISO 55000 Series: Asset Management— Vocabulary, Overview and Principles*, 2nd Edition. ISO, Geneva.

Khan, Z., Moselhi, O., and Zayed, T. (2015). "Identifying rehabilitation options for optimum improvement in municipal asset condition." *Journal of Infrastructure Systems*, ASCE, 21(2). https://ascelibrary.org/doi/10.1061/%28ASCE%29IS.1943-555X.0000220

Michele, D. and Daniela, L. (2011). "Decision-Support Tools for Municipal Infrastructure Maintenance Management." *World Conference on Information Technology (WCIT)*, vol. 3 (pp. 36–41), Oct 5–7, Elsevier, Istanbul.

Mirza, S. (2007). "Danger Ahead: The Coming Collapse of Canada's Municipal Infrastructure." FCM (Federation of Canadian Municipalities). https://www.fcm.ca/Documents/reports/Danger_Ahead_The_coming_collapse_of_Canadas_municipal_infrastructure_EN.pdf. June 12, 2016.

Mirza, S. (2009). "Canadian Infrastructure Crisis Still Critical." https://ascelibrary.org/doi/abs/10.1061/(ASCE)CO.1943-7862.0001402. Jan 18, 2017.

NAMS (NZ Asset Management Support). (2006). *Infrastructure Asset Valuation and Depreciation Guidelines*, 1st Edition. NAMS (New Zealand Asset Management Support): Wellington.

NAMS (NZ Asset Management Support). (2007). *Developing Levels of Service and Performance Management Guidelines*, 2nd Edition. NAMS (New Zealand Asset Management Support): Wellington.

NAMS (NZ Asset Management Support). (2014). *Optimised Decision-Making Guidelines*, 1st Edition. NAMS (New Zealand Asset Management Support): Wellington.

NAMS (NZ Asset Management Support). (2015). *International Infrastructure Management Manual*, 5th Edition. NAMS (New Zealand Asset Management Support): Wellington.

Shahata, K. and Zayed, T. (2010). "Integrated Decision-Support Framework for Municipal Infrastructure Asset." *Conference on Pipelines: Climbing New Peaks to Infrastructure Reliability: Renew, Rehab, and Reinvest* (pp. 1492–1502), ASCE, Keystone, CO.

Shahata, K. and Zayed, T. (2016). "Integrated risk-assessment framework for municipal infrastructure." *Journal of Construction Engineering and Management*, ASCE, 142(1), 1–13.

Shen, Y. and Spainhour, L. (2001). "IT: A potential solution for managing the infrastructure life cycle." *Journal of Computing in Civil Engineering*, ASCE, 15(1), 1–2.

Uddin, W., Hudson, W. R., and Haas, R. (2013). *Public Infrastructure Asset Management*, 2nd Edition. McGraw-Hill Education, Columbus, OH.

6 PBC-Based Contractual Scheme for Co-Located Assets

INTRODUCTION

PBC have become increasingly popular in recent years as a means of improving the quality and efficiency of public infrastructure maintenance. This type of contract is designed to incentivize contractors to achieve specific performance metrics by tying payment to measurable outcomes rather than simply completing a set of predetermined tasks. In the context of road, water, and sewer networks, PBC can be particularly effective in promoting cost-effective maintenance practices and improving service quality. By focusing on performance outcomes such as the road condition or water quality, PBC encourages contractors to use innovative technologies and strategies to achieve better results with limited resources. There are various benefits to using PBC for managing municipal infrastructure:

- **Improved Service Quality**: PBC incentivize contractors to focus on achieving measurable performance outcomes such as road condition or water quality.
- **Cost-Effective Maintenance Practices**: Contractors are encouraged to use innovative technologies and strategies to achieve better results with limited resources, leading to more cost-effective maintenance practices.
- **Reduced Risk for Public Agencies**: PBC shifts some of the risk from public agencies to contractors, who are responsible for achieving performance outcomes.
- **Greater Accountability**: Contractors are held accountable to achieve specific performance outcomes.

To fully exploit the benefits of PBC, municipalities and asset owners shall pay further attention to the following key aspects while designing and implementing the PBC:

- **Set Clear Performance Metrics**: Performance metrics should be clearly defined, measurable, and tied to specific outcomes. Public agencies should also establish benchmarks for acceptable performance.
- **Align Incentives**: Payment should be tied to achieving performance outcomes, and penalties should be established for failing to meet performance metrics. Public agencies should also consider offering bonuses for exceeding performance metrics.

DOI: 10.1201/9781003327509-8

- **Monitor and Evaluate Performance**: Public agencies should closely monitor contractor performance and evaluate outcomes to ensure that performance metrics are being achieved.
- **Foster Collaboration**: Public agencies should work closely with contractors to establish shared goals and promote collaboration. This can help build trust and ensure that both parties are invested in achieving performance outcomes.
- **Consider the Long-Term**: Public agencies should carefully consider the long-term sustainability of infrastructure maintenance practices when designing PBC. This may involve balancing short-term performance outcomes with long-term goals such as sustainability and resilience.

CONTRACTUAL SCHEME FOR INTEGRATED ASSET MANAGEMENT UNDER PBC

The integrated PBC contractual scheme revolves through three contractual models to fit the multi-asset nature and guarantee proper risk allocation among the parties involved. In this case, the parties involved are divided into two categories: (1) municipal departments, and (2) maintenance contractors. Due to the differences in how those co-located assets evolve throughout their lifecycles (e.g., service life, deterioration pattern, management scale, level of service, construction/intervention actions, etc.), asset management specialists are appointed to manage each asset type. However, given the fact that those assets are spatially interdependent, decisions shall not be solely taken and shall be coordinated. Thus, consortium, joint venture, and PBC were integrated, as shown in Figure 6.1, to guide the relation between municipal departments and maintenance contractors (Abudamra 2019). The first contractual model was the "consortium model among the municipal departments". The consortia should:

- Define common PBC asset-based performance metrics.
- Identify the asset-based thresholds.
- Set the annual intervention budget.

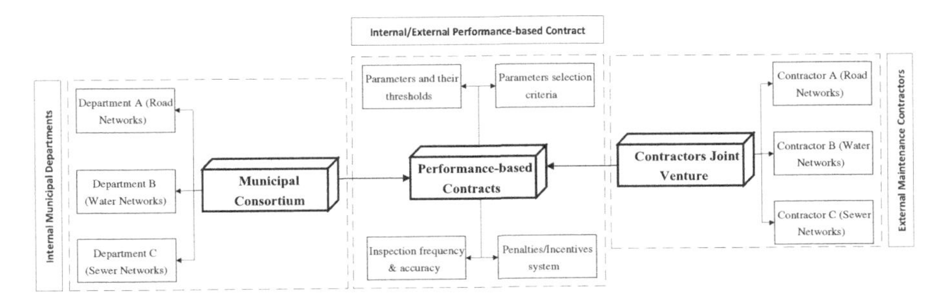

FIGURE 6.1 Contractual relationships (Integrated PBC, consortium and joint venture).

- Specify the inspection frequency and techniques for assessing the performance metrics.
- Classify the assets interdependencies.
- Outline the performance metrics and their associated penalties and incentives.
- Determine the optimal contractual period.

After completing this phase, the consortium should be able to draft a PBC that contains a set of performance metrics along with their thresholds, associated penalties and incentives, and prequalification documents for the maintenance contractors. Hence after, the prequalification phase takes place to check the capability of the proponents, in that case could be a general contractor or joint venture among various specialized contractors to handle the job, based on the prequalification criteria set earlier. Finally, the bidding process takes place and the proponent with the highest score is awarded the contract.

On the other hand, the maintenance contractors are bound together through a joint venture contract, which is "*a business entity created by two or more parties, generally characterized by shared ownership, shared returns and risks, and shared governance.*" (Roos et al. 2019). The objective of choosing the joint venture contractual model is that the risks and returns are shared among the parties, which guarantees more diligence for enhancing the cooperation to maximize the interventions' efficiency and share the profits and losses.

REFERENCES

Abusamra, S. (2019). "Coordination and Multi-Objective Optimization Framework for Managing Municipal Infrastructure under Performance-Based Contracts." Doctoral of Philosophy Thesis, Concordia University, Montreal.

Roos, A., Khanna, D., Verma, S., Lang, N., Dolya, A., Nath, G., and Hammoud, T. (2009). "Getting More Value from Joint Ventures." Transaction Advisors, Boston Consulting Group (BCG). https://www.bcg.com/publications/2014/m-a-divestitures-more-value-joint-ventures. May 25, 2022.

Part 3

Key Performance Indicators (KPIs): Measurement and Modeling

7 Criteria for KPIs Selection

KPIs SELECTION CRITERIA

Ensuring the effective delivery of public services relies heavily on choosing the appropriate KPIs. These KPIs serve as measurable benchmarks to evaluate performance and track progress against the desired goals. Given the fact that we are dealing with multi-asset with different characteristics (e.g., deterioration rates, useful lives, installation years, intervention costs, etc.), establishing a unified comparison criteria among the assets to assess their performance throughout the planning horizon is crucial. Furthermore, the KPIs shall be indicative and SMART to predict their annual performance from economic, financial, physical, and social perspectives, before and after applying different intervention plans. Thus, upon reviewing literature review and best practices, a set of rules were set to define the KPIs, as detailed in Table 7.1.

Based on those rules, the KPIs were categorized into the following: (1) financial, (2) temporal, (3) spatial, (4) physical, (5) risk, (6) resilience preparedness; (7) intervention efficiency, and (8) intervention effectiveness. Those indicators account for the multi-asset nature and consider the different stakeholders' preferences. Let's start by the temporal category, it represents the time needed to undertake

TABLE 7.1
KPIs' Rules (Abu-Samra et al. 2018)

Indicator Category	Indicative Rules
Accuracy	• The indicator should consider the precision level needed while measuring • The indicator should consider the difficulty level, represented either by the frequency or easiness of measurement
Frequency	• The indicator should frequently (e.g., annually) track the assets' performance throughout the planning horizon • The indicator's rate of change shall be considered (e.g., the KPI should experience a periodical difference in the asset state)
Financial	• The indicator should consider the costs needed for frequently measuring and controlling the asset
Ownership	• The indicator should have an owner, who is held liable/responsible for
Portability	• The indicator should fit multi-assets with different features and attributes such as deterioration rates, useful lives, construction years
Subjectivity	• The indicator should be objective-oriented and should include a pre-defined set of rules for measuring an asset attribute to guarantee a consensus agreement among different parties
Understandability	• The indicator should consider the easiness of understanding and tracing the triggers behind a sudden rise/fall throughout the assets' lifecycle

DOI: 10.1201/9781003327509-10

an intervention or the service disruption duration. The outcome of the temporal category feeds both the spatial and financial categories. Moving on to the financial category, which represents the impact of decreasing ownership costs and increasing operating and maintenance costs of the asset. It computes the LCC including all the direct and indirect costs associated with undertaking an intervention or service disruption (e.g., rehabilitation, user costs, replacement, etc.). The disruption/intervention duration and spatial extent are utilized to compute the indirect costs within the financial category. The spatio-temporal category computes the disruption/intervention area for the different intervention scenarios. The outcome of the temporal category, represented through the disruption/intervention duration, was utilized in the spatio-temporal model as the spatial extent was not enough to represent the savings. The physical category represents the corridor condition state, which is necessary for taking timely intervention decisions. The physical category includes future prediction models for the right-of-way corridor assets to compute the physical performance of the assets and take prompt corrective actions. The condition of the asset is represented as a percentage, such that "0%" represents an asset in a failing condition whereas "100%" represents an asset in an excellent condition. The risk category represents the probability and consequences of failure for any assets within the corridor. It represents the corridors, including all the right-of-way assets, with a risk index to assist asset managers in their decision-making process. The resilience preparedness category represents the preparedness of the asset to adapt with demand changes taking place due to the urbanization (e.g., population growth, land use change) and climate change (e.g., increased rainfall intensity and frequency). The resilience preparedness of the asset is represented by a percentage between the demand and supply, such that the demand is computed from an urban hydrological model whereas the supply is simply based on the pipe diameter and the capacity of the water supply station or wastewater treatment plant, depending on the asset under investigation. For instance, a ratio of "90%" implies that the current supply is barely meeting the demand, which indicates the need to replace the existing with larger diameters to meet the demand. However, a ratio of "110%" implies that the demand exceeds the supply, which is troublesome in both sewer or stormwater systems and even more complicated in the combined sewer and stormwater systems because of the overflooding that will occur around the pipes' surrounding areas. The intervention efficiency category represents the performance of the intervention plan in terms of resources' utilization. Thus, it reflects the percentage of time consumed to undertake the intervention activities over the total disruption period to restore all the right-of-way assets. Finally, the intervention effectiveness represents the quality/performance of the intervention. In simpler terms, it reflects the amount of operating time expected until the next major intervention, without undertaking any disruptive intervention activities after the intervention program is accomplished. Table 7.2 summarizes the KPIs' categories along with their performance measures and their associated description and ranges.

TABLE 7.2
KPIs' Categories and Performance Measures (Abusamra 2019)

Performance Measure	KPI	Description	Value/Range
Financial	LIF	Potential cost savings that can be attained due to integrating common activities while undertaking the intervention	No savings: LIF = 1 Savings possible: LIF > 1
Temporal	NCR	Potential time savings that can be attained due to integrating common and parallel activities while undertaking the intervention	No savings: NCR = 1 Savings possible: NCR > 1
Spatial	STIF	Time and space savings that can be attained while undertaking a fully coordinated intervention as opposed to several conventional interventions	No savings: STIF = 1 Savings possible: STIF > 1
Physical	CIF	Condition improvement due to undertaking fully coordinated intervention as opposed to several conventional interventions	No savings: CIF = 1 Improvement possible: CIF < 1
Risk	RIF	Risk improvement due to undertaking fully coordinated intervention as opposed to several conventional interventions	No savings: RIF = 1 Improvement possible: RIF > 1
Resilience Preparedness	RPIF	Resilience preparedness improvement due to expanding the network capacity to meet the increasing demand	No savings: RPIF = 1 Improvement possible: RPIF > 1
Efficiency	IEF	Proxy for the nuisance caused by repeated interventions executed for the same corridor within a short time span	Highest efficiency: IEF = 1 Lower efficiency: IEF < 1
Effectiveness	IFF	Proxy of time without any disruption once the intervention program is completed	Higher effectiveness: IFF<< Lower effectiveness: IFF>>

REFERENCES

Abu-Samra, S., Ahmed, M., Hammad, A., and Zayed, T. (2018). "Multiobjective framework for managing municipal integrated infrastructure." *Journal of Construction Engineering and Management*, ASCE, 144(1).

Abusamra, S. (2019). "Coordination and Multi-Objective Optimization Framework for Managing Municipal Infrastructure Under Performance-Based Contracts." Doctoral of Philosophy Thesis, Concordia University, Montreal.

8 Condition and Resilience Preparedness KPIs

CONDITION KPI

The objective of this KPI is to track the condition state/reliability of each asset/system within the corridor. To compute the condition/reliability, an integrated deterioration model is developed for the corridor, including all the n_s systems, as shown in Figure 8.1. It features n_s deterioration models for all the n_s systems and compiles their outcomes to a corridor condition state (CCS) based on the weights of

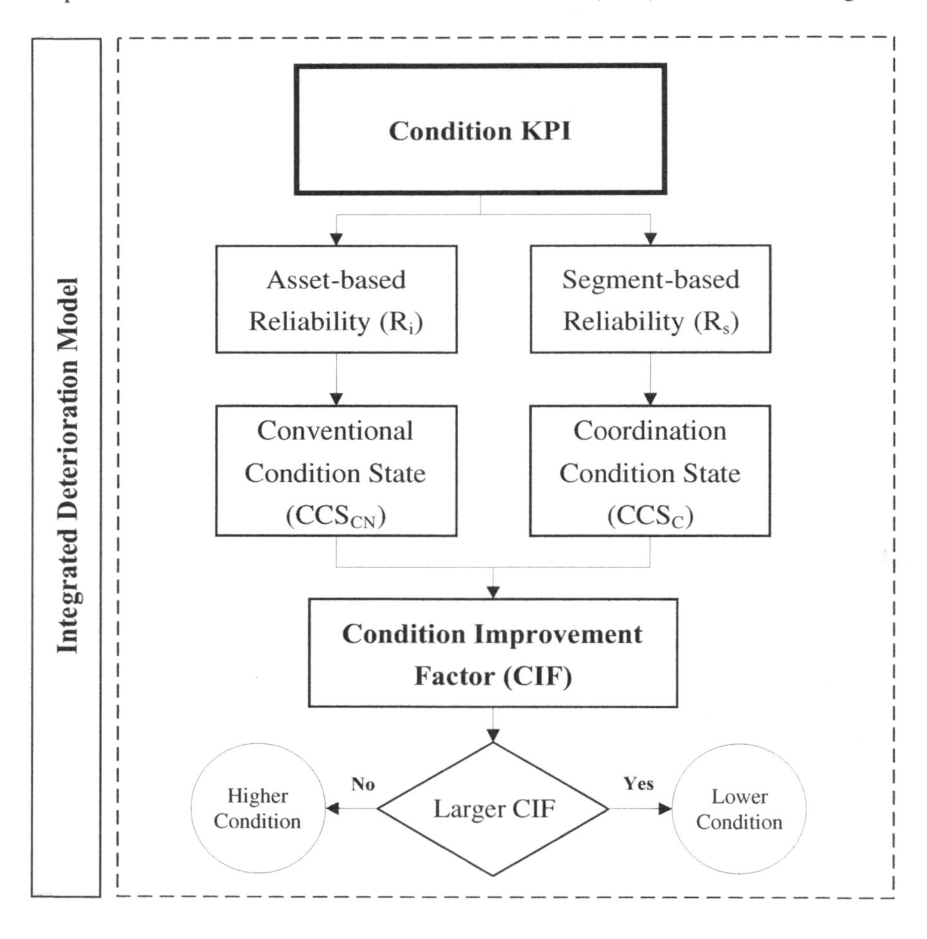

FIGURE 8.1 Integrated deterioration model flowchart (Abusamra 2019).

DOI: 10.1201/9781003327509-11

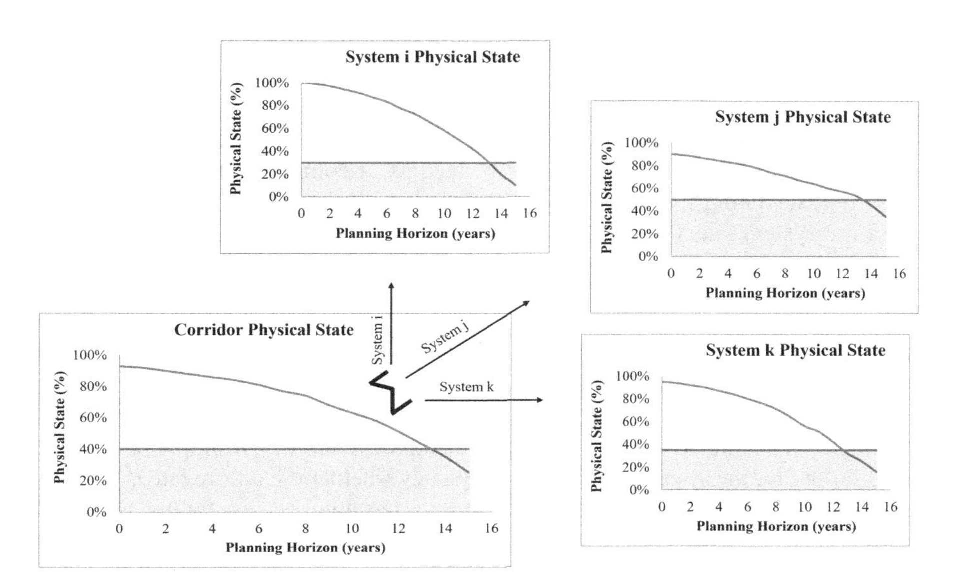

FIGURE 8.2 Corridor-based integrated deterioration curves.

the importance of each system, as shown in Figure 8.2. Due to their different service lives, deterioration patterns, surrounding conditions, etc., various deterioration models were built for the n_s systems. Based on the outcome of each deterioration model, the condition/reliability of each system (R_i) at a certain point of time (t) is known for each coordination scenario while considering the intervention actions' effect on the condition/reliability. Accordingly, the systems' condition/reliability are compiled based on the systems' weights of importance and the CCS is computed. The deterioration models were split into two categories: (1) super-corridor; and (2) sub-corridor. The super-corridor represents all the systems that are above the ground level (e.g., roads) whereas the sub-corridor represents all the systems that are below the ground level (e.g., water and sewer networks). Thus, the super-corridor deterioration model was built to represent the roads' condition evolution across the planning horizon while considering the negative impacts (e.g., aging, traffic, freeze and thaw, extreme weather conditions, etc.) and positive impacts (e.g., maintenance, rehabilitation, reconstruction). Given the fact that (1) the service lives of the roads are relatively short as opposed to the water and sewer networks; and (2) inspection and condition assessment is easier and less costly, historical condition records of various road types with different traffic levels were available, the super-corridor deterioration pattern was adopted from the town of Kindersley records (Amador and Magnuson 2011). A sample deterioration equation of the super-corridor is displayed in Equation 8.1. On the other hand, the sub-corridor deterioration model was built to represent the condition evolution of the water and sewer networks across the planning horizon while considering both the negative (e.g., aging, pipe break) and positive impacts (e.g., leak repair, rehabilitation, replacement). Given the fact that (1) the service lives of the water and sewer networks are long (e.g., 60–100 years); and (2) inspection and condition assessment is difficult and costly, Weibull-based deterioration model was used to reflect the deterioration pattern of the pipes across the planning horizon. To build a Weibull-based deterioration model, the initial date of installation, estimated service life, alpha, and beta distribution parameters need to be present. Weibull analysis is a widely used technique to analyze and predict failures and malfunctions for different types of assets (Jardine and Tsang 2013). It aims at computing the systems' reliability by calculating the probability density and cumulative distribution functions across the system's service life as displayed in Equations 8.2 and 8.3 respectively. Thenceforth, the system's reliability is computed through Equation 8.4. To account for different pipe materials and diameters, a probability distribution function along with its' distribution function parameters is assigned to each pipe category to account for the different pipe failure curves. The key to plotting the cumulative distribution function as well as the reliability function is properly estimating the shape, scale, and location parameters. The shape parameter, sometimes referred to as Beta (β), is the slope of the cumulative distribution curve and the reliability. It simply reflects the rate of failure for the system such that it designates whether the failure rate is increasing, constant or decreasing. For $\beta < 1$, the system has a decreasing failure rate. This scenario is typical of infant mortality and indicates that the system is failing during its initial burn-in period. For $\beta = 1$, the system has a constant failure rate. It typically reflects the systems that have survived the initial burn-in period as they will subsequently exhibit a constant failure rate. For $\beta > 1$, the system has an increasing failure rate, which represents the systems in their wearing out period. The scale parameter, sometimes referred to as Alpha (α), is the Weibull attribute life or service life adjustment factor. In other words, it is a measure of the range or spread in the distribution of data. The location parameter, sometimes referred to as Gamma (γ), represents the distribution along the planning horizon (time). For $\gamma = 0$, the distribution starts at $t = 0$ (origin). However, the distribution slides to the left or right for $\gamma < 0$ or $\gamma > 0$ respectively. In this study, the location parameter was set to zero and the reliability function was computed according to Equation 8.5.

$$R_{i_{o_t}} = \left[0.033 \times (t_{i_o})^2\right] - \left[2.688 \times t_{i_o}\right] + R_{i_{o\,t0}} \tag{8.1}$$

$$d_{i_{o_t}} = \frac{\beta}{\alpha} \times \left(\frac{t_{i_o} - \gamma}{\alpha}\right)^{\beta - 1} \times e^{-\left(\frac{t_{i_o} - \gamma}{\alpha}\right)^{\beta}} \tag{8.2}$$

$$D_{i_{o_t}} = 1 - e^{-\left(\frac{t_{i_o} - \gamma}{\alpha}\right)^{\beta}} \tag{8.3}$$

$$R_{i_{o_t}} = 1 - D_{i_{o_t}} = e^{-\left(\frac{t_{i_o} - \gamma}{\alpha}\right)^{\beta}}; \text{ for } \gamma > 0 \tag{8.4}$$

$$R_{i_{o_t}} = R_{i_{o\,t0}} \times e^{-\left(\frac{t_{i_o}}{\alpha}\right)^{\beta}}; \text{ for } \gamma = 0 \tag{8.5}$$

where t_{i_o} is the age of the system i within corridor o (years); $R_{i_{ot0}}$ is the reliability of system i within corridor o at the initial point of time t_0; $d_{i_{ot}}$ is the probability density function (deterioration) of system i within corridor o at point of time t (%); β is the shape parameter (>0); γ is the location parameter (>0); α is the scale parameter (years); $D_{i_{ot}}$ is the cumulative distribution function (deterioration) of system i within corridor o at point of time t (%); and $R_{i_{ot}}$ is the reliability of system i within corridor o at point of time t (%).

Random sudden failure may happen to a ratio of systems due to extreme conditions (Mohammed et al. 2017). This phenomenon of sudden failure happens randomly to numerous systems at random locations in almost every city where the system experiences a sharp change from a working state to a failing state. However, it is more likely to happen with older systems with lower condition states than others. This type of abrupt service/technical failure may not be adequately represented by the typical deterioration models. Hence, to account for those sudden failures, an extreme event random generator module was developed to enforce sudden failure of 10% of the network under study. This module randomly generates sudden failure to numerous corridors at different points of time. Thenceforth, that information is integrated with the deterioration models of each system and their negative impact is computed through Equation 8.6. This dual deterioration modes' representation would provide decision-makers with more informed decisions that account for real-life cases and enhances the overall network resilience on the long-term accordingly.

$$R_{i_{ot_e}} = \begin{array}{ll} 0\% & \text{Water pipe break} \\ 0\% & \text{Sewer pipe break} \\ R_{i_{ob}} \times 50\% & \text{Road extreme weather} \end{array} \tag{8.6}$$

where $R_{i_{ot_e}}$ is the negative impact of the extreme conditions on the reliability of system i within corridor o at a point of time t (%); and $R_{i_{ob}}$ is the reliability of system i within corridor o at a point of time t before the occurrence of an extreme event (%).

Given spatial interdependency among the systems, there are several cases for representing the systems' reliability improvement for each intervention scenario. For instance, conventional intervention for the roads will only improve the reliability of the roads. However, a conventional intervention for the water or sewer pipe, assuming open-cut replacement, will partially improve the road's reliability, given that the road had to be demolished and reconstructed to access the underneath pipe and replace it with a new one. For the partially coordinated program, the case when the roads and water or sewer network are integrated will improve the reliability of both systems and will not have any impact on the third system. However, the partially coordinated water and sewer intervention, assuming open-cut replacement, will improve the reliability of the three systems, given that the road will have to be demolished and reconstructed to access the underneath pipes and replace them with new ones. The full coordinated intervention scenario of the roads, water, and sewer network will improve the reliability of the three systems. For simplicity purposes, Table 8.1 summarizes the different coordination scenarios without taking into consideration the different intervention activities (e.g., minor repairs will not return the system to its pristine condition; etc.). Table 8.2 summarizes the impact of intervention activities on different systems. Furthermore, the

TABLE 8.1
Intervention-Asset Impact Matrix

Intervention Scenario	Assets	Road	Water	Sewer
Do Nothing	Do Nothing	×	×	×
Conventional (1 system)	Roads	✓	×	×
	Water	✓	✓	×
	Sewer	✓	×	✓
Partially coordinated (2 systems)	Roads and Water	✓	✓	×
	Roads and Sewer	✓	×	✓
	Water and Sewer	✓	✓	✓
Full coordination (3 systems)	Full Coordination	✓	✓	✓

TABLE 8.2
Intervention Reliability Impact Matrix

Intervention Description			Intervention Quantitative Effect (ΔR_i)		
Intervention Scenario	**Assets**	**Intervention Action**	**Roads (Reliability %)**	**Water (Age - Years)**	**Sewer (Age - Years)**
Do Nothing	Do Nothing	Do Nothing	0%	0	0
Conventional (1 system)	Roads	Surface overlay	Varies*	×	×
		Resurfacing	100%	×	×
	Water	Leaks repair	*Varies***	Varies*	×
		Pipe replacement	*Varies***	0	×
	Sewer	Leaks repair	*Varies***	×	Varies*
		Pipe replacement	*Varies***	×	0
Partially coordinated (2 systems)	Roads and Water	Pipe replacement and road resurfacing	100%	0	×
	Roads and Sewer	Pipe replacement and road resurfacing	100%	×	0
	Water and Sewer	Pipe replacement	*Varies***	0	0
Full coordination (3 systems)	Full Coordination	Pipe replacement and road resurfacing	100%	0	0

0% represents road segments in failing condition states.
100% represents road segments in pristine condition state.
Varies* represents a varying reliability improvement based on the reliability of the system right before undertaking the intervention.
Varies** represents a varying reliability impact of an intervention depending on the reconstruction area.
0 represents a water or sewer pipe in a pristine condition state.
× represents an asset that has not been affected by the corresponding intervention action.

impact of the intervention activities on the sub-corridor and super-corridor could be graphically displayed in Figures 8.3 and 8.4. As displayed in those figures, the initial reliability of the system is referred to as R_{INI} and it occurs at the start of the planning horizon T_0. The reliability of the system before and after undertaking the intervention is

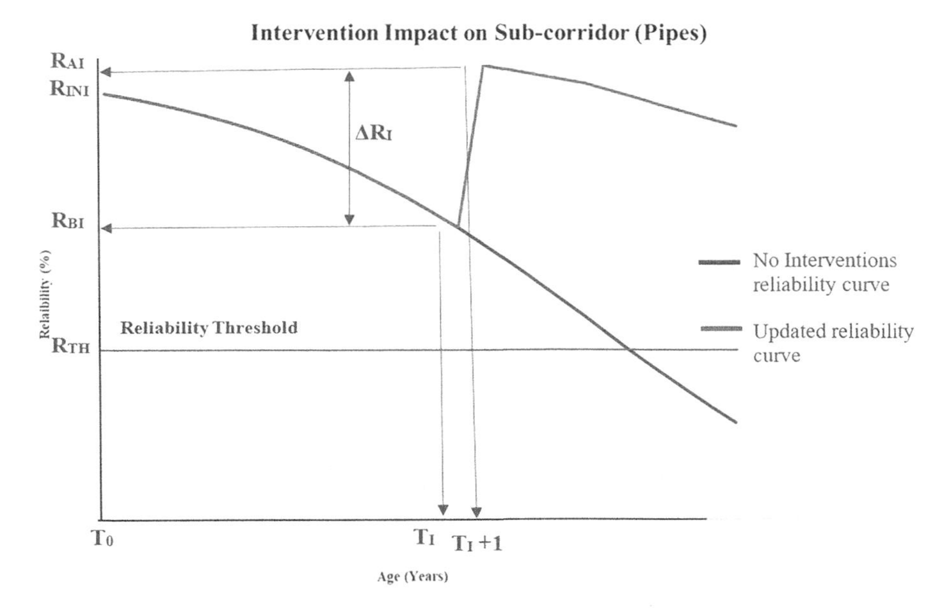

FIGURE 8.3 Impact of an intervention on sub-corridor (pipes).

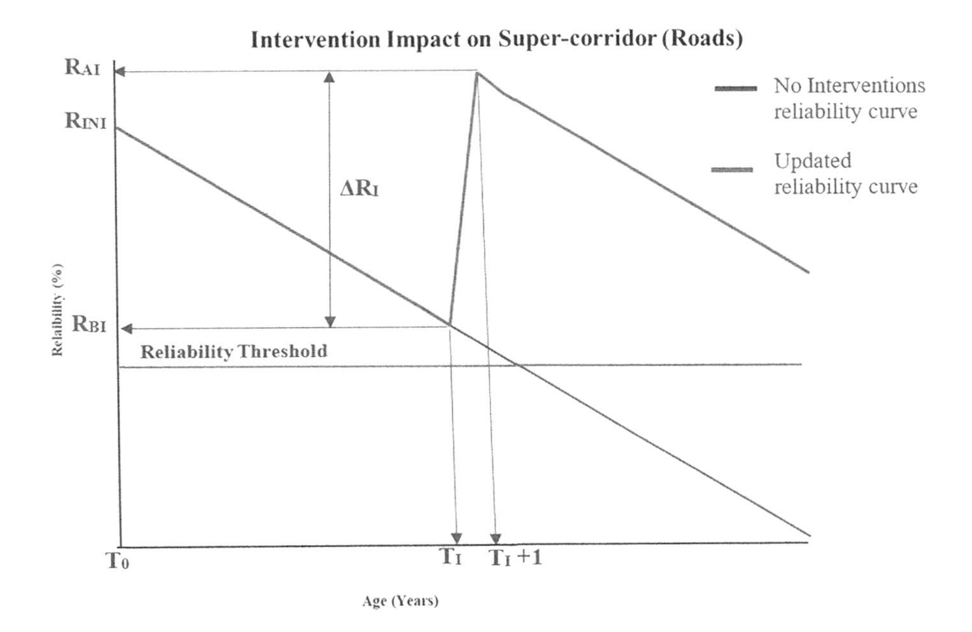

FIGURE 8.4 Impact of an intervention on super-corridor (roads).

referred to as R_{BI} and R_{AI} respectively. The impact of the intervention action is referred to as ΔR_I and it takes place on time T_I and the impact takes place at time T_I+1. The reliability threshold is referred to as R_{TH}. In both the sub-corridor and super-corridor cases, the impact of the intervention on the systems' reliability (R_{AI}) is represented through an improvement ($t_{i_{o\,imp}}$) in their age as displayed in Equation 8.7. However, given the fact that the deterioration pattern and models vary, the mathematical representation of the improvement (ΔR_I) varies. For the sub-corridor case, the mathematical computations of the intervention impact on the Weibull deterioration could be displayed in Equations 8.8 and 8.9. For the super-corridor case, the mathematical computations of the intervention impact on the regression deterioration could be displayed in Equations 8.10 and 8.11.

$$R_{AI_{i_{ot+1}}} = R_{BI_{i_{ot}}} + \Delta R_{I_{i_{ot}}} \tag{8.7}$$

$$R_{AI_{i_{o\,t+1}}} = R_{i_{ot0}} \times e^{-\left(\frac{t_{i_{o\,imp}}}{\alpha}\right)^{\beta}} \times IA_{o_{i_{t_r}}} \tag{8.8}$$

$$\Delta R_{I_{i_{ot}}} = \left[R_{i_{ot0}} \times e^{-\left(\frac{t_{i_{o\,imp}}}{\alpha}\right)^{\beta}} \times IA_{o_{i_{t_r}}} \right] - \left[R_{i_{ot0}} \times e^{-\left(\frac{t_{i_o}}{\alpha}\right)^{\beta}} \right] \tag{8.9}$$

$$R_{AI_{i_{ot+1}}} = \left[\left[0.033 \times (t_{i_{o\,imp}})^2\right] - \left[2.688 \times t_{i_{o\,imp}}\right] + R_{i_{ot0}}\right] \times IA_{o_{i_{t_r}}} \tag{8.10}$$

$$\begin{aligned} \Delta R_{I_{i_{ot}}} = {} & \left[\left[0.033 \times (t_{i_{o\,imp}})^2\right] - \left[2.688 \times t_{i_{o\,imp}}\right] + R_{i_{ot0}}\right] \\ & \times IA_{o_{i_{t_r}}} - \left[\left[0.033 \times (t_{i_o})^2\right] - \left[2.688 \times t_{i_o}\right] + R_{i_{ot0}}\right] \end{aligned} \tag{8.11}$$

where $R_{AI_{i_{ot+1}}}$ is the reliability of system i within corridor o at point of time $t+1$ after undertaking intervention r (%); $R_{BI_{i_{ot}}}$ is the reliability of system i within corridor o at point of time t before undertaking intervention r (%); $\Delta R_{I_{i_{ot}}}$ is the reliability improvement of system i within corridor o at point of time t after undertaking intervention r (%); and $t_{i_{o\,imp}}$ is the improved age after undertaking an intervention r in system i within corridor o at point of time t (years).

Condition Impact Factor

The deterioration models have pre-set condition thresholds that alert the decision-makers in case the condition state of any system reaches a value below the threshold to take rapid intervention decisions and avoid experiencing an increased probability of failure. The corridor condition is computed for the conventional, partially coordinated, and full coordination intervention scenarios and is represented by CCS_{CN}, CCS_{PC}, and CCS_C, as shown in Equations 8.12, 8.13, and 8.14 respectively. Finally, the

Condition Improvement Factor (CIF) was computed to compare the partially coordinated and full coordination intervention scenarios with the conventional one in terms of condition improvement, as shown in Equations 8.15 and 8.16 respectively. If the CIF is less than 1 (CIF < 1), the considered intervention scenario displays better condition as opposed to the conventional scenario. However, if the CIF is more than 1 (CIF > 1), the considered intervention scenario displays worse condition state as opposed to the conventional scenario. For instance, a CIF of "1.2" indicates that the considered intervention scenario displays 20% less condition compared to the conventional one.

$$\mathrm{CCS}_{\mathrm{CN}_o} = \sum_{i=1}^{n_s} \left(W_i \times \overline{R_{i\,\mathrm{CN}_{ot}}} \right). \tag{8.12}$$

$$\mathrm{CCS}_{\mathrm{PC}_o} = \sum_{i=1}^{n_s} \left(W_i \times \overline{R_{i\,\mathrm{PC}_{ot}}} \right) \tag{8.13}$$

$$\mathrm{CCS}_{\mathrm{C}_o} = \sum_{i=1}^{n_s} \left(W_i \times \overline{R_{i\,\mathrm{C}_{ot}}} \right) \tag{8.14}$$

$$\mathrm{CIF}_{\mathrm{PC}} \left(I_6^{+} \right) = \sum_{o=1}^{O} \left(\frac{L_o}{\sum_{o=1}^{O} L_o} \times \frac{\mathrm{CCS}_{\mathrm{CN}_o}}{\mathrm{CCS}_{\mathrm{PC}_o}} \right) \tag{8.15}$$

$$\mathrm{CIF}_{\mathrm{C}} \left(I_6^{+} \right) = \sum_{o=1}^{O} \left(\frac{L_o}{\sum_{o=1}^{O} L_o} \times \frac{\mathrm{CCS}_{\mathrm{CN}_o}}{\mathrm{CCS}_{\mathrm{C}_o}} \right) \tag{8.16}$$

where $\mathrm{CCS}_{\mathrm{CN}_o}$ is the corridor condition state of corridor o in the conventional intervention scenario (%); W_i is the weight of importance assigned to system i (%); L_o represents corridor o length (m); $\overline{R_{i\,\mathrm{CN}_{ot}}}$ is the average reliability of system i within corridor o across the planning horizon (t) in the conventional intervention scenario (%); $\mathrm{CCS}_{\mathrm{PC}_o}$ is the corridor condition state of corridor o in the partially coordinated intervention scenario (%); $\overline{R_{i\,\mathrm{PC}_{ot}}}$ is the average reliability of system i within corridor o across the planning horizon (t) in the partially coordinated intervention scenario (%); $\mathrm{CCS}_{\mathrm{C}_o}$ is the corridor condition state of corridor o in the full coordination intervention scenario (%); $\overline{R_{i\,\mathrm{C}_{ot}}}$ is the average reliability of system i within corridor o across the planning horizon (t) in the full coordination intervention scenario (%); $\mathrm{CIF}_{\mathrm{PC}}$ is the condition impact factor that compares the partially coordinated and conventional network intervention scenarios in terms of condition improvement (%); and $\mathrm{CIF}_{\mathrm{C}}$ is the condition impact factor that compares the full coordination and conventional network intervention scenarios in terms of condition improvement (%).

RESILIENCE PREPAREDNESS KPI

The resilience preparedness framework supports the transition from a "Reactive" approach, where assets are fixed after failure, to a "Proactive" approach, where asset management plans are developed to prevent assets from failure and prolong the assets' service lives. It computes the corridors' resiliency with respect to climate change and urbanization. The resilience preparedness model focuses on the water and sewer pipes' replacement given their long service lives and lengthy public disruptions. Furthermore, the resiliency model computes the impact of urbanization, represented through land-use change and population growth; and climate change, represented through the rainfall intensity and frequency increase, on the water and combined sewer and stormwater systems. It computes the corridor resilience preparedness of the applied intervention plan on the full coordination, partially coordinated, and conventional intervention scenarios as shown in Figure 8.5.

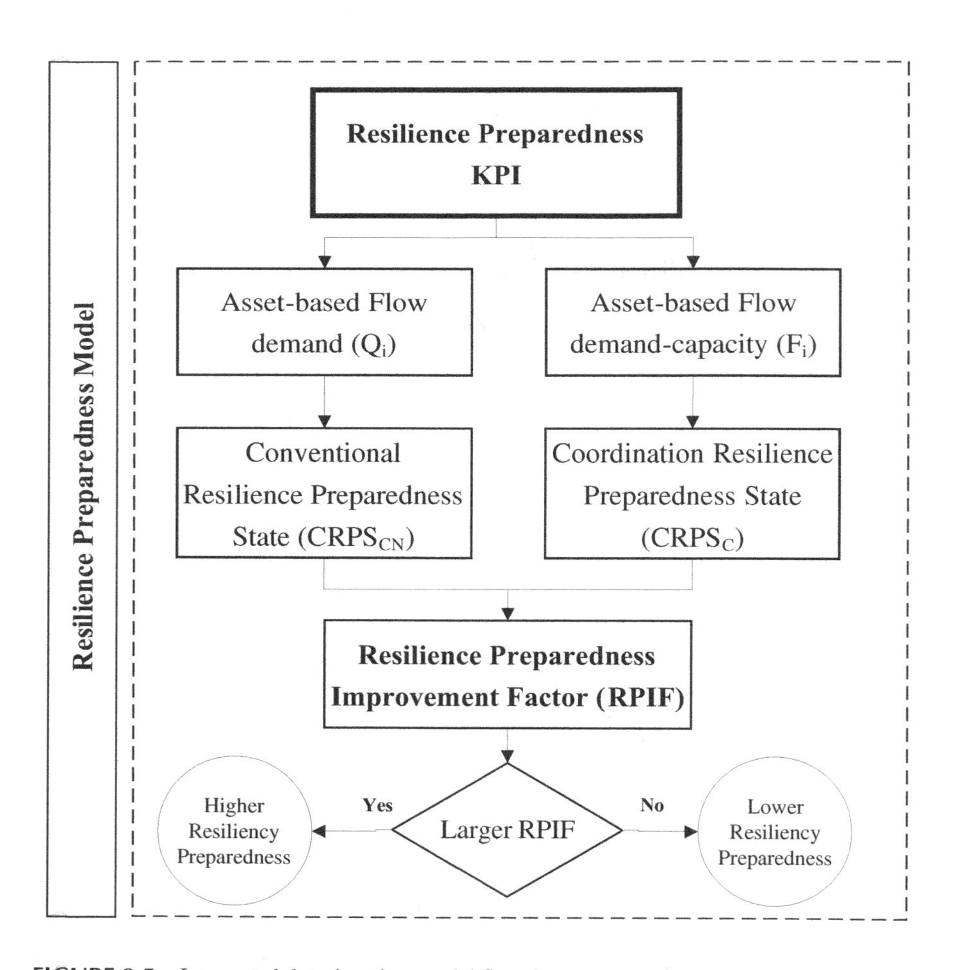

FIGURE 8.5 Integrated deterioration model flowchart.

The model revolves through four integrated models as shown in Figure 8.6: (1) urban and climate change models feed into an urban hydrological model that computes the runoff coefficient and rainfall intensity for each sub-catchment area; (2) capacity performance model that predicts the flow demand-capacity ratios according to the population growth, land-use changes for water and combined sewer and stormwater pipes; and climate change, only for combined sewer and stormwater pipes; (3) deterioration model that computes the condition state of the pipes throughout their lifetime, as highlighted in the previous section of this Chapter; and (4) financial model that computes the pipes' replacement costs, as will be highlighted in Chapter 9. The urban

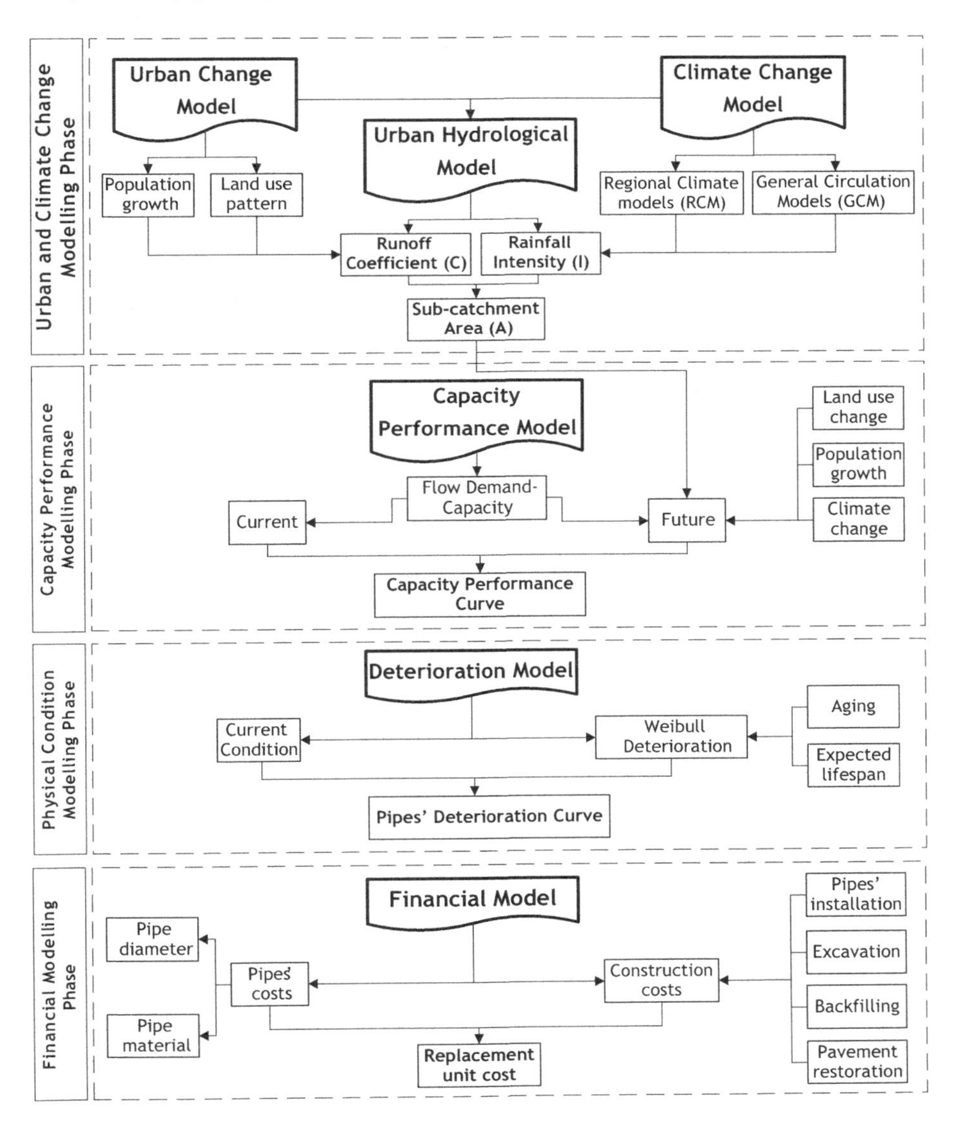

FIGURE 8.6 Resilience preparedness computational models' flowchart.

hydrological model integrates the outputs of climate change and urban change models to simulate the impact of rainfall intensity and land-use change/population growth on the flow of the study areas. For the water pipes, the impact of land-use change and population growth are considered to compute the demand flow increase, and the increased rainfall intensity will not be considered given that it does not impact the water pipes. However, in the case of combined sewer and stormwater pipes, the impact of climate change, represented through the increased rainfall intensity, are added to the impact of the land-use change and population growth. The result of combining those impacts is the increased demand flow. Thenceforth, the capacity performance model computes the current and future flow demand-capacity ratios based on future prediction models' of the population growth, land-use changes, and climate change as will be highlighted later in the upcoming sub-sections. The output of this model is a flow demand-capacity performance curve over the study planning horizon. Accordingly, a demand-capacity ratio greater than 1 ($F_{i_t} > 1$) represents the case when the demand flow exceeds the existing capacity. In that case, the existing pipes need to be replaced with bigger diameter ones to meet the increasing demand flow. Hence after, the deterioration model computes the reliability of the pipes across the planning horizon. Thenceforth, the financial model computes the pipes' replacement costs over the study horizon. The only difference is that additional alternatives are added to reflect the fact that a pipe could be replaced with a bigger diameter pipe, in case the demand-capacity ratio exceeds 1. In that case, the pipe replacement decisions for both water and sewer networks will be as follows: (1) replacing the pipe with the same diameter/hydraulic capacity in case the current diameter is enough to operate over its' lifetime and the only trigger to replace the pipe was the deteriorating condition state; and (2) increase the hydraulic capacity through installing a larger diameter pipe to account for growing population, increased rainfall intensity, and pipe condition. In the case of larger diameter replacement, the replacement decision trigger will be either (1) operational; where the hydraulic capacity is no longer sufficient to operate; or (2) physical and operational; where both the condition is deteriorating, and the hydraulic capacity is no longer enough for operation. The impact of replacing the pipe with a larger diameter one is considered in the affected models (Amador et al. 2019).

Computational Models

Urban Hydrological Models

The urban hydrological model computes the rainfall intensity and runoff coefficients of each sub-catchment area to calculate the increase in the demand flow. The runoff coefficient is computed through an urban change model and the rainfall intensity is computed through a climate change model. For the water pipes, the demand flow will be only affected by the runoff coefficient given the fact that they are not affected by the rainfall intensity. However, for the combined sewer and stormwater pipes, both the rainfall intensity and runoff coefficients are considered while computing the demand flow increase. The approach recommended by most drainage manuals is computing the hydrologic response of each sub-catchment to the design storms associated with different return periods. From a drainage perspective, the most dominant characteristic of the urban landscape is the high degree of impervious ground cover. Population growth and changes in urban land-use affect the extent of imperviousness of urban watersheds, leading to a rapid rate of increase on rainfall runoff. These factors result in more significant changes to the hydrologic regime compared with changes due to drainage works in rural and non-developed areas. Furthermore, the volume and rate of stormwater runoff directly rely on the magnitude of precipitation. Statistical frequency analysis of Canadian global climate models' series has shown that rainfall events' frequency and intensity will, most likely, increase over the next years due to the climate change (Environment Canada 2014). Accordingly, statistical downscaling models must be employed to downscale the general circulation models (GCMs) based rainfall projections.

The adopted urban change model is centered on the rational method (Dooge 1957). It computes the runoff coefficient and tributary area (A) that are affected by current and future land-use patterns, which respond to urban growth development strategies. Given the fact that there are different land uses for each sub-catchment area, the runoff coefficient of each pipe i is estimated through computing the individual runoff coefficient with respect to each land-use type area (A_i). Furthermore, the climate model estimates the changes across time of impervious areas, runoff and flows to the pipe system for an entire catchment based on remotely sensed data and GIS technologies (Thanapura et al. 2007; Savary et al. 2009; Gupta et al. 2012). The output of the climate model is rainfall intensity (I), which is estimated by regional climate models (RCMs) and intensity-duration-frequency (IDF). The mathematical formulation of the demand for both runoff coefficient and rainfall intensity could be displayed in Equations 8.17 and 8.18. Thenceforth, the demand of the water and combined sewer and stormwater pipes could be mathematically formulated in Equations 8.19 and 8.20 respectively.

$$\mathrm{RR}_{i_{o_t}} = \frac{\sum_{\mathrm{pc}=1}^{\mathrm{PC}} A_{\mathrm{pc}\,i_o} \times \mathrm{RR}_{\mathrm{pc}\,i_{o_t}}}{A_{i_o}} \tag{8.17}$$

$$A_{i_o} = \sum_{\mathrm{pc}=1}^{\mathrm{PC}} A_{\mathrm{pc}\,i_o} \tag{8.18}$$

$$Q_{i_{o_t}} = \mathrm{RR}_{i_{o_t}} \times A_{i_o} \tag{8.19}$$

$$Q_{i_{o_t}} = I_t \times \mathrm{RR}_{i_{o_t}} \times A_{i_o} \tag{8.20}$$

where $Q_{i_{o_t}}$ is the design discharge for the recurrence interval of pipe i within corridor o at point of time t (m^3/day); t is the analysis point of time throughout the planning horizon (years); i is the pipes counter; $\mathrm{RR}_{i_{o_t}}$ is the rational runoff coefficient of pipe i within corridor o at point of time t; I_t is the rainfall intensity at point of time t (mm-h); A_{i_o} is the catchment area of pipe i within corridor o (m^2); $A_{\mathrm{pc}\,i_o}$ is a fraction of pipe i area (A_{i_o}) covered within corridor o (m^2); and pc and PC are the counter and total number of components (pc) within pipe i area (A_{i_o}) respectively.

After predicting the future impact of climate change and urbanization, a sub-catchment future prediction model is built to estimate the future discharging of each sub-catchment based on projections of rainfall intensity-duration and land use, only for the combined sewer and stormwater pipes. Similarly, a future demand prediction model is built to estimate the future demand of each water pipe based on land-use change as well as population growth. Thenceforth, the flow demand is periodically computed for each pipe over its lifecycle. Different pipes feature different demand curves based on their spatial location. This difference impacts rainfall intensity, population growth, and land-use change. For instance, the population growth in the downtown area is much higher than in residential areas. Similarly, the rainfall intensity differs from one area to another.

Capacity Performance Model

The capacity performance model aims at computing the flow demand-capacity ratios (F). This ratio (F) can be used to characterize the system resiliency where it estimates the flow over the capacity ratio of each pipe over its lifecycle to ensure that the flow demand is met by the given pipe diameter. For instance, a ratio above 100% indicates a pipe facing fw demand superior to its capacity. In that case, the model alerts the decision-makers that the current pipe either (1) will experience overflow, in case of combined sewer and stormwater, or (2) will not fit the demand, in case of water. In both cases, it needs to be replaced with a larger diameter pipe to meet the flow demand, as shown in Equation 8.21. As mentioned earlier, different pipes feature different demand-capacity ratios based on their spatial location. Thus, the pipes were categorized into three groups based on their demand (e.g., low, medium, and high). A sample of the combined sewer and stormwater pipes' groups could be displayed in Figure 8.7. Similarly, the demand-flow capacity curves of the water pipes have been constructed.

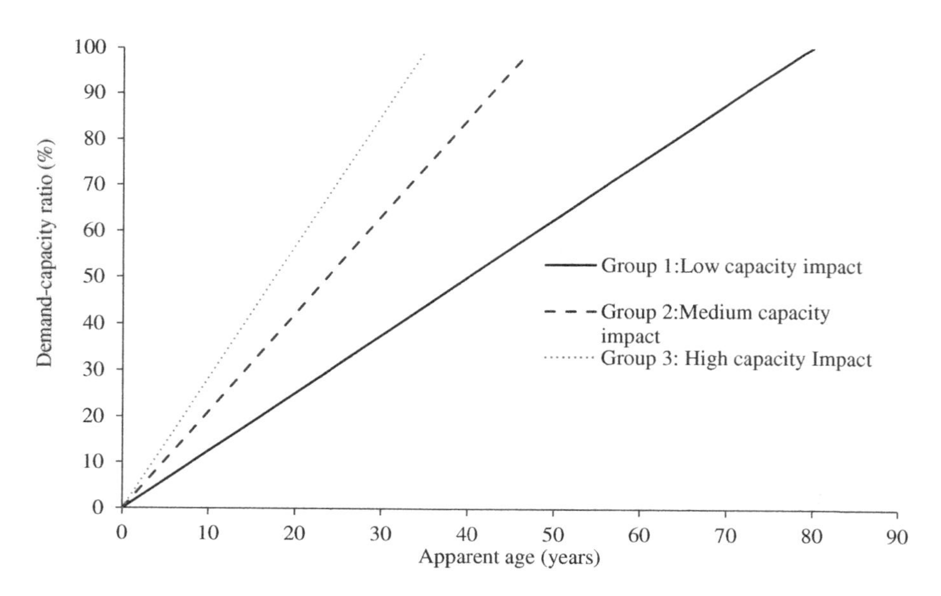

FIGURE 8.7 Demand-capacity ratio of different combined sewer and stormwater pipe groups.

$$F_{i_{ot}} = \begin{bmatrix} \text{Do Nothing or replacement with same diameter} & \dfrac{Q_{i_{ot}}}{\text{CP}i_{o_t}} \\ \text{Replacement with larger diameter} & \dfrac{Q_{i_{ot}}}{\text{CP}i_{\text{new}_{ot}}} \end{bmatrix} \tag{8.21}$$

where $F_{i_{o_t}}$ is the flow demand-capacity ratio of pipe i within corridor o at point of time t (%); $\text{CP}i_t$ is the capacity of pipe i within corridor o at point of time t (m^3/day); $\text{CP}i_{\text{new}_{ot}}$ is the capacity of new pipe i with a larger diameter within corridor o at point of time t (m^3/day).

Deterioration Model

The deterioration model predicts the reliability of the pipes across their service lives. A detailed description of the Weibull pipe's deterioration was provided in the previous section. However, it is worth noting that the impact of replacing the pipe with a same or larger diameter is the same as returning the system to a pristine condition state as displayed in Equation 8.22.

$$R_{\text{AI}_{i_{ot+1}}} = \begin{bmatrix} \text{Do Nothing} & R_{\text{BI}_{i_{ot}}} - d_{i_{ot}} \\ \text{Replacement with same or larger diameter} & R_{i_{ot0}} \end{bmatrix} \tag{8.22}$$

Financial Model

The financial model calculates the replacement costs of each intervention scenario. Given the diversity of pipe diameters, depths, and materials, replacement costs are estimated at a pipe level. The replacement costs could even vary within the same pipe at different periods of time given the fact that some pipes might require replacement with larger diameters to account for the increased capacity, resulting from increased rainfall intensity-duration-frequency or land-use change or population growth. For instance, a 300 mm diameter pipe could be replaced either by the same diameter pipe or a larger one (e.g., 375 mm) depending on future demand. Thus, a flow demand-capacity replacement threshold of 50% has been defined to guarantee a safety margin of 25 years without overflooding or operational-triggered replacement that makes the current pipe diameter no longer sufficient to meet the increasing demand. For instance, a deteriorating pipe with flow demand-capacity less than 50% would be replaced with the same diameter and a deteriorating pipe with flow demand-capacity more than 50% will be replaced with a larger pipe diameter given that their hydraulic capacity will not be enough to meet the increasing future demand (Abusamra 2019).

Resilience Preparedness Impact Factor

The resilience preparedness models have pre-set flow demand-capacity thresholds that alert the decision-makers in case the flow demand-capacity of any system reaches a value above the threshold to take rapid intervention decisions and avoid experiencing either an overflood, in case of combined sewer and stormwater system, or unmet demand, in case of the water system. The corridor resiliency preparedness state (CRPS) is computed for water and combined sewer and stormwater systems in the conventional, partially coordinated, and full coordination intervention scenarios and is represented by $CRPS_{CN}$, $CRPS_{PC}$, and $CRPS_{C}$, as shown in Equations 8.23, 8.24, and 8.25 respectively. Finally, the Resilience Preparedness Improvement Factor (RPIF) was computed to compare the partially coordinated and full coordination intervention scenarios with the conventional one in terms of resilience preparedness improvement, as shown in Equations 8.26 and 8.27 respectively. If the RPIF is less than 1 ($RPIF < 1$), the considered intervention scenario displays better resiliency preparedness as opposed to the conventional scenario. However, if the RPIF is more than 1 ($RPIF > 1$), the considered intervention scenario displays worse resiliency preparedness as opposed to the conventional scenario. For instance, an RPIF of "1.2" indicates that the considered intervention scenario displays 20% better resiliency preparedness compared to the conventional one.

$$CRPS_{CN_o} = \sum_{i=1}^{n_S} \left(W_i \times \overline{F_{i\,CN_{ot}}} \right) \quad (8.23)$$

$$CRPS_{PC_o} = \sum_{i=1}^{n_S} \left(W_i \times \overline{F_{i\,PC_{ot}}} \right) \quad (8.24)$$

$$CRPS_{C_o} = \sum_{i=1}^{n_S} \left(W_i \times \overline{F_{i\,C_{ot}}} \right) \quad (8.25)$$

$$RPIF_{PC}\left(I_7^{+}\right) = \sum_{o=1}^{O} \left(\frac{L_o}{\sum_{o=1}^{O} L_o} \times \frac{CRPS_{CN_o}}{CRPS_{PC_o}} \right) \quad (8.26)$$

$$RPIF_{C}\left(I_7^{+}\right) = \sum_{o=1}^{O} \left(\frac{L_o}{\sum_{o=1}^{O} L_o} \times \frac{CRPS_{CN_o}}{CRPS_{C_o}} \right) \quad (8.27)$$

where $CRPS_{CN_o}$ is the corridor resiliency preparedness state of corridor o in the conventional intervention scenario (%); W_i is the weight of importance assigned to system i (%); L_o represents corridor o length (m); $\overline{F_{i\,CN_{ot}}}$ is the average flow demand-capacity ratio of system i within corridor o across the planning horizon (t) in the conventional intervention scenario (%); $CRPS_{PC_o}$ is the corridor resiliency preparedness state of corridor o in the partially coordinated intervention scenario (%); $\overline{F_{i\,PC_{ot}}}$ is the average flow demand-capacity ratio of system i within corridor o across the planning horizon (t) in the partially coordinated intervention scenario (%); $CRPS_{C_o}$ is the corridor resiliency preparedness state of corridor o in the full coordination intervention scenario (%); $\overline{F_{iC_{ot}}}$ is the average flow demand-capacity ratio of system i within corridor o across the planning horizon (t) in the full coordination intervention scenario (%); $RPIF_{PC}$ is the resilience preparedness impact factor that compares the partially coordinated and conventional network intervention scenarios in terms of resiliency preparedness improvement (%); and $RPIF_{C}$ is the resilience preparedness impact factor that compares the full coordination and conventional network intervention scenarios in terms of resiliency preparedness improvement (%).

REFERENCES

Abusamra, S. (2019). "Coordination and Multi-Objective Optimization Framework for Managing Municipal Infrastructure under Performance-Based Contracts." Doctoral of Philosophy Thesis, Concordia University, Montreal.

Amador, L. and Magnuson, S. (2011). "Adjacency modeling for coordination of investments in infrastructure asset management: Case study of Kindersley, Saskatchewan, Canada." *Transportation Research Board (TRB)*, 2246, 8–15.

Amador, L., Mohammadi, A., Abu-Samra, S., and Maghsoudi, R. (2019). "Resilient storm pipes: A multi-stage decision support system." *Structure and Infrastructure Engineering*, Taylor & Francis, 16(1), 1–13.

Dooge, J. C. (1957). "The rational method for estimating flood peaks." *Engineering*, 184(1), 311–313.

Environment Canada. (2014). "Engineering Climate Datasets." https://climate.weather.gc.ca/prods_servs/engineering_e.html. Aug 22, 2014.

Gupta, P. K., Punalekar, S., Panigrahy, S., Sonakia, A., and Parihar, J. S. (2012). "Runoff modeling in an agro-forested watershed using remote sensing and GIS." *Journal of Hydrologic Engineering*, 17(11), 1255–1267.

Jardine, A. K. S. and Tsang, A. H. C. (2013). *Maintenance, Replacement, and Reliability: Theory and Applications*. Boca Raton, FL: CRC Press, Taylor & Francis.

Mohammed, A., Abu-Samra, S., and Zayed, T. (2017). "Resilience Assessment Framework for Municipal Infrastructure." *MAIREINFRA-The International Conference on Maintenance and Rehabilitation of Constructed Infrastructure Facilities,* July 19–21, Seoul.

Thanapura, P., Helder, D. L., Burckhard, S., Warmath, E., O'Neill, M., and Galster, D. (2007). "Mapping urban land cover using QuickBird NDVI and GIS spatial modeling for runoff coefficient determination." *Photogrammetric Engineering and Remote Sensing*, 73(1), 57–65.

Savary, S., Rousseau, A. N., and Quilbe, R. (2009). "Assessing the effects of historical land cover changes on runoff and low flows using remote sensing and hydrological modeling." *Journal of Hydrologic Engineering*, 14(6), 575–587.

9 Temporal, Spatial, and Financial KPIs

TEMPORAL KPI

The objective of this KPI is to compute the intervention duration and potential savings, arising from coordinating the interventions compared to undertaking them in silos. The duration savings model dynamically computes the duration of the fully coordinated, partially coordinated, and conventional interventions, based on the categorized activities and their production rates as shown in Figure 9.1. The benefit

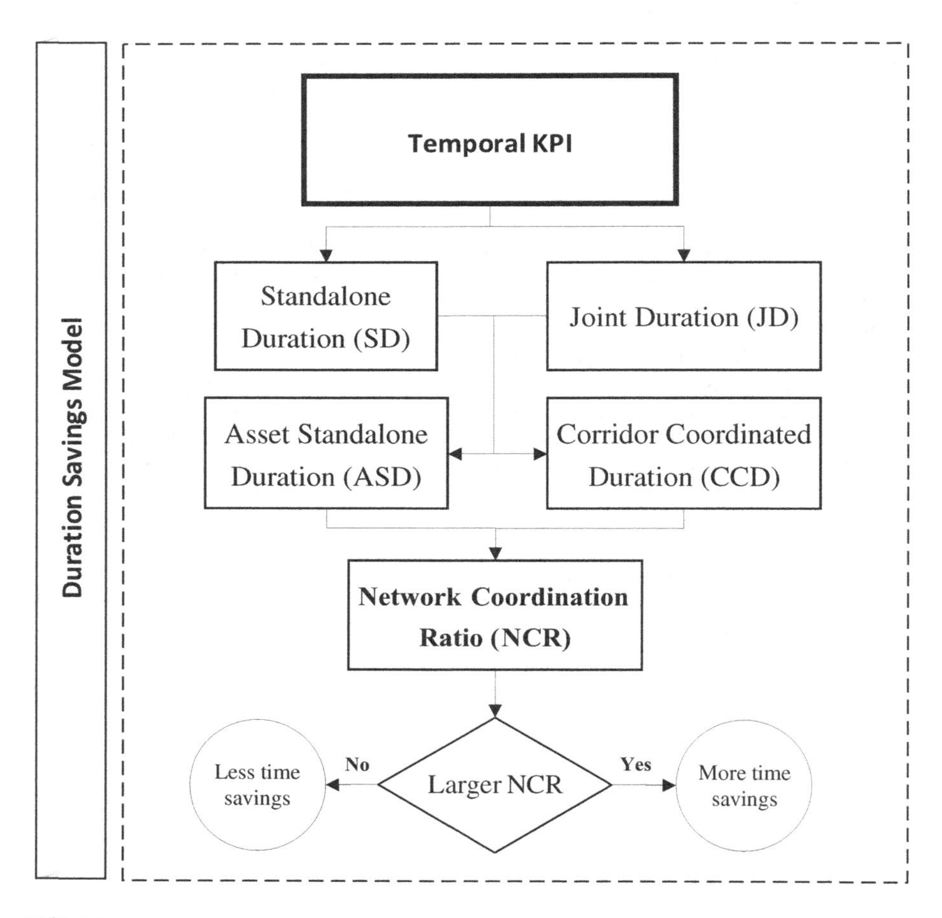

FIGURE 9.1 Duration savings model flowchart (Abusamra 2019).

DOI: 10.1201/9781003327509-12

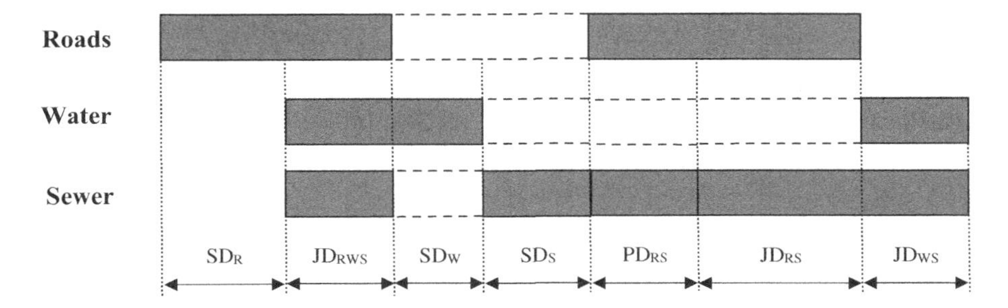

FIGURE 9.2 Assets' durations' categories and relationships.

of coordinating the intervention actions is generating time savings in the corridor intervention duration compared to the conventional approach. Those time savings take place because of the existence of joint activities that are shared among the three systems as well as the possibility of undertaking parallel activities rather than series ones, in case proper coordination takes place (e.g., road resurfacing can occur concurrently while working on reinstating sewer laterals). As such, these activities can be undertaken only once, in the case of the fully coordinated approach, rather than n_a or n_s, in the case of the partially coordinated or the conventional approach, where n is the number of standalone interventions and a and s are the number of systems in cases of partially coordinated and conventional intervention scenarios, respectively. Accordingly, those overlaps can be globalized through the basis of standalone duration (SD), parallel duration (PD), and joint duration (JD), as shown in Figure 9.2.

To better understand the theory, let SD_i represent the duration of the intervention activities required only for system i and no other work can take place concurrently (e.g., installation of new sewer manholes). Furthermore, let PD_{ijk} represent the duration of the intervention activities that can take place concurrently. Furthermore, let JD_{ijk} represent the duration of the intervention actions required for two or more systems. This duration represents the activities that can take place between two or more systems concurrently (e.g., excavation of entrance and exit pits for water and sewer systems is an example of trenchless rehabilitation for both systems, traffic control devices, excavation and backfilling of common areas, traffic control systems set up, residents' notification, and site reinstatement works). Table 9.1 lists the intervention activities among the three systems along with their duration categories (e.g., SD_i, PD_{ijk}, and JD_{ijk}). Those durations are formulated as the summation of the activities' durations under the same category, as shown in Equations 9.1–9.4:

$$SD_{i_{ot}} = \sum_{m=1}^{M} UR_{m_t} \times Q_{m_{ot}} \times IA_{o_{i_{tr}}} \tag{9.1}$$

$$PD_{ijk_{ot}} = \sum_{y=1}^{Y} \left(\max.\left[UR_{y_t} \times Q_{y_{ot}} \times IA_{o_{i_{tr}}} \right] \right) \tag{9.2}$$

TABLE 9.1
List of intervention activities along with their categories and units of measurement

Intervention activities	Roads (R)	Water (W)	Sewer (S)	Category	Parallel	Unit
Site reinstatement	■	■	■	JD_{RWS}	Yes	m′
Joint excavation and shuttering		■	■	JD_{WS}	No	m^3
Sewer excavation and shuttering			■	SD_S	No	m^3
Joint backfilling and compaction		■	■	SD_{WS}	No	m^3
Sewer backfilling and compaction			■	SD_S	No	m^3
Reinstating sewer laterals			■	SD_S	Yes	No.
Traffic control systems	■	■	■	JD_{RWS}	Yes	day
Resident notification	■	■	■	JD_{RWS}	Yes	No.
Excavation of entrance and exit pits		■	■	JD_{WS}	Yes	m^3
Installation of sewer manholes			■	SD_S	Yes	No.
Water pipe repair/ installation		■		SD_W	Yes	m′
Sewer pipe repair/ installation			■	SD_S	Yes	m′
Surface overlay	■			SD_R	No	m^2
Road resurfacing	■			SD_R	No	m^2
Water pipe bedding		■		SD_W	Yes	m′
Sewer pipe bedding			■	SD_S	Yes	m′
Water main pipe leak repair		■		SD_W	Yes	leak
Sewer main pipe leak repair		■		SD_S	Yes	leak
Water pipe installation (Trenchless)		■		SD_W	Yes	m′
Sewer pipe installation (Trenchless)			■	SD_S	Yes	m′

■ Intervention
No intervention

$$JD_{ij_{o_t}} = \sum_{e=1}^{E} UR_{e_t} \times Q_{e_{o_t}} \times IA_{o_{i_{t_r}}} \quad (9.3)$$

$$JD_{ijk_{o_t}} = \sum_{q=1}^{Q} UR_{q_t} \times Q_{q_{o_t}} \times IA_{o_{i_{t_r}}} \quad (9.4)$$

where $SD_{i_{o_t}}$ is the SD of system *i* in corridor *o* at point of time *t* (hours); *o* is the corridors' counter (number); *t* is the point of time in the planning horizon *T* (years); $IA_{o_{i_{t_r}}}$ is a binary intervention applicability index for intervention *r* at time *t* for system *i* within corridor *o* (0 or 1); *m*, *y*, *e*, and *q* are the counters of the standalone, parallel, and joint activities for system *i*, *j*, and *k* (number), respectively; *M*, *Y*, *E*, and *Q* are the number of standalone, parallel, and joint activities for systems *i*, *j*, and *k* (number), respectively; UR_{m_t} is the unit rate for activity *m* at point of time *t* (hours/unit); $Q_{m_{o_t}}$ is the quantity of work needed to complete activity m in corridor *o* at point of time *t* (varies according to the units of measurement of the work); $PD_{ijk_{o_t}}$ is the PD for intervention actions of systems *i*, *j*, and *k* in corridor *o* at point of time *t* (hours); $JD_{ij_{o_t}}$ is the JD required for intervention actions of systems *i* and *j* of corridor *o* at point of time *t* (hours); and $JD_{ijk_{o_t}}$ is the JD required for intervention activities in all the three systems *i*, *j*, and *k* of corridor *o* at point of time *t* (hours).

The model development went through the following four phases: (1) activity breakdown, (2) activity categorization, (3) duration computation, and (4) network coordination ratio computation. The first phase is breaking down the intervention cost centers into tangible activities along with their estimated unit rates, taking into consideration the varying attributes among different corridors (e.g., corridor length, number of residents, soil type, pipe type, and diameter). Then, the second phase is categorizing those activities and identifying the parallel activities for each coordination scenario, as detailed in the corridor intervention activities displayed in Table 9.1. This list represents the main intervention activities that are considered in the system along with their categories, applicability to be undertaken in parallel with other intervention activities, and units of measurements. The list of intervention was identified after conducting an exhaustive literature review on the existing Bill of Quantities (BOQs) and tender documents that include detailed activity list along with their unit rates and costs. However, municipalities and maintenance contractors could input their own list of intervention activities, identify the parallel activities, and undertake the computations accordingly.

According to the predefined corridor intervention activity list, the unit rates associated with those intervention activities were adopted from several Canadian BOQs and tender documents and validated by experts from various Canadian municipalities, as shown in Tables 9.2 and 9.3. However, given that municipalities' and maintenance contractors' productivity rates might vary due to several factors such as maintenance technology, available crew, crew size, number of projects, project scale and extent, availability of experienced crews, etc., the user can change those unit rates and undertake the computations.

TABLE 9.2
Pipes' intervention repair time per street category per leak (Hachey 2018)

Road category	Repair time (hours)
Local roads	4
Main roads	10
Arterial roads	16

TABLE 9.3
Intervention activity unit rates

Intervention activities	Unit	Unit rate (hour/unit)
Site reinstatement	m′	25
Joint excavation and shuttering	m^3	84.82
Sewer excavation and shuttering	m^3	84.82
Joint backfilling and compaction	m^3	19.22
Sewer backfilling and compaction	m^3	19.22
Reinstating sewer laterals	No.	6
Traffic control systems	Day	13.77
Residents notification	No.	250
Excavation of entrance and exit pits	m^3	60
Installation of sewer manholes	No.	10
Water pipe repair/installation	m′	11
Sewer pipe repair/installation	m′	12
Surface overlay	m^2	1
Road resurfacing	m^2	13
Water pipe bedding	m′	15
Sewer pipe bedding	m′	16
Water main pipe leak repair	Leak	17
Sewer main pipe leak repair	Leak	18
Water pipe installation (Trenchless)	m′	10
Sewer pipe installation (Trenchless)	m′	8

Network Coordination Ratio

The third phase is computing the durations for three intervention scenarios. Let asset standalone duration (ASD_i) represent the duration of all the intervention activities required for system i without interruptions, assuming no coordination takes place; and corridor coordinated duration (CCD) represents the total duration of the entire project, assuming either partial or fully coordinated scenarios. Finally, the fourth phase is computing the network coordination ratio (NCR), which reflects the potential time savings that could be attained from coordinating the intervention activities, either partially or fully, during the execution phase. The less the NCR is, the less the extent of time savings resulting from the coordination. A ratio of 100% represents no possible time savings due to the absence of either joint activities or activities that can be undertaken in parallel. They could be mathematically formulated as follows:

$$\text{ASD}_{i_{o_t}} = \text{SD}_{i_{o_t}} + \text{JD}_{ijk_{o_t}} + \sum_{i=1}^{n_s} \text{JD}_{ij_{o_t}} \ (i \neq j) \tag{9.5}$$

$$\text{CCD}_{o_t} = \sum_{i=1}^{n_s} \text{SD}_{i_{o_t}} + \sum_{i=1}^{n_s} \text{PD}_{ijk_{o_t}} + \left\{\text{JD}_{ijk_{o_t}} \times n_a\right\} + \sum_{i=1}^{n_s}\sum_{j=1}^{n_s} \text{JD}_{ij_{o_t}} \ (i \neq j) \tag{9.6}$$

$$\text{NCR}_t = \sum_{o=1}^{O} \left(\frac{\sum_{i=1}^{n_s} \text{ASD}_{i_{o_t}}}{\text{CCD}_{o_t}} \right) \tag{9.7}$$

$$\text{NCR}\left(I_1^-\right) = \overline{\text{NCR}_t} \tag{9.8}$$

where $\text{ASD}_{i_{o_t}}$ is the asset SD for all the systems n_s in corridor o at point of time t (hours); i is the counter for the systems (number); n_s is the total number of systems (number); CCD_{o_t} is the corridor coordination duration for all the systems n_s in corridor o at point of time t (hours); n_a is the number of intervention actions that occurred at the same corridor (number); j is the counter for the systems (number); NCR_t is the network coordination ratio at point of time t (%); O is the total number of corridors (number); and NCR and $\overline{\text{NCR}_t}$ are the average network coordination ratio across the planning horizon T (%).

SPATIAL KPI

The objective of this KPI is to compute the spatial considerations and potential spatiotemporal savings arising from coordinating the interventions compared to undertaking them in silos. The spatial intervention savings model considers the amount of space needed to be occupied while undertaking any intervention. The spatial interdependency savings model flowchart is shown in Figure 9.3.

Based on the lane rental approach (Scott et al. 2006), which is applied on the roads for expediting their rehabilitation works, asset managers aim at minimizing the space, time, and disruption caused by maintenance contractors while undertaking the interventions. To better understand the theory, let us assume that A_i is the amount of space needed to be utilized during the rehabilitation for system i in the case of no coordination. Therefore, the total area required in case of no coordination will be the sum of

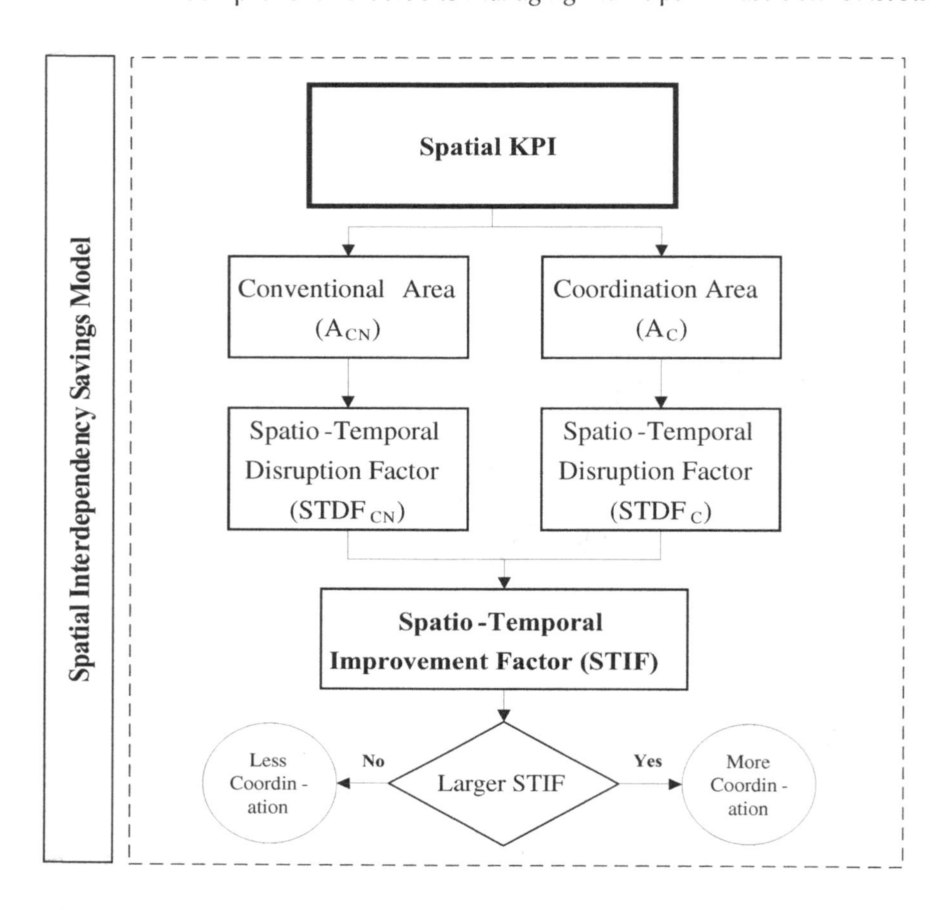

FIGURE 9.3 Spatial interdependency savings model flowchart.

the rehabilitation areas of the three right-of-way assets ($A_R + A_W + A_S$), representing the area of roads, water, and sewer, respectively, as shown in Equation 9.9. On the contrary, due to the spatial overlap among the systems sharing the same right-of-way, the total area required to undertake the rehabilitation for both the partially coordinated and fully coordinated intervention scenarios could be referred to as A_{PC} and A_C, and they can be mathematically calculated, as shown in Equations 9.10 and 9.11, respectively. To build the model, the extent and required areas for each system have been separately identified, and spatial interdependencies in partially coordinated and fully coordinated intervention scenarios have been identified to compute the above-mentioned areas (A_i, A_j, and A_k). Hence after, the duration model outcomes, ASD_{i_o} and CCD_o, have been used to represent the time a specific area will be occupied for undertaking the rehabilitation. As such, the spatiotemporal disruption factor (STDF) integrates the spatial and temporal dimensions for conventional, partially coordinated, and fully coordinated intervention scenarios, as shown in Equations 9.12, 9.13, and 9.14, respectively.

$$A_{\mathrm{CN}_{ot}} = \sum_{i=1}^{n_s} A_{i_t} \times \mathrm{IA}_{o_{i_{t_r}}} \tag{9.9}$$

$$A_{\mathrm{PC}_{ot}} = \left[A_{i_t} + \left(a \times A_{j_t}\right) + A_{k_t}\right] \times \mathrm{IA}_{o_{i_{t_r}}} \tag{9.10}$$

$$A_{\mathrm{C}_{ot}} = \left[A_{i_t} + \left(a \times A_{j_t}\right) + \left(b \times A_{k_t}\right)\right] \times \mathrm{IA}_{o_{i_{t_r}}} \tag{9.11}$$

$$\mathrm{STDF}_{\mathrm{CN}_t} = \sum_{o=1}^{O} \sum_{i=1}^{n_s} \mathrm{ASD}_{i_{ot}} \times A_{\mathrm{CN}_{ot}} \tag{9.12}$$

$$\mathrm{STDF}_{\mathrm{PC}_t} = \sum_{o=1}^{O} \mathrm{CCD}_{o\mathrm{PC}_t} \times A_{\mathrm{PC}_{ot}} \tag{9.13}$$

$$\mathrm{STDF}_{\mathrm{C}_t} = \sum_{o=1}^{O} \mathrm{CCD}_{o\mathrm{C}_t} \times A_{\mathrm{C}_{ot}} \tag{9.14}$$

where $A_{\mathrm{CN}_{ot}}$ is the maintenance/rehabilitation area of all the systems n_s in corridor o at point of time t for the conventional intervention scenario (m^2); A_{i_t} is the maintenance/rehabilitation area of system i (m^2); $A_{\mathrm{PC}_{ot}}$ is the maintenance/rehabilitation area of all systems n_s in corridor o at point of time t for the partially coordinated intervention scenario (m^2); a is the percentage of independent area of system j needed to undertake the maintenance/rehabilitation works in the cases of partially coordinated and fully coordinated intervention scenarios ($1< a <0$) (%); A_{j_t} is the maintenance/rehabilitation area of system j for both partially coordinated and fully coordinated intervention scenarios (m^2); A_{k_t} is the maintenance/rehabilitation area of system k in the case of the fully coordinated intervention scenario (m^2); $A_{\mathrm{C}_{ot}}$ is the maintenance/rehabilitation area of all systems n_s in corridor o at point of time t for the fully coordinated intervention scenario (m^2); b is the percentage of independent area of system k needed to undertake the maintenance/rehabilitation works in case of partially coordinated and fully coordinated intervention scenario ($1<b<0$) (%); $\mathrm{STDF}_{\mathrm{CN}_t}$ is the STDF in the case of conventional network intervention scenario at point of time t (%); $\mathrm{STDF}_{\mathrm{PC}_t}$ is the STDF in the case of partially coordinated network intervention scenario at point of time t (%); and $\mathrm{STDF}_{\mathrm{C}_t}$ is the STDF in the case of fully coordinated network intervention scenario at point of time t (%).

Spatiotemporal Improvement Factor (STIF)

The potential spatial savings of the fully coordinated and partially coordinated intervention scenarios over the conventional intervention scenario could be visualized in Figure 9.4. It is obvious that the fully coordinated scenario will consume less area

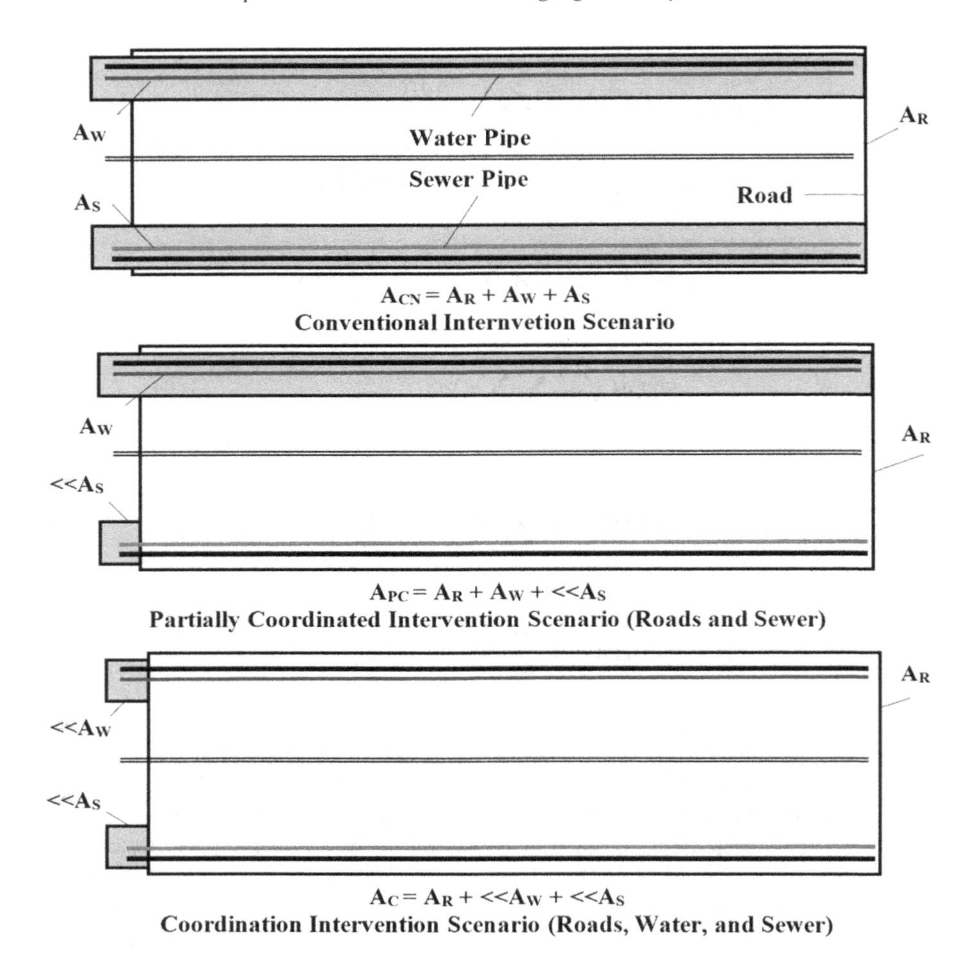

FIGURE 9.4 Area savings for different coordination intervention scenarios.

as opposed to the partially coordinated and conventional scenarios ($A_C < A_{PC} < A_{CN}$). This is simply because of the spatial interdependency among the co-located assets. For instance, a water or sewer pipe rehabilitation requires demolishing and reconstructing the above road section, causing duplication of work within relatively short time spans and accordingly extra nuisance to the public. Finally, a STIF is computed to compare the fully coordinated and partially coordinated intervention scenarios with the conventional one in terms of space and time savings, as shown in Equations 9.15, 9.16, 9.17, and 9.18, respectively. For instance, an STIF of “2” indicates that the considered intervention scenario consumes two times less time and space compared to the conventional intervention scenario.

$$\mathrm{STIF}_{\mathrm{PC}_t} = \frac{\mathrm{STDF}_{\mathrm{CN}_t}}{\mathrm{STDF}_{\mathrm{PC}_t}} \tag{9.15}$$

$$\text{STIF}_{\text{C}_t} = \frac{\text{STDF}_{\text{CN}_t}}{\text{STDF}_{\text{C}_t}} \tag{9.16}$$

$$\text{STIF}_{\text{PC}}\left(I_2^{-}\right) = \overline{\text{STIF}_{\text{PC}_t}} \tag{9.17}$$

$$\text{STIF}_{\text{C}}\left(I_2^{-}\right) = \overline{\text{STIF}_{\text{C}_t}} \tag{9.18}$$

where $\text{STIF}_{\text{PC}_t}$ is the spatiotemporal impact factor that compares the partially coordinated network intervention scenario with the conventional intervention one in terms of space and time at point of time t (%); STIF_{C_t} is the spatiotemporal impact factor that compares the fully coordinated network intervention scenario with the conventional intervention one in terms of space and time at point of time t (%); STIF_{PC} and $\overline{\text{STIF}_{\text{PC}_t}}$ are the average spatiotemporal impact factors that compares the partially coordinated network intervention scenario with the conventional intervention one in terms of space and time across the planning horizon T (%); and STIF_{C} and $\overline{\text{STIF}_{\text{C}_t}}$ are the average spatiotemporal impact factor that compares the fully coordinated network intervention scenario with the conventional intervention one in terms of space and time across the planning horizon T (%).

FINANCIAL KPI

The objective of this KPI is to compute the intervention costs and the potential savings arising from coordinating the interventions compared to undertaking them in silos. The financial savings model calculates the direct and indirect ownership and operational costs of the systems, as displayed in Figure 9.5.

The direct costs represent the costs of the intervention activities needed to be undertaken throughout the planning horizon to deliver the services in an "acceptable" manner without interruption (e.g., site preparation, residents' notification, installation of traffic control systems, road reconstruction, and pipe laying). On the contrary, the indirect costs, sometimes referred to as "Social" or "User" costs, reflect all the costs that are not directly related to the intervention (e.g., traffic disruption, vehicles or properties repair, business loss, noise disturbance, dirt and dust, and environmental or health and safety issues). Those costs are "*the estimated daily cost to the traveling public resulting from the construction work being performed*" (Daniels et al. 2000). They are also "*the consideration of opportunity cost of time for drivers when inconvenienced due to infrastructure overtime*". They refer to the lost time caused by several factors including detours and rerouting that add to travel time; reduced roadway capacity that slows the travel speed and increases the travel time; and delays the opening of a new or improved facility that prevents users from gaining travel time benefits. Accordingly, they were calculated based on the following factors: (1) average annual daily traffic (AADT) for each corridor; (2) user costs per vehicle per hour; (3) passenger cars vs truck ratio; (4) corridor speed limit; and (5) intervention time per unit length. Those costs are subjective and rely on probabilistic approaches for predicting their amounts over the systems' service lives (Qin and Cutler 2014; TDOT 2015). The calculations of the indirect costs were based on the output of the duration savings and spatial models to

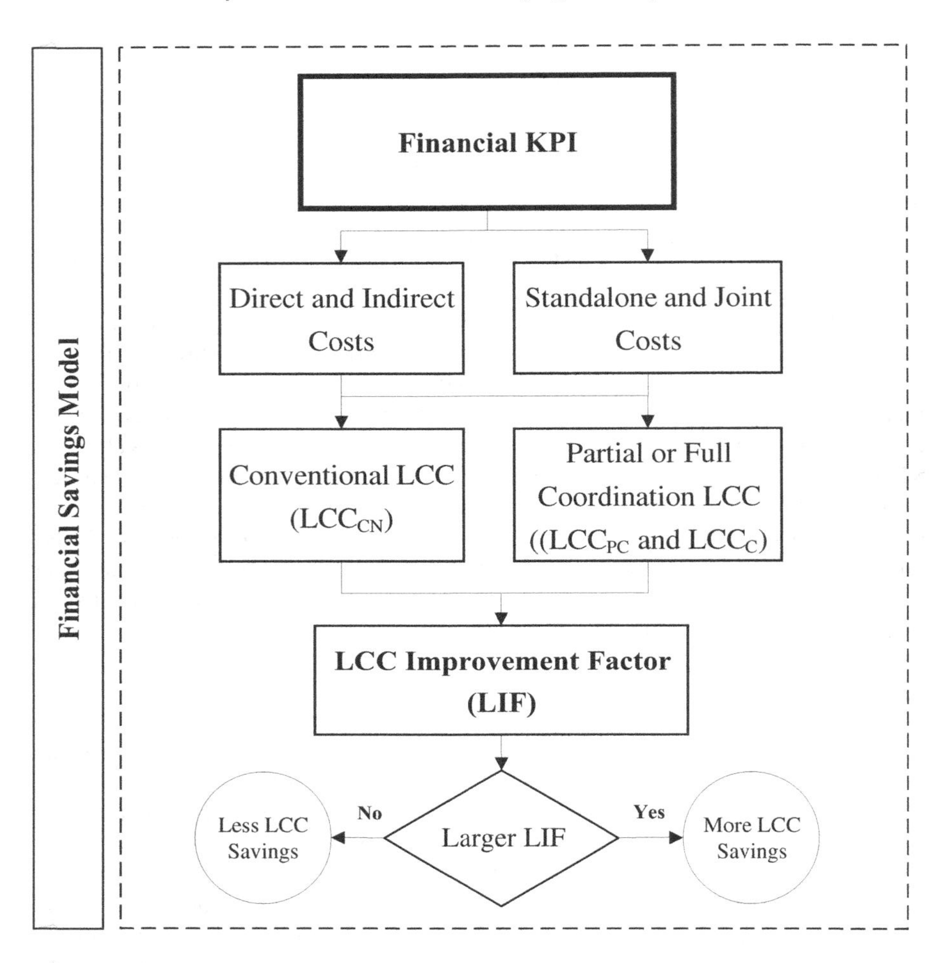

FIGURE 9.5 Financial savings model flowchart.

consider the temporal and spatial extent of disruption. To compute the life-cycle costs (LCC) for each intervention scenario, the cost centers were divided into the following three categories: (1) standalone direct and indirect costs for system i (SDC_i and SIC_i), (2) Joint Direct and Indirect cost centers between systems i and j (JDC_{ij} and JIC_{ij}), and (3) joint direct and indirect cost centers among systems i, j, and k (JDC_{ijk} and JIC_{ijk}). The mathematical computations of the direct and indirect costs for the three categories are shown in the below equations:

$$\mathrm{SDC}_{i_{ot}} = \sum_{m=1}^{M} Q_{m_{ot}} \times \mathrm{UC}_{m_t} \times \mathrm{IA}_{o_{i_{t_r}}} \tag{9.19}$$

$$\mathrm{SIC}_{i_{ot}} = \mathrm{ASD}_{i_t} \times \frac{A_{i_{ot}}}{\sum_{i=1}^{n_s} A_{i_{ot}}} \times \left[\left((1-T_p)\times \mathrm{AADT}\times \mathrm{UUC}_T\right) + \left(T_p \times \mathrm{AADT}\times \mathrm{UUC}_p\right)\right] \tag{9.20}$$

$$\text{JDC}_{ij_{o_t}} = \sum_{y=1}^{Y} Q_{y_{o_t}} \times \text{UC}_{y_t} \times \text{IA}_{o_{i_{t_r}}} \tag{9.21}$$

$$\text{JIC}_{ij_{o_t}} = \text{CCD}_{o_{\text{PC}_t}} \times \frac{A_{\text{PC}_{o_t}}}{\sum_{i=1}^{n_s} A_{i_{o_t}}} \times \left[\left(\left(1-T_p\right) \times \text{AADT} \times \text{UUC}_T\right) + \left(T_p \times \text{AADT} \times \text{UUC}_p\right)\right] \tag{9.22}$$

$$\text{JDC}_{ijk_{o_t}} = \sum_{e=1}^{E} Q_{e_{o_t}} \times \text{UC}_{e_t} \times \text{IA}_{o_{i_{t_r}}} \tag{9.23}$$

$$\text{JIC}_{ijk_{o_t}} = \text{CCD}_{o_{C_t}} \times \frac{A_{C_{o_t}}}{\sum_{i=1}^{n_s} A_{i_{o_t}}} \times \left[\left(\left(1-T_p\right) \times \text{AADT} \times \text{UUC}_T\right) + \left(T_p \times \text{AADT} \times \text{UUC}_p\right)\right] \tag{9.24}$$

where $\text{SDC}_{i_{o_t}}$ is the total direct costs for the standalone activities of system i in corridor o at point of time t ($); UC_{m_t} is the unit cost for each standalone activity in system i at point of time t ($); $\text{SIC}_{i_{o_t}}$ is the total indirect costs for the standalone activities of system i in corridor o at point of time t ($); $A_{i_{o_t}}$ is the maintenance/rehabilitation area of system i at point of time t (m^2); T_p is the percentage of trucks (%); AADT is the average annual daily traffic representing the average number of daily vehicles (vehicles); UUC_p is the unit user cost for the passenger cars ($); UUC_T is the unit user cost for the trucks ($); $\text{JDC}_{ij_{o_t}}$ is the total direct costs for the joint activities between systems i and j in corridor o at point of time t ($); $\text{JIC}_{ij_{o_t}}$ is the total indirect costs for the joint activities between systems i and j in corridor o at point of time t ($); $\text{CCD}_{o_{\text{PC}_t}}$ is the corridor coordination duration of corridor o for the partially coordinated intervention scenario at point of time t (days); JDC_{ijk_t} is the total direct costs for the joint activities among systems i, j, and k in corridor o at point of time t ($); $\text{JIC}_{ijk_{o_t}}$ is the total indirect costs for the joint activities among systems i, j, and k in corridor o at point of time t ($); and $\text{CCD}_{o_{C_t}}$ is the corridor coordinated duration of corridor o for the fully coordinated intervention scenario at point of time t (days).

According to the predefined corridor intervention activities list, outlined in Table 9.3, the unit costs associated with those intervention activities were adopted from recent Canadian recent BOQs and tender documents, taking the time value of money into consideration, as shown in Tables 9.4 and 9.5 (Abusamra 2019). However, given that activities' unit costs could vary due to several factors such as location, technology, available crews, crew size, number of projects, project scale and extent, and availability of experienced crews, they can be changed, and computations will be undertaken accordingly.

TABLE 9.4
Intervention activity unit costs

Intervention activities	Unit	Unit cost ($/unit)
Site reinstatement	m′	$660
Joint excavation and shuttering	m^3	$459
Sewer excavation and shuttering	m^3	$459
Joint backfilling and compaction	m^3	$303
Sewer backfilling and compaction	m^3	$303
Reinstating sewer laterals	No	Varies**
Traffic control systems	Day	$190
Residents notification	No	$3
Excavation of entrance and exit pits	m^3	Varies**
Installation of sewer manholes	No	$950
Water pipe repair/installation	m′	Varies**
Sewer pipe repair/installation	m′	Varies**
Surface overlay	m^2	$25
Road resurfacing	m^2	$65
Water pipe bedding	m′	$157
Sewer pipe bedding	m′	$157
Water main pipe leak repair	Leak	$825
Sewer main pipe leak repair	Leak	$938
Water pipe installation (Trenchless)	m′	Varies**
Sewer pipe installation (Trenchless)	m′	Varies**

Varies** represents a varying intervention unit cost depending on the pipe diameter and material.

TABLE 9.5
Water intervention repair costs per street category per leak (Hachey 2018)

Road category	Repair costs ($)
Local roads	$8,000
Main roads	$10,000
Arterial roads	$12,000

Lifecycle Cost Improvement Factor

Once the unit costs are established, the LCC could be calculated for the three intervention scenarios. The conventional intervention scenario will result in the highest cost given that all the joint direct and indirect cost centers, either between two systems or among the three systems, will be applied n_s times, drastically increasing the direct and indirect costs. However, the partially coordinated intervention scenario will experience n_a repetitions for the joint activities as there have been some potential activities that were not coordinated. Thenceforth, the fully coordinated intervention scenario will not experience any repetitions as the systems were fully coordinated and all the potentially coordinated activities were applied only once, decreasing the overall costs and the amount/extent of disruption across the planning horizon. Finally, the LCC Improvement Factor (LIF) was calculated to compare the partially coordinated or fully coordinated intervention scenarios with the conventional intervention scenario and quantify their potential cost savings. For instance, a LIF of "2" indicates that the fully coordinated intervention scenario utilizes two times less cost compared to the conventional intervention scenario. The mathematical computations of the LCC and LIF for the three categories are shown in the below equations:

$$C_{\mathrm{CN}_{ot}} = \left[\sum_{i=1}^{n_s}\left(\mathrm{SDC}_{i_{ot}} + \mathrm{SIC}_{i_{ot}}\right)\right] + \left[\sum_{i=1}^{n_s}\sum_{j=1}^{n_s}\left(\mathrm{JDC}_{ij_{ot}} + \mathrm{JIC}_{ij_{ot}}\right)\times n_s\right] + \left[\sum_{i=1}^{n_s}\sum_{j=1}^{n_s}\sum_{k=1}^{n_s}\left(\mathrm{JDC}_{ijk_{ot}} + \mathrm{JIC}_{ijk_{ot}}\right)\times n_s\right] + \left[P_{O_t} - I_{O_t}\right] \tag{9.25}$$

$$C_{\mathrm{PC}_{ot}} = \left[\sum_{i=1}^{n_s}\left(\mathrm{SDC}_{i_{ot}} + \mathrm{SIC}_{i_{ot}}\right)\right] + \left[\sum_{i=1}^{n_s}\sum_{j=1}^{n_s}\left(\mathrm{JDC}_{ij_{ot}} + \mathrm{JIC}_{ij_{ot}}\right)\times n_a\right] + \left[\sum_{i=1}^{n_s}\sum_{j=1}^{n_s}\sum_{k=1}^{n_s}\left(\mathrm{JDC}_{ijk_{ot}} + \mathrm{JIC}_{ijk_{ot}}\right)\times n_a\right] + \left[P_{O_t} - I_{O_t}\right] \tag{9.26}$$

$$C_{\mathrm{C}_{ot}} = \left[\sum_{i=1}^{n_s}\left(\mathrm{SDC}_{i_{ot}} + \mathrm{SIC}_{i_{ot}}\right)\right] + \left[\sum_{i=1}^{n_s}\sum_{j=1}^{n_s}\left(\mathrm{JDC}_{ij_{ot}} + \mathrm{JIC}_{ij_{ot}}\right)\right] + \left[\sum_{i=1}^{n_s}\sum_{j=1}^{n_s}\sum_{k=1}^{n_s}\left(\mathrm{JDC}_{ijk_{ot}} + \mathrm{JIC}_{ijk_{ot}}\right)\right] + \left[P_{O_t} - I_{O_t}\right] \tag{9.27}$$

$$\mathrm{LCC}_{\mathrm{CN}_o} = \sum_{t=1}^{T} C_{\mathrm{CN}_{ot}} \tag{9.28}$$

$$\mathrm{LCC}_{\mathrm{PC}_o} = \sum_{t=1}^{T} C_{\mathrm{PC}_{ot}} \tag{9.29}$$

$$\mathrm{LCC}_{\mathrm{C}_o} = \sum_{t=1}^{T} C_{\mathrm{C}_{ot}} \tag{9.30}$$

$$\mathrm{LIF}_{\mathrm{PC}}\left(I_3^-\right) = \sum_{o=1}^{O} \left(\frac{\mathrm{LCC}_{\mathrm{CN}_o}}{\mathrm{LCC}_{\mathrm{PC}_o}} \right) \tag{9.31}$$

$$\mathrm{LIF}_{\mathrm{C}}\left(I_3^-\right) = \sum_{o=1}^{O} \left(\frac{\mathrm{LCC}_{\mathrm{CN}_o}}{\mathrm{LCC}_{\mathrm{C}_o}} \right) \tag{9.32}$$

where $C_{\mathrm{CN}_{ot}}$ is the maintenance and rehabilitation costs of corridor o for the conventional intervention scenario at point of time t ($); P_{o_t} is the financial penalty applied on corridor o for all systems n_s over the planning horizon ($); I_{o_t} is the financial incentive applied on corridor o for all systems n_s over the planning horizon ($); $C_{\mathrm{PC}_{ot}}$ is the maintenance and rehabilitation of corridor o for the partially coordinated intervention scenario at point of time t ($); $C_{\mathrm{C}_{ot}}$ is the maintenance and rehabilitation of corridor o for the fully coordinated intervention scenario at point of time t ($); $\mathrm{LCC}_{\mathrm{CN}_o}$ is the lifecycle costs of corridor o for the conventional intervention scenario ($); $\mathrm{LCC}_{\mathrm{PC}_o}$ is the lifecycle costs of corridor o for the partially coordinated intervention scenario ($); $\mathrm{LCC}_{\mathrm{C}_o}$ is the lifecycle costs of corridor o for the fully coordinated intervention scenario ($); $\mathrm{LIF}_{\mathrm{PC}}$ is the lifecycle cost impact factor of the partially coordinated network intervention scenario over the conventional intervention one (%); and $\mathrm{LIF}_{\mathrm{C}}$ is the lifecycle cost impact factor of the fully coordinated network intervention scenario over the conventional intervention one (%).

REFERENCES

Abusamra, S. (2019). “Coordination and Multi-Objective Optimization Framework for Managing Municipal Infrastructure under Performance-Based Contracts.” Doctoral of Philosophy Thesis, Concordia University, Montreal.

Daniels, G., Stockton, W., and Hundley, R. (2000). “Estimating Road User Costs Associated with Highway Construction Projects: Simplified Method.” Project # 10–61, *Transportation Research Record: Journal of the Transportation Research Board*, 1732(1), 70–79.

Hachey, N. (2018). “B/C Analysis Project for Montréal Water Network.” Interview, Sep 10, 2018, Montreal.

Scott, S., Molenaar, K., Gransberg, D., and Smith, N. (2006). *Best-Value Procurement Methods for Highway Construction Projects.* Project # 10–61, National Cooperative Highway Research Program, Transportation Research Board (TRB) and National Research Council (NRC), Washington, DC.

TDOT (Texas Department of Transportation). (2015). “Value of Time and User Costs.” https://www.researchgate.net/publication/362429765_Calculating_Road_User_Cost_for_Specific_Sections_of_Highway_for_Use_in_Alternative_Contracting_Project. Apr 25, 2023.

Qin, X. and Cutler, C. (2014). *Review of Road User Costs and Methods.* South Dakota University, Vermillion, SD.

10 Efficiency and Effectiveness KPIs

INTERVENTION EFFICIENCY AND EFFECTIVENESS MODELS

The intervention efficiency (utility cut) and effectiveness (free of maintenance) models support the transition from an "asset stewardship" approach that focuses on cost and condition while undertaking intervention decisions to an "asset serviceability" approach that is centered around the cost, performance, and risk exposure (Osman 2015). It is aligned with the concepts of the performance-based contracts (PBC) such that the assessment is based on the performance and not the level of exerted efforts. In that context, the models introduce efficiency and effectiveness measures to assess the intervention performance over the contractual period, as shown in Figure 10.1.

INTERVENTION EFFICIENCY

The efficiency represents the percentage of time consumed to undertake all the systems' intervention activities over the total disruption period to restore all the n_s systems. The disruption of the surrounding communities depends on the spacing between the intervention activities. For instance, the disruptive roadway intervention activities that are undertaken immediately after one another will cause more discomfort to the surrounding communities compared to the ones spaced over time. Consequently, many municipalities have taken steps toward implementing measures such as pavement moratoriums and pavement cutting fees to discourage any excavation within 3–6 years of a road's reconstruction (Wilde et al. 2003). To better understand the idea, let us assume the time to undertake the intervention as "disruption time (DT)", the time between one intervention and another as "inter-disruption time (IDT)", and the "total disruption time (TDT)" as the total time to undertake all the interventions for systems n_s, as shown in Figure 10.2. The larger the IDT between the undertaken interventions, the lower the corridor intervention efficiency is, which increases the nuisance to the surrounding environment and the residents will be wondering why those interventions were not coordinated together given the small-time frame they were undertaken at. Accordingly, an intervention efficiency factor (IEF) was computed to assess the overall efficiency of the intervention program for all n_s systems in terms of inter-disruption time extent for the conventional, partially coordinated, and fully coordinated intervention scenarios, as shown in Equations 10.1, 10.2, and 10.3, respectively. For instance, an IEF of "1" indicates a superior efficiency of the fully coordinated intervention scenario in which the IDT is equal to "0" and the TDT is equal to the DT of the fully coordinated intervention. However, an IEF of "0.5" indicates a 50% DT from the TDT, resulting in discomfort and nuisance to the surrounding environment and residents.

DOI: 10.1201/9781003327509-13

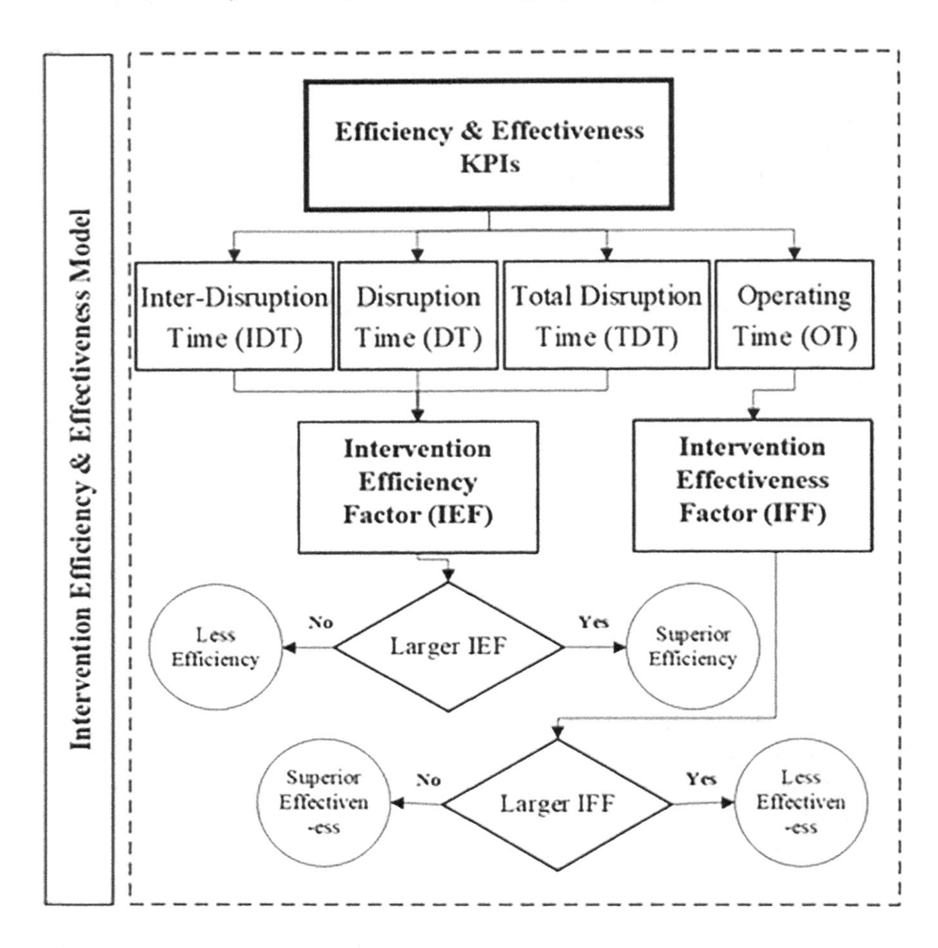

FIGURE 10.1 Intervention efficiency and effectiveness model flowchart (Abusamra 2019).

$$\text{IEF}_{\text{CN}}\left(I_4^-\right) = \sum_{o=1}^{O}\left(\frac{\sum_{i=1}^{n_s}\text{DT}_{i_o}}{\text{TDT}_o}\right) \tag{10.1}$$

$$\text{IEF}_{\text{PC}}\left(I_4^-\right) = \sum_{o=1}^{O}\left(\frac{\sum_{i=1}^{n_s}\sum_{j=1}^{n_s}\text{DT}_{ij_o} + \text{DT}_{k_o}}{\text{TDT}_o}\right) \tag{10.2}$$

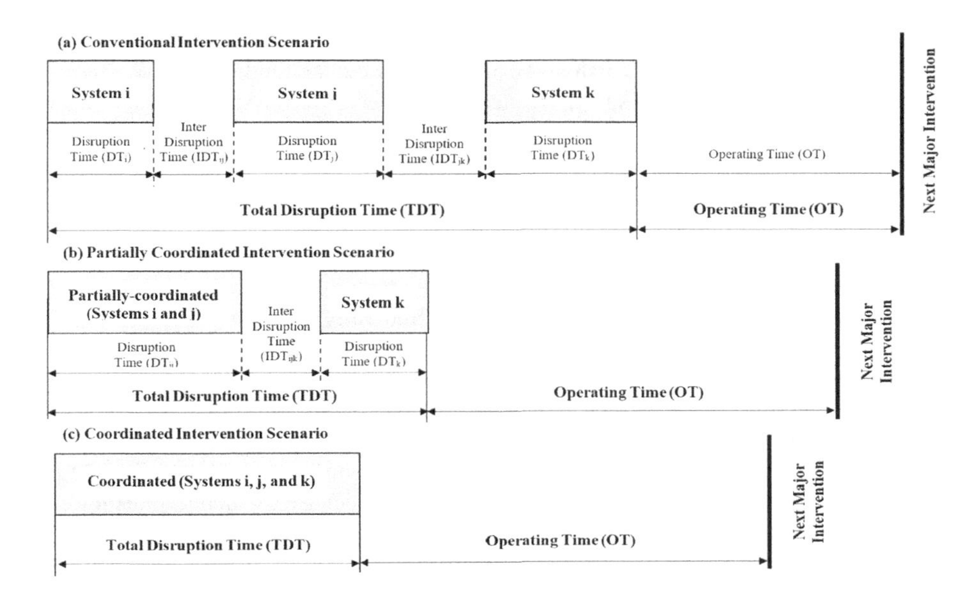

FIGURE 10.2 Disruption vs operating time for different intervention scenarios.

$$\text{IEF}_{\text{C}}\left(I_4^{-}\right)=\sum_{o=1}^{O}\left(\frac{\sum_{i=1}^{n_s}\sum_{j=1}^{n_s}\sum_{k=1}^{n_s}\text{DT}_{ijk_o}}{\text{TDT}_o}\right) \tag{10.3}$$

where IEF_{CN} is the intervention efficiency factor of the conventional network intervention program for systems n_s in terms of inter-disruption time extent (%); DT_{i_o} is the disruption time for system i in corridor o (days); TDT_o is the total disruption time to undertake all the interventions for systems n_s in corridor o (days); IEF_{PC} is the intervention efficiency factor of the partially coordinated network intervention program for systems n_s in terms of inter-disruption time extent (%); DT_{ij_o} is the disruption time for systems' i and j coordinated intervention in corridor o (days); IEF_{C} is the intervention efficiency factor of the fully coordinated network intervention program for systems n_s in terms of inter-disruption time extent (%); k is the counter for the systems (number); and DT_{ijk_o} is the disruption time for systems' i, j, and k coordinated intervention in corridor o (days).

Intervention Effectiveness

The effectiveness represents the performance/quality of the intervention. In simpler terms, it reflects the amount of operating time expected until the next major intervention, without undertaking any disruptive intervention activities, after the intervention program is accomplished. The time of the next major intervention relies on the following: (1) system's service life and deterioration rate; (2) occurrence of any unforeseeable risk event (e.g., earthquake and severe weather conditions); (3) compliance with new safety measures (e.g., banned pipe material); (4) system obsolescence; and (5) effectiveness of the undertaken intervention (e.g., condition impact factor, which represents the extent to which the intervention enhances the system's condition state). To better understand the idea, let us assume the time between the corridor's last and next intervention as "operating time (OT)". The larger OT reflects higher intervention effectiveness, longer system's service life and less disruption time. Thus, an intervention effectiveness factor (IFF) was computed to compare the interventions' effectiveness of the partially coordinated and fully coordinated intervention scenarios with the conventional one, as shown in Equations 10.4 and 10.5, respectively. If the IFF is less than 1 ($IFF < 1$), the considered coordination scenario reveals improved intervention effectiveness. However, if the IFF is more than 1 ($IFF > 1$), the considered coordination scenario reveals worse intervention effectiveness. For instance, an IFF of "1.2" indicates that the considered coordination scenario has 20% shorter OT compared to the conventional intervention scenario, which implies lower intervention effectiveness. However, an IFF of "0.3" indicates that the considered coordination scenario has 70% longer OT compared to the conventional intervention scenario, which implies superior intervention effectiveness.

$$\mathrm{IFF_{PC}}\left(I_5^+\right) = \sum_{o=1}^{O}\left(\frac{\mathrm{OT}_{\mathrm{CN}_o}}{\mathrm{OT}_{\mathrm{PC}_o}}\right) \tag{10.4}$$

$$\mathrm{IFF_{C}}\left(I_5^+\right) = \sum_{o=1}^{O}\left(\frac{\mathrm{OT}_{\mathrm{CN}_o}}{\mathrm{OT}_{\mathrm{C}_o}}\right) \tag{10.5}$$

where $\mathrm{IFF_{PC}}$ is the intervention effectiveness factor that compares the partially coordinated network intervention effectiveness with the conventional one in terms of operating time (%); $\mathrm{OT}_{\mathrm{CN}_o}$ is the operating time of corridor o in the case of conventional intervention scenario (days); $\mathrm{OT}_{\mathrm{PC}_o}$ is the operating time of corridor o in the case of partially coordinated intervention scenario (days); $\mathrm{IFF_C}$ is the intervention effectiveness factor that compares the fully coordinated network intervention effectiveness with the conventional intervention one in terms of OT (%); and $\mathrm{OT}_{\mathrm{C}_o}$ is the operating time of corridor o in the case of the fully coordinated intervention scenario (days).

REFERENCES

Abusamra, S. (2019). "Coordination and Multi-Objective Optimization Framework for Managing Municipal Infrastructure under Performance-Based Contracts." Doctoral of Philosophy Thesis, Concordia University, Montreal.

Osman, H. (2015). "Coordination of urban infrastructure reconstruction projects." *Structure and Infrastructure Engineering*, Taylor & Francis, 12(1), 108–121.

Wilde, W., Grant, C., and White, C. (2003). *Manual for Controlling and Reducing the Frequency of Pavement Utility Cuts.* The Transtec Group Inc., Austin, TX.

11 Risk KPI

INTRODUCTION

"*Municipal infrastructure is at risk*", a statement widely spread among decision-makers and stakeholders concerning their municipal infrastructure assets (Shahata and Zayed 2016; Shahata 2013). Municipalities are faced with challenging decisions to plan the coordinated repair/renewal for the roads, water, and sewer networks as they are sharing the same right-of-way. It is time for renewal, the cumulative grade point average (GPA) for US corridor infrastructure is "D", water networks are about to end their useful lives, and 75% of the wastewater capital needs rehabilitation (ASCE 2021). In Canada, the corridor infrastructure is in fair condition state, 52% of the roads need further attention, and the corridors' replacement cost is estimated at $171.8 billion (BCG 2017). In addition, this is reflected by the growing budget deficits among most of the municipalities who suffer lack of collaboration and coordination among various internal and external groups, which results in unnecessary restoration works, duplication of efforts, etc. The risk impact model aims at computing the risk index (RI) for each corridor through both the probability of failure (POF) and consequences of failure (COF), as shown in Figure 11.1. Even though different infrastructure systems feature different failure modes and deterioration rates, the POF was independently calculated for each system, based on the current and future condition state, as discussed in deterioration models' section. On the contrary, the COF was computed for each system based on the repair costs, loss of revenue, loss of service, loss of life, injury, health and environmental impacts, damage to surrounding infrastructure or property, failure to meet safety regulations, third party losses, loss of image, pollution, contamination, etc. Those parameters were split into four categories, as displayed in Table 11.1, and could be summarized as follows:

- **Economic**: Economic effect of the system's failure on monetary resources.
- **Operational**: Operational effect of the system's failure on the surrounding society.
- **Social**: Social effect of the system's failure on the surrounding society.
- **Environmental**: Environmental effect of the system's failure on the surrounding environment.

RISK FACTORS' AND SUB-FACTORS' WEIGHTS

Given that the risk factors' assessment is either quantitative or qualitative, a risk scoring system was needed to define and unify the risk assessment criteria among the different factors. Upon thorough review of the literature and interviews with experts, the risk factors and sub-factors along with their associated weights were developed. The POF

DOI: 10.1201/9781003327509-14

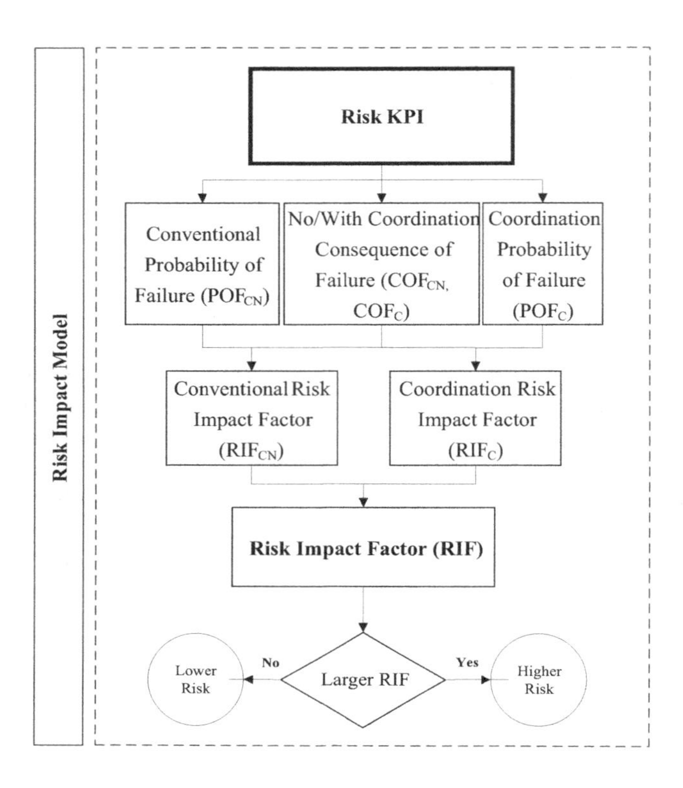

FIGURE 11.1 Risk model flowchart (Abusamra 2019).

TABLE.11.1
Risk categories along with their parameters

Effect category	Parameter
Economic	Repair costs
	Loss of revenue
Operational	Loss of production
	Damage to surrounding infrastructure or property
	Failure to meet safety regulations
Social	Loss of life
	Injury
	Loss of service
	Third party losses
	Loss of image
Environmental	Health and environmental impacts
	Pollution
	Contamination

was computed according to the output of the deterioration models for the n_s systems. The COF scoring (SW_{var_r}) ranges between 1 and 5 from insignificant to catastrophic, respectively, as shown in Table 11.2. The risk factors and sub-factors along with their weights of importance are defined in Table 11.3. Afterward, the Decomposed Weight (DW_{cr}) of each sub-factor was computed by multiplying the main factor (c) weight

TABLE 11.2
Risk consequences scoring description

Score	Consequence level	Description
1	Insignificant	No significant impact Little or no public exposure No impact on health risk Can be tolerated indefinitely
2	Minor	Limited public exposure Minor health risk Can be tolerated for an expected period of time
3	Moderate	Minor public exposure Health risk on a small part of the population Can be tolerated for a brief period of time (e.g., sufficient to plan and take action)
4	Major	Large part of the population at risk Requires expedient and/or emergency measures to address
5	Catastrophic	Major impact for a large part of the population at risk Complete failure of systems Requires extreme emergency measures

TABLE 11.3
Factors and sub-factors' weights and decomposed weights

Main factor	Sub-factor	Sub-factor weight (W_{var_r})(%)	Decomposed weight (DW_{cr}) (%)
Economic (39%)	Pipe size (diameter)	19	7.41
	Pipe depth	21	8.19
	Material (type of pipe)	16	6.24
	Land use	6	2.34
	Accessibility	28	10.92
	Road type	10	3.9
Operational (27%)	Business disruption – critical customer	33	8.91
	Hydraulic impact	18	4.86
	Pipe size (diameter)	16	4.32
	Damage to surrounding assets	33	8.91

(Continued)

TABLE 11.3 (*Continued*)
Factors and sub-factors' weights and decomposed weights

Main factor	Sub-factor	Sub-factor weight (W_{var_r})(%)	Decomposed weight (DW_{cr}) (%)
Environmental	Water body proximity	18	3.78
(21%)	Sensitive area	47	9.87
	Average daily traffic (road class)	24	5.04
	Type of soil	11	2.31
Social	No diversion	40	5.2
(13%)	Land use	10	1.3
	Transit route	20	2.6
	Average daily traffic (road class)	30	3.9

by its sub-factor weight (W_{var_r}) to represent the overall weight of the sub-factor (r), as displayed in Equation 11.1, and thus, the priority of each sub-factor (r) could be established. Each parameter has several attributes that do not have a similar effect on the COF. For instance, the pipe diameter sub-factor has various values: (1) less than or equal 300mm, (2) 300–450mm, (3) 450–750mm, (4) 750–1,200mm, and (5) greater or equal 1,200mm. The larger the pipe diameter is, the higher its COF score (SW_{var_r}) is, representing a higher financial impact in case of failure. The COF is calculated according to Equation 11.2. Similar criteria were established for the other sub-factors and for each asset class to compute the COF, as outlined in Tables 11.4, 11.5, and 11.6 for sewer mains, water mains, and roads, respectively.

$$DW_{cr} = W_{index_c} \times W_{var_r} \tag{11.1}$$

$$COF_o = \sum_{i}^{n_s} \sum_{c}^{C} \sum_{r}^{R} DW_{cr} \times SW_{var_{r\,i\,o}} \tag{11.2}$$

where DW_{cr} is the overall DW for each sub-factor (e.g., sub-factor r within factor c) (%); W_{index_c} is the weight of each factor (c) (e.g., economic, operational, environmental, and social) (%); W_{var_r} is the weight of each sub-factor r within the factor c (%); COF_o is the COF of corridor o including all the n_s systems along with their factors and sub-factors (0–5); c and r are the counters of the factors and sub-factors, respectively (number); C and R are the total number of factors and sub-factors, respectively (number); and $SW_{var_{r\,i\,o}}$ is the SW_{var} of sub-factor r in system i within corridor o (1–5).

On the contrary, the POF was calculated based on the future deterioration models, as discussed earlier in the deterioration models' section. The POF scale ranges between 0 and 1 representing "Rare" and "Almost certain" POF, respectively. Furthermore, the ranges of the POF vary according to the systems' nature of deterioration as well as their remaining service life as outlined in Table 11.7.

TABLE 11.4
Sewer main COF variable scoring

	Sewer COF variable scores			
	Factors	**COF score**	**Factors**	**COF score**
Economic parameters	**1.1 Pipe size (diameter)**		**1.4 Land use**	
	Less or equal 300 mm	1	Agricultural	1
	300–450 mm	2	Park / open space	2
	450–750 mm	3	Residential	3
	750–1,200 mm	4	Commercial	4
	Greater or equal 1,200 mm	5	Institutional	5
			Industrial	5
	1.2 Pipe depth			
	Less or equal 2.0 m	1	**1.5 Accessibility**	
	2.0–3.0 m	2	Good	1
	3.0–3.5 m	3	Marginal	3
	3.5–4.0 m	4	Low	5
	Greater or equal 4.0 m	5	**1.6 Material (type of pipe)**	
			Poly Vinyl Chloride (PVC)	1
	1.3 Road type			
	Local	1	Clay (CT, VC)	2
	Collector	2	Asbestos Cement (AC)	3
	Arterial	3	Corrugated Steel (CS)	3
	Custom (e.g., University)	4	Metallic (STL, DI, and CI)	4
	Expressway/highway	5	Concrete (RC)	5
Operational parameters	**2.1 Business disruption critical customer**		**2.4 Sewer main blockages**	
	Low	1	Low	1
	High (major users, hospitals, and health clinics)	5	Medium	3
			High	5
	2.2 Damage to surrounding assets			
	Low	1	**2.5 Pipe size (diameter)**	
	Medium	3	Less or equal 300 mm	1
	High	5	300–450 mm	2
			450–750 mm	3
	2.3 Hydraulic impact			
	$d/D \leq 0.5$	1	750–1,200 mm	4
	0.5–0.65	2	Greater or equal 1,200 mm	5
	0.65–0.75	3		
	0.75–0.85	4		
	$d/D \geq 0.85$	5		

(*Continued*)

TABLE 11.4 (*Continued*)
Sewer main COF variable scoring

Sewer COF variable scores				
	Factors	**COF score**	**Factors**	**COF score**
Environmental parameters	**3.1 Water body proximity**		**3.3 Sensitive area**	
	Greater or equal 200 m away	1	No	1
	101–200 m	2	Yes	5
	51–100 m	3	**3.4 Type of soil**	
	5–50 m	4	Nonaggressive	1
	Less or equal 5 m	5	Moderate	2
	3.2 Average daily traffic (road class)		Aggressive	3
	Low	1	Highly aggressive	5
	Moderate	3		
	Heavy	5		
Environmental parameters	**4.1 No diversion**		**4.3 Average daily traffic (road class)**	
	No	1	Low	1
	Yes	5	Moderate	3
	4.2 Land use		Heavy	5
	Agricultural	1	**4.4 Transit route**	
	Park / open space	2	No	1
	Residential	3	Yes	5
	Commercial	4		
	Institutional	5		
	Industrial	5		

RI

Risk management is an important driver in the decision-making process. It quantifies all the direct (labor, materials, and equipment) and indirect impacts (social, operational, and environmental) of failures and computes a RI for a corridor, inclusive of all its colocated assets. The corridor was divided into sub-corridor and super-corridor. The water and sewer networks represent the sub-corridor, and the roads represent the super-corridor. The POF of each asset class is computed from the deterioration models, as outlined in Equation 11.3. Thenceforth, the RI for the sub-corridor is computed through multiplying the POF and COF using the systems' weights of importance, as detailed in Equations 11.4, 11.5, and 11.6. For the super-corridor, the POF and COF for the super-corridor are equal to the POF and COF of the roads, and the RI is computed through multiplying the POF and COF as outlined in Equation 11.7. Finally, the RI for the corridor is computed at each point of time (t) across the planning horizon, taking into consideration the weights of the importance of the sub-corridor and super-corridor, as outlined in Equations 11.8 and 11.9.

TABLE 11.5
Water main COF variable scoring

	Water main COF variable scoresw			
	Factors	**COF score**	**Factors**	**COF score**
Economic parameters	**1.1 Pipe size (diameter)**		**1.5 Land use**	
	Less or equal 300 mm	1	Agricultural	1
	300–450 mm	2	Park/open space	2
	450–750 mm	3	Residential	3
	750–1,200 mm	4	Commercial	4
	Greater or equal 1,200 mm	5	Institutional	5
	1.2 Pipe depth		Industrial	5
	Less or equal 2.0 m	1	**1.6 Material (type of pipe)**	
	2.0–3.0 m	2	Galvanized Steel (GALV)	5
	3.0–3.5 m	3	Steel (ST)	4
	3.5–4.0 m	4	Pitted Cast Iron (CIP)	3
	Greater or equal 4.0 m	5	Spun Cast Iron (CIS)	3
	1.3 Road type		Ductile iron (DI)	4
	Local	1	Copper (CU)	5
	Collector	2	Concrete Pressure (CP)	5
	Arterial	3	Asbestos Cement (AC)	3
	Custom (e.g., University)	4	Poly Vinyl Chloride (PVC)	1
	Expressway/highway	5	High-Density Polyethylene (HDPE)	2
	1.4 Accessibility		Ductile Iron Hyprotech (DIHY)	2
	Good	1		
	Marginal	3		
	Low	5		
Operational parameters	**2.1 Business disruption critical customer**		**2.3 Hydraulic impact**	
	Low	1	Pass	1
	High (major users, hospitals, and health clinics)	5	Fail	5
	2.2 Damage to surrounding assets		**2.4 Pipe size (diameter)**	
	Low	1	Less or equal 300 mm	1
	Medium	3	300–450 mm	2
	High	5	450–750 mm	3
			750–1,200 mm	4
	Greater or equal 1,200 mm		5	

(*Continued*)

TABLE 11.5 (*Continued*)
Water main COF variable scoring

Water main COF variable scoresw				
	Factors	**COF score**	**Factors**	**COF score**
Environmental parameters	**3.1 Water body proximity**		**3.3 Sensitive area**	
	Greater or equal 200 m away	1	No	1
	101–200 m	2	Yes	5
	51–100 m	3	**3.4 Type of soil**	
	5–50 m	4	Nonaggressive	1
	Less or equal 5 m	5	Moderate	2
	3.2 Average daily traffic (road class)		Aggressive	3
	Low	1	Highly aggressive	5
	Moderate	3		
	Heavy	5		
Environmental parameters	**4.1 No diversion**		**4.3 Average daily traffic (road class)**	
	No	1	Low	1
	Yes	5	Moderate	3
	4.2 Land use		Heavy	5
	Agricultural	1	**4.4 Transit route**	
	Park / open space	2	No	1
	Residential	3	Yes	5
	Commercial	4		
	Institutional	5		
	Industrial	5		

TABLE 11.6
Road COF variable scoring

Road COF variable scores				
	Factors	**COF score**	**Factors**	**COF score**
Economic parameters	**1.1 Road size (#lanes)**		**1.3 Land use**	
	Local	1	Agricultural	1
	Collector—2 lane	1	Park / open space	2
	Collector—3 lane	2	Residential	3
	Arterial—2 lane	3	Commercial	4
	Arterial—3 lane	3	Institutional	5
	Arterial—4 lane	4	Industrial	5
	Arterial—5 lane	4	**1.4 Road width**	
	Arterial—6 lane	5	Less or equal 8.0 m	1
	Expressway—4 lane	5	8.0–12.0 m	2

(*Continued*)

TABLE 11.6 (*Continued*)
Road COF variable scoring

Road COF variable scores				
	Factors	**COF score**	**Factors**	**COF score**
			12.0–16.0 m	3
	1.2 Road type			
	Local	1	16.0–20.0 m	4
	Collector	2	Greater or equal 20.0 m	5
	Arterial	3	**1.5 Road material GST granular**	
	Custom (e.g., University)	4	Low-Class Bituminous	2
	Expressway/highway	5	High-Class Bituminous	3
			Asphalt over concrete	4
			Concrete 1	5
Operational parameters	**2.1 Business disruption critical customer**		**2.3 Road width**	
	Low	1	Less or equal 8.0 m	1
	High (major users, hospitals, and health clinics)	5	8.0–12.0 m	2
	2.2 Damage to surrounding assets		12.0–16.0 m	3
	Low	1	16.0–20.0 m	4
	Medium	3	Greater or equal 20.0 m	5
	High	5		
Environmental parameters	**3.1 Water body proximity**		**3.3 Sensitive area**	
	Greater or equal 200 m away	1	No	1
	101–200 m	2	Yes	5
	51–100 m	3	**3.4 Type of soil**	
	5–50 m	4	Nonaggressive	1
	Less or equal 5 m	5	Moderate	2
	3.2 Average daily traffic (road class)		Aggressive	3
	Low	1	Highly aggressive	5
	Moderate	3		
	Heavy	5		
Environmental parameters	**4.1 No diversion**		**4.3 Average daily traffic (road class)**	
	No	1	Low	1
	Yes	5	Moderate	3
	4.2 Land use		Heavy	5
	Agricultural	1	**4.4 Transit route**	
	Park / open space	2	No	1
	Residential	3	Yes	5
	Commercial	4		
	Institutional	5		
	Industrial	5		

TABLE 11.7
Probability of failure ranges and reliability

POF score range	POF description	System condition	Reliability values
0.9–1	Almost certain	Failing	Roads: $0 < R_r < 30$
			Water: $1 < R_w < 0.85$
			Sewer: $1 < R_s < 0.9$
0.7–0.9	Most likely	Poor	Roads: $30 < R_r < 45$
			Water: $0.85 < R_w < 0.65$
			Sewer: $0.9 < R_s < 0.7$
0.5–0.7	Likely	Fair	Roads: $45 < R_r < 60$
			Water: $0.65 < R_w < 0.35$
			Sewer: $0.7 < R_s < 0.4$
0.25–0.5	Unlikely	Good	Roads: $60 < R_r < 80$
			Water: $0.35 < R_w < 0.15$
			Sewer: $0.4 < R_s < 0.2$
0.01–0.25	Rare	Excellent	Roads: $80 < R_r < 100$
			Water: $0.15 < R_w < 0.01$
			Sewer: $0.2 < R_s < 0.01$

$$\text{POF}_{i_{ot}} = 1 - R_{i_{ot}} \tag{11.3}$$

$$\text{POF}_{\text{Sub}_{ot}} = \sum_{i=1}^{n_s} \left(W_i \times \text{POF}_{i_{ot}} \right) \tag{11.4}$$

$$\text{COF}_{\text{Sub}_{ot}} = \text{Max}\left(\text{COF}_{i_{ot}} \right) \tag{11.5}$$

$$\text{RI}_{\text{Sub}_{ot}} = \text{POF}_{\text{Sub}_{ot}} \times \text{COF}_{\text{Sub}_{ot}} \tag{11.6}$$

$$\text{RI}_{\text{Super}_{ot}} = \text{POF}_{\text{Super}_{ot}} \times \text{COF}_{\text{Super}_{ot}} \tag{11.7}$$

$$\text{RI}_{ot} = \left(W_{\text{Sub}} \times \text{RI}_{\text{Sub}_{ot}} \right) + \left(W_{\text{Super}} \times \text{RI}_{\text{Super}_{ot}} \right) \tag{11.8}$$

$$\text{RI}_o = \overline{\text{RI}_{ot}} \tag{11.9}$$

where $\text{POF}_{i_{ot}}$ is the probability of failure of system i in corridor o at point of time t (%); $R_{i_{ot}}$ is the condition state/reliability of system i in corridor o at point of time t (%); $\text{POF}_{\text{Sub}_{ot}}$ is the probability of failure of the sub-corridor of corridor o at point of time t (%); $\text{COF}_{\text{Sub}_{ot}}$ is the consequence of failure of the sub-corridor of corridor o at point of time t (0–5); $\text{COF}_{i_{ot}}$ is the consequence of failure of system i in corridor o at point of time t (0–5); $\text{RI}_{\text{Sub}_{ot}}$ is the RI of the sub-corridor in corridor o at point of time t (0–5); $\text{RI}_{\text{Super}_{ot}}$ is the RI of the super-corridor of corridor o at point of time t (0–5); $\text{POF}_{\text{Super}_{ot}}$ is the probability of failure of the super-corridor of corridor o at point of time t (%); $\text{COF}_{\text{Super}_{ot}}$ is the consequence of failure of the super-corridor of corridor o at point of time t (0–5); RI_{o_t} is the RI of corridor o at point of time t (0–5); W_{Sub} is the weight of importance assigned to the sub-corridor (%); W_{Super} is the weight of importance assigned to the super-corridor (%); and RI_o and $\overline{\mathrm{RI}_{o_t}}$ are the average RI for corridor o across the planning horizon (%).

The risk improvement factor (RIF) is computed to compare the partially coordinated and fully coordinated intervention scenarios with the conventional one in terms of risk improvement, given that the POF changes with time and according to the intervention actions. For instance, at the time of the first intervention action of the conventional intervention scenario, the POF of the two nonrehabilitated assets will be high as their reliability/condition states are low. However, at the time of the first intervention in the fully coordinated scenario, the POF of the three assets will be low as their reliability/condition states are high. Accordingly, the RIF computes the risk improvements of the partially coordinated and fully coordinated intervention scenarios compared to the conventional intervention scenario, as shown in Equations 11.10 and 11.11 respectively. If the RIF is more than 1 (RIF > 1), the considered coordination scenario reveals less risk exposure compared to the conventional scenario. However, if the RIF is less than 1 (RIF < 1), the considered coordination scenario reveals higher risk exposure compared to the conventional scenario. For instance, a RIF of "1.6" indicates that the considered intervention scenario has 60% less risk exposure compared to the conventional intervention scenario.

$$\mathrm{RIF}_{\mathrm{PC}}\left(I_8^-\right)=\sum_{o=1}^{O}\left(\frac{\mathrm{RI}_{\mathrm{CN}_o}}{\mathrm{RI}_{\mathrm{PC}_o}}\right) \tag{11.10}$$

$$\mathrm{RIF}_{\mathrm{C}}\left(I_8^-\right)=\sum_{o=1}^{O}\left(\frac{\mathrm{RI}_{\mathrm{CN}_o}}{\mathrm{RI}_{\mathrm{C}_o}}\right) \tag{11.11}$$

where $\mathrm{RIF}_{\mathrm{PC}}$ is the RIF of the partially coordinated network intervention scenario as opposed to the conventional intervention one (%); $\mathrm{RI}_{\mathrm{CN}_o}$ is the RI of corridor o in the conventional intervention scenario (0–5); $\mathrm{RI}_{\mathrm{PC}_o}$ is the RI of corridor o in the partially coordinated intervention scenario (0–5); $\mathrm{RIF}_{\mathrm{C}}$ is the RIF of the fully coordinated network intervention scenario as opposed to the conventional intervention one (%); and $\mathrm{RI}_{\mathrm{C}_o}$ is the RI of corridor o in the fully coordinated intervention scenario (0–5).

REFERENCES

Abusamra, S. (2019). "Coordination and Multi-Objective Optimization Framework for Managing Municipal Infrastructure under Performance-Based Contracts." Doctoral of Philosophy Thesis, Concordia University, Montreal.

ASCE (American Society of Civil Engineers). (2021). "ASCE Infrastructure Report Card." American Society of Civil Engineering. https://www.infrastructurereportcard.org/. June 16, 2022.

BCG (Boston Consulting Group). (2017). "15 Things to Know about Canadian Infrastructure." https://www.caninfra.ca/insights/. Dec 18, 2023.

Shahata, K. (2013). "Decision-Support Framework for Integrated Asset Management of Major Municipal Infrastructure." Doctoral of Philosophy Thesis, Concordia University, Montreal.

Shahata, K. and Zayed, T. (2016). "Integrated risk-assessment framework for municipal infrastructure." *Journal of Construction Engineering and Management*, ASCE, 142(1), 1–13.

12 Multidimensional Performance Assessment Saving Model

CORRIDOR PRIORITIZATION MODEL

To assist decision-makers in prioritizing the corridors for interventions based on the key performance indicators (KPIs), a multidimensional performance assessment savings model was developed. The model aims at providing the decision-makers with a framework for ranking the corridor interventions based on a priority index that integrates the outcome of all the KPIs, as outlined in Figure 12.1. The corridor priority index (CPI) represents the urgency of undertaking an intervention based on the KPIs (time, space, cost, efficiency, effectiveness, condition, resilience preparedness,

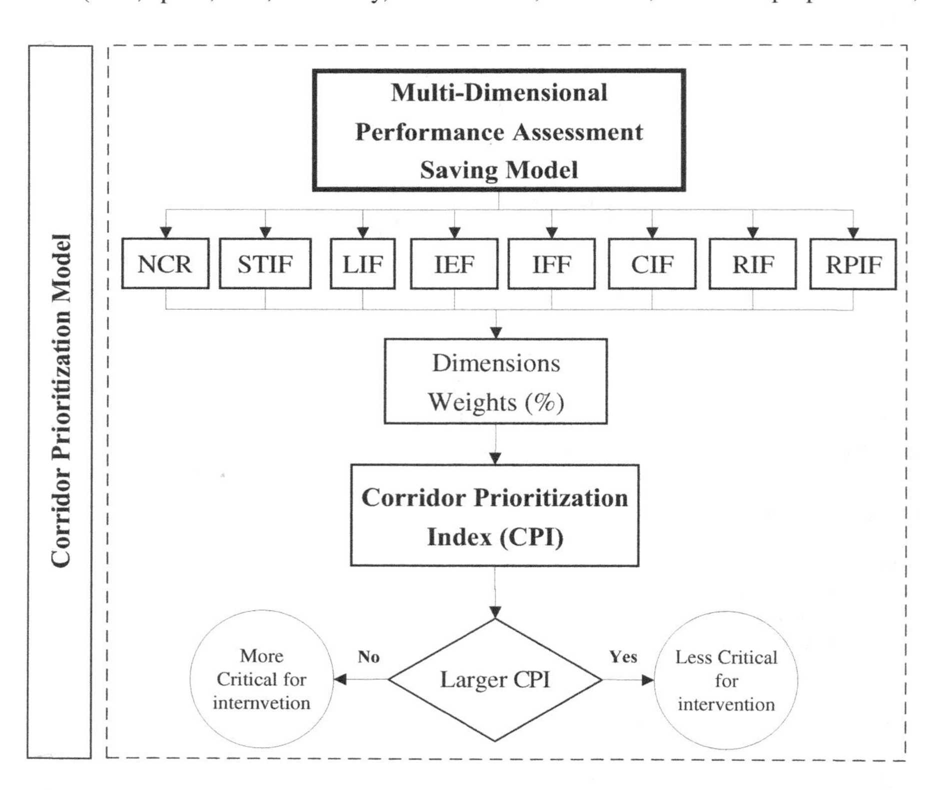

FIGURE 12.1 Corridor health prioritization model flowchart (Abusamra 2019).

DOI: 10.1201/9781003327509-15

TABLE 12.1
KPIs' Weights of Importance

KPI	Basis	Weights of Importance (%)	Illustrative Example
Time	Intervention duration	10	Critical areas (i.e., commercial areas)
Space	Intervention spatial and interdependency	10	Critical areas (i.e., commercial areas)
Cost	Life-cycle costs	20	Tough budgetary constraints
Efficiency	Intervention crew	5	Low intervention crew productivity issues
Effectiveness	Intervention quality	5	Low intervention quality and more frequent issues
Condition	Physical state, reliability, and LOS	20	Higher LOS (i.e., water and sewer mains; large pipes beside the water or sewer treatment plants, arterial roads, etc.)
Resilience preparedness	Demand and supply theory	10	Steep demand increasing area (i.e., land-use change from residential to industrial [water pipes], increasing rainfall frequency and intensity due to climate change [combined sewer and stormwater pipes], etc.)
Risk	Probability and consequences of failure	20	Critical area (i.e., mixed residential and commercial uses, industrial areas)

and risk). Each KPI was assigned a weight of importance that represents its contribution to the overall CPI. The weight of the importance of each KPI could vary from one municipality to another, and even from one area/corridor to another, according to their needs. For instance, a municipality that is facing tough budgetary constraint will assign a higher weight to the cost. Another case could be a critical area with mixed residential and commercial facilities, such that higher weights will be assigned to the time and space KPIs to minimize the disruption impacts on the surrounding communities. A corridor in an area that features numerous events will receive higher weights of importance for the time and space KPIs as longer disruption times will result in extremely high financial consequences of failure. Table 12.1 summarizes the different cases, their basis, assigned weights, and an illustrative example. In this book, the cost, condition, and risk were assigned the highest weights due to the tough budgetary constraints, arising from the increasing budget deficits, low assets' condition and the low assets' reliability.

Upon computing the KPIs at each point in time t, the CPI is calculated, as shown in Equation 12.1. A weighted sum method has been used to integrate all the KPIs in one indicator that reflects the vision of the management and develops an intervention plan that fits the municipality's priorities and preferences. The weights could vary from one municipality to another, and even from one area/corridor to another, according to their needs. The KPIs' score is adopted from the models to reflect the compliance percentage against predefined thresholds (i.e., LCC should be less than the budget, condition should not be less than the threshold, disruption in terms of time, space, and frequency should not exceed the thresholds to avoid extra indirect costs or contractual penalties, reliability should be higher than the threshold, etc.). The CPI is calculated for each corridor at a certain point of time (end of the planning horizon, annually), and asset managers will be able to visualize the detailed information for each KPI and trigger the main reason behind a low CPI. If the CPI is more than 1 (CPI > 1), the corridor requires further attention compared to other corridors with CPIs less than 1. Similar concept has been developed for ranking networks (a groups of corridors) as outlined in Equations 12.2 and 12.3.

$$\text{CPI}_{o_t} = \sum_{f}^{F} W_f \times S_{f_o} \tag{12.1}$$

$$\text{NPI}_t = \sum_{o}^{O} \frac{L_o}{\sum_{o}^{O} L_o} \times \text{CHI}_{o_t} \tag{12.2}$$

$$\text{NPI} = \overline{\text{NHI}_t} \tag{12.3}$$

where CPI_{o_t} is the corridor priority index of corridor o at point of time t (%); f is the counter of factors (number); F is the total number of factors (number); W_f is the weight of factor f (%); S_{f_o} is the score of factor f in corridor o (%); NPI_t is the network priority index at point of time t (%); and L_o is the length of corridor o (m); and NPI and $\overline{\text{NPI}_t}$ are the average network priority indexes across the planning horizon (%).

REFERENCES

Abusamra, S. (2019). "Coordination and Multi-Objective Optimization Framework for Managing Municipal Infrastructure under Performance-Based Contracts." Doctoral of Philosophy Thesis, Concordia University, Montreal.

Part 4

Optimization Framework for Municipal Infrastructure

13 An Overview of Decision-Making and Optimization for Asset Management

STATE-OF-THE-ART REVIEW

Introduction

Over the past decades, scholars have developed innovative funding and prioritization approaches for asset management. The relationships among factors affecting asset performance, deterioration processes, and service/physical failures are complex and nonlinear. Consequently, integrated planning, fund allocation, and prioritization of intervention projects across multiple colocated assets' life cycles are challenging. To address this complexity, various prioritization techniques and approaches have been investigated, considering operational, physical, and environmental factors.

Modeling and Decision-Making Techniques

Different multi-criteria decision-making methods require varying levels of knowledge and expertise to assign criteria values and combine their scores. Most methods assign values to alternatives for each criterion and multiply them by corresponding weights to obtain a final score (Huang et al. 2011). Within this domain, scholars have prioritized life-cycle funding by estimating asset failure using hierarchical models such as analytic hierarchy process (AHP) (Al-Barqawi and Zayed 2006b; Shahata and Zayed 2016; Sharma et al. 2008) or network models such as analytic network process (ANP) (Al-Barqawi and Zayed 2006a). AHP and ANP focus on modeling expert knowledge to determine the relative weights of factors and sub-factors. Scholars categorize criteria under specific factors chosen by domain experts. For example, Al-Barqawi and Zayed (2008) categorized deterioration factors into physical, operational, and environmental. Combining different sets of criteria under these relevant factors results in a structured decision model. Fares and Zayed (2009) integrated sixteen pre-failure and post-failure factors into a risk-based prioritization model. These models can be either crisp or fuzzy and are often integrated with multi-attribute utility theory (MAUT). Data ambiguity is typically modeled using fuzzy logic.

DOI: 10.1201/9781003327509-17

For deterioration and failure models, scholars have categorized them into two main types:

- **Statistical and Mathematical Survival Models**: For example, the Weibull model predicts the probability of failure (POF) of assets and/or network reliability (Abu-Samra et al. 2017a–c).
- **Data-Driven Approaches**: These use historical maintenance records to either: (1) Build empirical deterioration models using techniques such as regression models (Kimutai et al. 2015; Abu-Samra et al. 2017a–c). (2) Use model-free methods like artificial neural networks (ANNs) to identify internal relationships among decision factors and their impacts (Al-Barqawi and Zayed 2006a, 2008; Bakry et al. 2016; Osman and Ali 2012).

Other scholars have used simulation models to support decision-making when historical data is available (Shahata and Zayed 2010, 2013; Francisque et al. 2011; Sadiq et al. 2014; Gharaibeh et al. 2006; Ganjidoost et al. 2015; Duenas-Osorio et al. 2007; Scaparra and Church 2008; Roelich et al. 2015; Goodall et al. 2015; Too and Too 2010; Hansen and Neale 2014).

To evaluate alternative investment options more effectively, some scholars have utilized spatial modeling to coordinate municipal corridor interventions using Geographic Information System (GIS) and dynamic neighborhood methodologies (Osman and El-Diraby 2006; Kielhauser et al. 2017; Halfawy et al. 2000, 2002). Others have used lifecycle cost (LCC) analysis, considering all direct and indirect cost categories such as planning, design, acquisition, maintenance, ownership, and operation of the asset (Ammar et al. 2012b; Farran and Zayed 2012; Moselhi et al. 2009; Faust et al. 2013, 2015; Matthews et al. 2011). For instance, Kleiner et al. (2010) used dynamic programming to calculate the LCC for rehabilitation projects. Additionally, Ismaeel and Zayed (2018) developed a budget allocation model using a fuzzy approach for water mains. The use of LCC techniques has increased the potential for cost-effectiveness and savings (Moselhi et al. 2005, 2009). LCC modeling approaches can be either deterministic, where decisions are typically based on the net present value of intervention alternatives, or probabilistic, where different costs are associated with relevant probabilities (Shahata and Zayed 2013; Oh et al. 2011; Michele and Daniela 2011; Lambert et al. 2012; Karvetski et al. 2009; Elbeltagi and Tantawy 2008, Abu-Samra 2017a-b).

Optimization

Single-objective optimization has been widely used over the past decade, but it has several limitations that need addressing:

- **Single Objective Constraints**: Relying on a single objective and placing others as constraints do not yield near-optimal solutions for the constrained objectives.
- **Subjectivity in Weighting**: Integrating multiple objectives into one indicator increases subjectivity and biases results toward higher-weighted objectives.

- **Lack of Trade-Off Analysis and Pareto Optimization**: Missing opportunities to compare different solutions' fitness and balance conflicting objectives.
- **Ignoring Lifecycle Costs (LCCs) and Time**: Not considering LCC and the dimension of time in decision-making processes.
- **Computational Complexity and Scalability Issues**: Extended planning horizons, multiple assets, and large search spaces increase running time and scalability challenges.
- **Fragmented Optimization and Deterministic Modeling**: Separating project and network-level decisions and failing to account for uncertainties in asset deterioration and costs.

To address these limitations, a more integrated and dynamic multi-objective optimization approach is needed, considering the diverse requirements of decision-makers, uncertainties, and the holistic nature of infrastructure systems. Multi-objective optimization is extensively used in the infrastructure sector, particularly for budget allocation, efficient expenditure utilization, performance enhancement, and intervention planning and scheduling. The complexity of these tasks is heightened by uncertainties related to deterioration patterns, economic conditions (e.g., inflation rates, available budgets), and political views (Frangopol et al. 2017). Additionally, the involvement of multiple stakeholders (e.g., municipalities, asset managers, end users, politicians) with conflicting preferences further complicates reaching a consensus (Saad and Hegazy 2015a). These preferences can vary significantly between assets and stakeholders, including goals such as maximizing network performance, minimizing risk consequences, maximizing intervention efficiency and effectiveness, and minimizing LCCs and social/user costs.

Balancing these conflicting objectives at both the network and project levels is another challenge (Barco 1994; Uddin et al. 2013). Network-level decisions involve determining which assets require intervention and when these interventions should occur, representing strategic and tactical management levels. Project-level decisions focus on what type of intervention is needed for each asset, representing the operational management level, where factors like asset condition guide the final decision.

The problem at hand includes:

- Multiple spatially located assets (e.g., roads, water, sewer networks) with varying deterioration patterns and service lives.
- Limited budgets with many corridors needing interventions.
- Various intervention types and strategies for each asset (e.g., preventive maintenance, minor repair, major rehabilitation, reconstruction, replacement).
- Trade-offs between in-house sub-contracting and private partnerships for interventions.
- Interdependencies among the assets studied.
- Varying performance expectations from end users.

Saad et al. (2017)

Due to increasing infrastructure deficits, additional challenges have been introduced to the decision-making process, such as optimal budget utilization, prioritization of municipal projects, enhancement of network performance, and reduction of service failure and disruption risks. Moreover, planning and executing corridor infrastructure intervention projects have significant social, economic, and environmental impacts on society. As a result, asset managers are continuously seeking efficient, effective, and near-optimal approaches that maximize benefits and savings while minimizing losses and LCCs. This inherently makes it a multi-objective optimization problem due to the presence of multiple conflicting objectives.

According to Hegazy et al. (2004), there is no single global solution to a multi-objective problem. Multi-objective optimization can generate a set of Pareto near-optimal solutions in one run, whereas single-objective optimization would require multiple runs to obtain the same level of information. Zeleny (2011) argued that maximizing a single criterion is not true optimization. In single-objective optimization, decision-makers must express preferences in advance, such as the order or priority of goals (e.g., maximizing performance first, then using it as a constraint to optimize another indicator like LCC). Preferences include assigning relative weights of importance. However, in multi-objective optimization, preferences are expressed after running the model (Savic 2002; Hutto 2016). Users can explore various scenarios with different relative weights after the optimization engine runs.

Scholars have used goal optimization to balance multiple competing objectives, such as LCC, risk, level of service (LOS), user costs, and economic losses (Osman 2015; Abu-Samra et al. 2018), by combining all weighted deviations from thresholds along with their relative weights to form an overall deviational goal.

The application of multi-objective optimization in infrastructure asset management has garnered significant attention from researchers. For example, Rashedi and Hegazy (2014) compared segmented genetic algorithms (GAs) and exact numerical optimization methods (GAMS/CPLEX) in capital renewal planning for large infrastructure systems, concluding that numerical methods are superior. Additionally, many scholars have studied multi-objective techniques, including linear programming and integer programming. For instance, De la Garza et al. (2011) selected the optimal pavement intervention plan at a network level. Hegazy and Elhakeem (2011) designed a framework for solving large-scale combinatorial bi-level optimization problems that include discrete, integer, and two-level decisions. This framework was applied to the building sector to determine the optimal repair timing for various building components.

Abu-Samra et al. (2018) developed a multi-objective optimization model for corridor infrastructure. To address the issue of a vast search space, the model employed a phased-network approach, where project-level decisions were made separately for each asset and then used as input for network-level trade-off analysis. The system integrated goal optimization, dynamic programming, integer programming, and GAs to prioritize corridors for renewal, rehabilitation, and preventive maintenance over the planning horizon while adhering to preset performance thresholds and the available annual budget.

Goal optimization was used to model conflicting objectives through a percentile ranking approach. Dynamic programming addressed the extended planning horizon issue by dividing it into smaller segments of 5 years. Bi-level dimensional integer programming was used to model project and network-level decision variables.

Similarly, Abu-Samra et al. (2017a) developed a preemptive goal optimization-based multi-objective model for municipal colocated infrastructure. This system fragmented the multi-objective problem into single-objective ones based on the ranking of objectives. Once the desired outcome for the first objective was achieved, the system automatically moved to the second objective, treating the first as a constraint.

Other scholars developed integrated non-preemptive goal optimization and GA models for corridor infrastructure to minimize financial, temporal, and performance deviations from the preset budget, resources, and performance thresholds (Abu-Samra et al. 2018; Osman 2015). Additionally, scholars created bi-level goal optimization models for transportation networks using penalty and compromise methods to minimize financial and performance deviations. These models were formulated in the general algebraic modeling system (GAMS) modeling environment using the CPLEX solver.

The penalty method uses the weighted sum technique to combine multiple objectives into a single function by assigning different weights to each objective, creating a set of Pareto-optimal solutions. The compromise method, based on the ε-constraint technique, maximizes or minimizes one objective while setting the remaining objectives as constraints. These remaining objectives are mathematically formulated as equalities, meeting certain limits identified by separately running individual optimizations for each objective to determine the most efficient value (Frangopol and Liu 2007; Mavrotas 2009; Saad et al. 2016, 2017; Colson et al. 2007).

Saad and Hegazy (2017) developed an enhanced benefit–cost analysis optimization method to achieve fair and equitable allocations among all co–located asset categories. The study used the CPLEX solver to maximize the network benefit–cost ratio but applied deterministic deterioration only to roads and bridges. El-Anwar et al. (2016a, b) developed a mixed integer-linear programming and Pareto optimization model for scheduling post-disaster reconstruction plans for transportation networks. The study aimed to minimize network recovery time and public expenditures while prioritizing post-disaster reconstruction projects, focusing only on extreme events like hurricanes, earthquakes, and tsunamis and not on typical aging deterioration.

CGI (2015) emphasized the importance of a holistic asset management approach and the role of optimization in making balanced decisions that consider costs, risks, opportunities, and performance. Elsawah et al. (2016) developed a decision-support system with a dynamic weighting system to prioritize repairs of water and sewer networks based on the risk index (RI), represented by the POF and consequence of failure (COF). However, this study did not consider LCCs in prioritizing repairs.

Shahata and Zayed (2016) established an integrated risk-assessment framework for municipal infrastructure to aid decision-making in corridor rehabilitation projects. The study used risk to prioritize rehabilitation using mixed-Delphi and AHP with unsupervised K-means clustering and proposed cost-effective risk mitigation measures for each risk category. However, it did not consider LCC and assumed equal importance for POF and COF.

While many scholars have investigated multi-objective optimization for modeling individual municipal networks like water or roads, few have attempted to integrate two or more networks. Integrated asset management, especially with multi-objective optimization, remains relatively limited in literature. For example, Osman (2015) developed a framework for temporal coordination of colocated infrastructure systems, considering financial, risk, and LOS triggers while planning corridor interventions. The trade-off between delaying and advancing intervention activities was conducted using a goal optimization approach based on pre-defined goal constraints.

Similarly, Tscheikner-Gratl et al. (2015) developed a strategic asset management approach that prioritizes street sections and underlying infrastructure (sewer and water supply networks) for rehabilitation. The study created a priority model accounting for pipe deterioration in water supply networks and urban flooding in sewer networks. Carey and Lueke (2013) developed an asset management plan for municipal right-of-way infrastructure networks, considering economic outcomes and monetary savings. Atef and Moselhi (2014) used local spatial vulnerability of failure to assign priorities. Other approaches have been developed to integrate varying municipal perspectives (Rashedi and Hegazy 2014; Khan et al. 2015) or to integrate multiple components within the same network or across two networks. El Hakea and Fakhr (2023) provided a comprehensive and up-to-date bibliometric analysis on recent computer vision applications, trends, modalities, prominent models and image datasets for pavement management, including distress detection and condition assessment.

Table 13.1 summarizes some of these research efforts around decision-making and single- and multi-objective optimization for the colocated infrastructure systems.

KEY FINDINGS

Despite the development of various asset management frameworks and models by previous scholars, several limitations remain:

- **Propagation of System Disruption**: Many models fail to account for the propagation of disruptions due to spatial interdependencies among colocated infrastructure systems.
- **Holistic Intervention Planning**: Most interventions are planned independently based on the condition of individual assets, neglecting the condition of colocated assets.
- **Time Dimension**: Decision-making processes often overlook the dimension of time, focusing on current asset conditions without forecasting future states.
- **Integrated Contractual and Asset Management System**: There is a lack of systems that link KPIs and P/I applications with the decision-making process.

Efforts have primarily focused on developing decision-support systems for single assets. However, there is a scarcity of literature on integrated asset management within the broader context of optimization and decision-making. While several scholars have developed single-objective optimization models for integrated asset management, few have created dynamic multi-objective optimization models that incorporate conflicting perspectives and plan corridor interventions accordingly.

TABLE 13.1
Summary of Research on Decision-Making and Optimization

Research	Domain of Application	Scale of Application	Optimization Type	Optimization Tool	Objective (s)	Category
Abu-Samra et al. (2017c)	Roads	Network level	Multi-objective	Integrated goal optimization and GAs	Minimize deviations from the KPIs' thresholds	Evolutionary algorithms
Osman et al. (2017)	Water	Network level	Multi-objective	Integrated discrete event simulation and GAs	Minimize the repair time, cost, and pipe break impact	Evolutionary algorithms
Saad et al. (2017)	Roads	Network level	Multi-objective	Bi-level goal optimization with Pareto (penalty and compromise methods)	Minimize deviations from the pre-defined targets	Mathematical/Linear optimization
Saad et al. (2016)	Roads	Network level	Multi-objective	Bi-level goal optimization using GAMS/CPLEX	Minimize deviations from the pre-defined targets	Mathematical/Linear optimization
Abu-Samra (2015)	Roads	Phased project and network level	Multi-objective	Integrated goal optimization and GAs	Minimize deviations from the KPIs' thresholds	Evolutionary algorithms
Matar et al. (2017)	Water	Project level	Multi-objective	Systems engineering and System of Systems (SoS)	Maximize the project sustainability index	Decision-making
Elhadidy et al. (2015)	Roads	Network level	Multi-objective	GAs with Pareto optimization	Maximize the condition and minimize the cost	Evolutionary algorithms
Marzouk and Omar (2013)	Sewer	Network level	Multi-objective	GAs	Maximize condition and minimize costs	Evolutionary algorithms
Salman et al. (2013)	Water	Network level	Multi-objective	Integrated Unsupervised Neural Networks (UNN) and Mixed Integer Nonlinear Programming (MINLP)	Minimize the repair time and maximize reliability	Mathematical/Linear optimization
Sitzabee and Harnly (2013)	Roads	Network level	Multi-objective	Goal optimization	Maximize the priority index	Evolutionary algorithms

(Continued)

TABLE 13.1 (*Continued*)
Summary of Research on Decision-Making and Optimization

Research	Domain of Application	Scale of Application	Optimization Type	Optimization Tool	Objective (s)	Category
Ward and Savic (2013)	Sewer	Project level	Multi-objective	Integrated AHP and MAUT	Maximize structural condition and minimize costs and risk	Decision-making
Orabi and El-Rayes (2012)	Roads	Network level	Multi-objective	GAs	Minimize reconstruction costs and network disruption	Evolutionary algorithms
Shehab-Eldeen and Moselhi (2011)	Sewer	Network level	Multi-objective	MAUT	Minimize the cost and time	Decision-making
Alvisi and Franchini (2009)	Water	Network level	Multi-objective	GAs with Pareto optimization	Minimize repair costs and water losses	Evolutionary algorithms
Orabi et al. (2009)	Roads	Project level	Multi-objective	GAs	Minimize reconstruction costs	Evolutionary algorithms
Scheinberg and Anastasopoulos (2009)	Roads	Phased project and network level	Multi-objective	Mathematical optimization and mixed integer programming	Minimize costs and maximize condition	Mathematical/Linear optimization
Muschalla (2008)	Sewer	Network level	Multi-objective	GAs with Pareto optimization	Maximize condition and minimize costs	Evolutionary algorithms
Wu and Flintsch (2008)	Roads	Project level	Multi-objective	GAs	Maximize the level of service and minimize the preservation costs	Evolutionary algorithms
Alvisi and Franchini (2007)	Water	Network level	Multi-objective	GAs	Minimize cost and maximize performance	Evolutionary algorithms
Guistolisi et al. (2006)	Water	Network level	Multi-objective	Benefit/Cost analysis	Maximize benefit/cost ratio	Decision-making

(*Continued*)

TABLE 13.1 (*Continued*)
Summary of Research on Decision-Making and Optimization

Research	Domain of Application	Scale of Application	Optimization Type	Optimization Tool	Objective (s)	Category
Fwa et al. (2000)	Roads	Project level	Multi-objective	GAs	Minimize maintenance costs, maximize condition and maintenance efficiency	Evolutionary algorithms
Hegazy et al. (1994)	Roads	Network level	Multi-objective	Enhanced backpropagation for neural networks using heuristics	Maximize the network performance	Evolutionary algorithms
Abu-Samra et al. (2018)	Roads, water, and sewer	Phased-network level	Multi-objective	Integrated goal optimization, dynamic, and integer programming, and GAs	Minimize deviations from the budget and performance targets	Evolutionary algorithms
Abu-Samra et al. (2017a)	Roads, water, and sewer	Network level	Multi-objective	Preemptive goal optimization	Maximize reliability, minimize LCCs, minimize economic losses	Evolutionary algorithms
Abu-Samra and Ahmed (2017)	Roads and water	Network level	Multi-objective	Non-preemptive goal optimization	Minimize financial, temporal, and condition deviations	Evolutionary algorithms
Abu-Samra et al. (2017b)	Roads, water, and sewer	Network level	Multi-objective	Non-preemptive goal optimization	Minimize deviations from annual budget and performance target	Evolutionary algorithms
Rashedi and Hegazy (2016)	Roads, water, and sewer	Network level	Multi-objective	Casual loop diagrams and system dynamics	Maximize performance and minimize costs	Decision-making
CGI (2015)	Roads, water, and sewer	Network level	Multi-objective	Mathematical optimization	Minimize risks and maximize return of investment	Mathematical/Linear optimization
Osman (2015)	Roads, water, and sewer	Network level	Multi-objective	Non-preemptive goal optimization	Minimize deviations from the pre-defined targets	Evolutionary algorithms

(*Continued*)

TABLE 13.1 *(Continued)*
Summary of Research on Decision-Making and Optimization

Research	Domain of Application	Scale of Application	Optimization Type	Optimization Tool	Objective (s)	Category
Tscheikner-Gratl et al. (2015)	Roads, water, and sewer	Network level	Multi-objective	Decision tree	Maximize the street priority index	Decision-making
Elsawah et al. (2014)	Roads, water, and sewer	Network level	Multi-objective	Decision tree	Minimize risk consequences and maximize condition	Decision-making
Farran and Zayed (2015)	Roads, water, and sewer	Network level	Multi-objective	Integrated Markov chains and GAs	Minimize LCCs and maximize performance	Evolutionary algorithms
Osman et al. (2012)	Water and sewer	Phased project and network level	Multi-objective	GAs with Pareto optimization	Minimize risk exposure and condition assessment cost	Evolutionary algorithms
Ali et al. (2012)	Water and sewer	Project and network level	Multi-objective	Integrated Markov chains and GAs	Minimize inspection costs	Evolutionary algorithms
Amador and Magnuson (2011)	Roads, water, and sewer	Network level	Multi-objective	Integrated classical time-space adjacency modeling, heuristic simulation, and mathematical optimization	Minimize the LCCs and service disruption	Mathematical/Linear optimization
Atef et al. (2011)	Water and sewer	Network level	Multi-objective	Integrated Markov chains and GAs	Maximize condition and minimize costs	Evolutionary algorithms
Atef et al. (2010)	Water and sewer	Network level	Multi-objective	Partially observable Markov decision process	Minimize the cost and maximize the reliability	Decision-making
Osman (2005)	Roads, water, and sewer	Network level	Multi-objective	Monte Carlo simulation	Minimize risks and maximize return of investment	Decision-making

(Continued)

TABLE 13.1 (*Continued*)
Summary of Research on Decision-Making and Optimization

Research	Domain of Application	Scale of Application	Optimization Type	Optimization Tool	Objective (s)	Category
Halfawy et al. (2002)	Roads, water, and sewer	Network level	Multi-objective	Integrated GIS and mathematical optimization	Minimize cost and maximize condition	Mathematical/Linear optimization
Ismaeel and Zayed (2018)	Water	Network level	Single objective	GAs	Maximize the network performance	Evolutionary algorithms
Kaddoura et al. (2018)	Sewer	Project level	Single objective	MAUT	Prioritize the corridors for rehabilitation based on the aggregated condition index	Decision-making
Al-Zahab et al. (2017)	Water	Network level	Single objective	GAs	Maximize the benefit/cost ratio for prioritizing the leak repairs	Evolutionary algorithms
Marzouk and Hamid (2017)	Water	Network level	Single objective	Integrated simo procedure and decision tree	Prioritize the corridors for repair	Decision-making
Mohammed et al. (2017)	Roads	Network level	Single objective	GAs	Maximize resilience index	Evolutionary algorithms
Abu-Samra et al. (2016)	Water	Project level	Single objective	Integrated discrete event simulation and GAs	Minimize the risk index represented by consequences of failure and leak severity	Evolutionary algorithms
Hawari et al. (2017)	Sewer	Project level	Single objective	Integrated FANP and Monte Carlo simulation	Prioritize the corridors for rehabilitation	Decision-making
Ismaeel and Zayed (2016)	Water	Network level	Single objective	GAs	Maximize the network performance	Evolutionary algorithms
Kaddoura et al. (2016)	Sewer	Project level	Single objective	MAUT	Prioritize the corridors for rehabilitation based on the aggregated condition index	Decision-making

(*Continued*)

TABLE 13.1 *(Continued)*
Summary of Research on Decision-Making and Optimization

Research	Domain of Application	Scale of Application	Optimization Type	Optimization Tool	Objective (s)	Category
Marzouk et al. (2015)	Water	Network level	Single objective	Integrated GIS, simo procedure and decision tree	Prioritize the corridors for repair	Decision-making
Saad and Hegazy (2015b)	Roads	Network level	Single objective	Loss-aversion	Maximize the gain within the limited budget	Mathematical/Linear optimization
Zdenko et al. (2015)	Water	Network level	Single objective	Decision tree	Maximize network performance	Decision-making
Khan et al. (2015)	Water	Network level	Single objective	Decision tree	Prioritize the corridors for repair	Decision-making
Azeez et al. (2013)	Sewer	Network level	Single objective	Fuzzy and simulation-based ranking	Minimize the LCCs	Decision-making
Elsayed and Zayed (2013)	Water	Network level	Single objective	Integrated AHP and MAUT	Prioritize the water main rehabilitation projects	Decision-making
Mohamed and Zayed (2013)	Water	Network level	Single objective	Integrated MAUT and AHP	Prioritize the corridors for fund allocation	Decision-making
Zayed and Mohamed (2013)	Water	Network level	Single objective	Integrated MAUT and AHP	Prioritize the corridors' repair based on the budget priority index	Decision-making
Zhang et al. (2013)	Roads	Network level	Single objective	Dynamic programming	Minimize the LCCs	Decision-making
Adey and Hajdin (2011) and Adey et al. (2012)	Roads	Project level	Single objective	Mathematical optimization	Maximize net benefits	Mathematical/Linear optimization
Ammar et al. (2012a)	Water	Network level	Single objective	Integrated Day–Stout–Warren (DSW) algorithm and the vertex method	Minimize the LCCs	Evolutionary algorithms

(Continued)

TABLE 13.1 (*Continued*)
Summary of Research on Decision-Making and Optimization

Research	Domain of Application	Scale of Application	Optimization Type	Optimization Tool	Objective (s)	Category
Fares et al. (2012)	Roads	Project level	Single objective	Gas	Minimize costs	Evolutionary algorithms
Hegazy et al. (2012)	Roads	Network level	Single objective	Heuristic approach	Minimize the LCCs and prioritize the corridors for repair	Mathematical/Linear optimization
De la Garza et al. (2011)	Roads	Network level	Single objective	Mathematical optimization	Maximize network performance	Mathematical/Linear optimization
Shahata and Zayed (2011)	Water	Project level	Single objective	Simulation-based LCCs and decision tree	Minimize the LCCs	Decision-making
Ammar et al. (2017)	Water	Network level	Single objective	Integrated DSW algorithm and fuzzy set theory	Minimize the LCCs	Evolutionary algorithms
Moselhi et al. (2010)	Water	Network level	Single objective	AHP decision-making	Maximize the level of service within the available budget	Evolutionary algorithms
Shahata and Zayed (2009)	Water	Network level	Single objective	Monte Carlo simulation	Minimize the LCCs	Evolutionary algorithms
Dridi et al. (2008)	Water	Network level	Single objective	GAs	Minimize the LCCs	Evolutionary algorithms
Al-Barqawi and Zayed (2006a)	Water	Network level	Single objective	Integrated AHP and ANN	Maximize the network performance	Decision-making
Chootinan et al. (2006)	Roads	Project level	Single objective	Stochastic simulation and GAs	Maximize the level of service	Evolutionary algorithms
Hegazy (2005)	Roads	Network level	Single objective	GAs	Minimize the LCCs	Evolutionary algorithms

(*Continued*)

TABLE 13.1 (*Continued*)
Summary of Research on Decision-Making and Optimization

Research	Domain of Application	Scale of Application	Optimization Type	Optimization Tool	Objective (s)	Category
Abaza et al. (2004)	Roads	Network level	Single objective	Markovian nonlinear programming	Maximize the network condition within the budget	Evolutionary algorithms
Hegazy (2006)	Roads	Network level	Single objective	GAs	Maximize the network performance	Evolutionary algorithms
Liu and Wang (1996)	Roads	Network level	Single objective	Linear programming	Maximize the network performance	Mathematical/Linear optimization
Elsawah et al. (2016)	Water and sewer	Network level	Single objective	Ranking method and dynamic weighting system	Prioritize the corridors for repair based on the risk index	Decision-making
Shahata and Zayed (2016)	Roads, water, and sewer	Network level	Single objective	Mixed-Delphi and AHP and K-means clustering	Minimize risk index	Decision-making
Fathy et al. (2015)	Water and sewer	Project level	Single objective	Integrated hierarchical ANN and GAs	Minimize the ANN training error to select the best rehabilitation strategy	Evolutionary algorithms
Shahata and Zayed (2010)	Roads, water, and sewer	Network level	Single objective	GAs	Minimize repair costs	Evolutionary algorithms
Elhakeem and Hegazy (2005)	Roads, water, and sewer	Network level	Single objective	Nomographs	Minimize the cost to allocate the manpower resources	Evolutionary algorithms

PBC has been recognized as a potential solution for transferring operational risks to the private sector while maintaining high service levels. However, scholars have not developed integrated contractual and asset management models to help municipalities and maintenance contractors set performance thresholds, P/I systems, and maintenance plans. PBC has mainly been applied to roads and transportation projects, with limited applications in water and sewer rehabilitation.

Therefore, in Chapter 14, the optimization model of an integrated performance-based multi-objective asset management system for colocated infrastructure will be presented. The system accounts for contractual performance indicators while selecting the optimal intervention plan and considers P/I in financial computations to estimate the LCCs associated with corridor infrastructure interventions.

REFERENCES

Abaza, K. A., Ashur, S. A., and Al-Khatib, I. (2004). "Integrated pavement management system with a Markovian prediction model." *Journal of Transportation Engineering*, ASCE, 130(1), 24–33.

Abu-Samra, S. (2015). *Integrated Asset Management System for Highways Infrastructure*. LAPLAMBERT Academic Publishing, Saarbrücken.

Abu-Samra, S. (2017a). *B/C Analysis Study on Expanding the Coverage of Noise Loggers Leak Detection Equipment: Case Study on the City of Montreal*. Amazon and Kindle Publishing.

Abu-Samra, S. (2017b). *Montréal Festival City Risk Triggered Economic Feasibility Study*. Amazon and Kindle Publishing.

Abu-Samra, S. and Ahmed, M. (2017). *Integrated Asset Management for Corridor Infrastructure*. Universal Publishers, Boca Raton, FL.

Abu-Samra, S., Ahmed, M., Hammad, A., and Zayed, T. (2018). "Multiobjective framework for managing municipal integrated infrastructure." *Journal of Construction Engineering and Management*, ASCE, 144(1). https://ascelibrary.org/doi/abs/10.1061/(ASCE)CO.1943-7862.0001402

Abu-Samra, S., Al-Zahab, S., Ahmed, M., and Zayed, T. (2017a). "Multi-Objective Preemptive Goal Optimization Framework for Corridor Infrastructure." *MAIREINFRA-The International Conference on Maintenance and Rehabilitation of Constructed Infrastructure Facilities,* July 19–21, Seoul.

Abu-Samra, S., Al-Zahab, S., and Zayed, T. (2016). "Risk-Based Leaks Repair Prioritization Model." *Canadian Water and Wastewater Association (CWWA 2016)*, Nov 16–18, Toronto.

Abu-Samra, S., Mohammed, A., and Zayed, T. (2017b). "Decision-Making Framework for Integrated Asset Management." *Canadian Society of Civil Engineers (CSCE) General International Conference,* May 31–June 3, Vancouver.

Abu-Samra, S., Zayed, T., and Tabra, W. (2017c). "Pavement condition rating using Multiattribute utility theory." *Journal of Transportation Engineering, Part B: Pavements*, ASCE, 143(3).

Adey, B. T. and Hajdin, R. (2011). "Methodology for determination of financial needs of gradually deteriorating bridges with only structure level data." *Structure and Infrastructure Engineering*, Taylor & Francis, 7(7–8), 645–660.

Adey, B. T., Herrmann, T., Tsafatinos, K., Lüking, J., Schindele, N., and Hajdin, R. (2012). "Methodology and base cost models to determine the total benefits of preservation interventions on road sections in Switzerland." *Structure and Infrastructure Engineering*, Taylor & Francis, 8(7), 639–654.

Al-Barqawi, H. and Tarek, Z. (2008). "Infrastructure management: Integrated AHP/ANN model to evaluate municipal water mains' performance." *Journal of Infrastructure Systems*, ASCE, 14(4), 305–318.

Al-Barqawi, H. and Zayed, T. (2006a). "Condition rating model for underground infrastructure sustainable water mains." *Journal of Performance of Constructed Facilities*, ASCE, 20(2), 126–135.

Al-Barqawi, H. and Zayed, T. (2006b). "Assessment Model of Water Main Conditions." *Pipeline Division Specialty Conference 2006,* July 30–Aug 2, Chicago, IL.

Ali, M. R. H., Osman, H., Marzouk, M., and Ibrahim, M. (2012). "Valuation of Minimum Revenue Guarantees for PPP Wastewater Treatment Plants." *Construction Research Congress 2012*, May 21–23, ASCE, West Lafayette, IN.

Alvisi, S. and Franchini, M. (2007). "Near-optimal rehabilitation scheduling of water distribution systems based on a multi-objective genetic algorithm." *Civil Engineering and Environmental Systems*, Taylor & Francis, 23(3), 143–160.

Alvisi, S. and Franchini, M. (2009). "Multiobjective optimization of rehabilitation and leakage detection scheduling in water distribution systems." *Journal of Water Resources Planning and Management,* ASCE, 135(6), 426–439.

Al-Zahab, S., Abu-Samra, S., and Zayed, T. (2017). "An Optimized Approach to Leak Repair Prioritization." *MAIREINFRA-The International Conference on Maintenance and Rehabilitation of Constructed Infrastructure Facilities,* July 19–21, Seoul.

Amador, L. and Magnuson, S. (2011). "Adjacency modeling for coordination of investments in infrastructure asset management: Case study of Kindersley, Saskatchewan, Canada." *Transportation Research Board (TRB)*, 2246, 8–15.

Ammar, M., Moselhi, O., and Zayed, T. (2012a). "Decision-support model for selection of rehabilitation methods of water mains." *Structure and Infrastructure Engineering*, Taylor & Francis, 8(9), 847–855.

Ammar, M., Zayed, T., and Moselhi, O. (2012b). "Fuzzy-based life-cycle cost model for decision making under subjectivity." *Journal of Construction Engineering and Management*, ASCE, 139(5), 556–563.

Ammar, M., El-Said, M., and Osman, H. (2017). "Optimal scheduling of water network repair crews considering multiple objectives." *Journal of Civil Engineering and Management*, ASCE, 23(1), 28–36.

Atef, A. and Moselhi, O. (2014). "Modeling Spatial and Functional Interdependencies of Civil Infrastructure Networks." *Pipelines 2014*, Aug 3–6, ASCE, Portland, OR.

Atef, A., Osman, H., and Moselhi, O. (2010). "Towards Optimum Condition Assessment Policies for Water and Sewer Networks." *Construction Research Congress 2010*, May 8–10, ASCE, Banff.

Atef, A., Osman, H., and Moselhi, O. (2011). "Optimizing Budget Allocation for Condition Assessment of Water and Sewer Infrastructures." *3rd International and 9th Construction Specialty Conference,* June 8–10, CSCE, Ottawa.

Azeez, K., Zayed, T., and Ammar, M. (2013). "Fuzzy versus simulation-based life cycle cost for sewer rehabilitation alternatives." *Journal of Performance of Constructed facilities*, ASCE, 27(5), 656–665.

Bakry, I., Elsawah, H., and Moselhi, O. (2016). "Multi-Tiered Database Schema for Integrated Municipal Asset Management." *Construction Research Congress 2016* (pp. 1567–1576), May 31–June 2, ASCE, San Juan, PR.

Barco, A. L. (1994). "Budgeting for facility repair and maintenance." *Journal of Management in Engineering*, ASCE, 15(4), 28–34.

Carey, B. and Lueke, J. (2013). "Optimized holistic municipal right-of-way capital improvement planning." *Canadian Journal of Civil Engineering*, 40(12), 1244–1251.

CGI. (2015). "The Value of Optimization in Asset Management." https://www.cgi.com/sites/default/files/white-papers/cgi-the-value-optimization-asset-management-wp.pdf. Jan 16, 2017.

Chootinan, P., Chen, A., Horrocks, M. R., and Bolling, D. (2006). "A multiyear pavement maintenance program using a stochastic simulation based genetic algorithm approach." *Transportation Research Part A: Policy and Practice*, Elsevier, 40(9), 725–743.

Colson, B., Marcotte, P., and Savard, G. (2007). "An overview of bilevel optimization." *Annals of Operations Research*, 153(1), 235–256.

De la Garza, J., Akyildiz, S., Bish, D., and Krueger, D. (2011). "Network- level optimization of pavement maintenance renewal strategies." *Advanced Engineering Informatics,* Elsevier, 25(4), 699–712.

Dridi, L., Parizeau, M., Mailhot, A., and Villeneuve, J. (2008). "Using evolutionary optimisation techniques for scheduling water pipe renewal considering a short planning horizon." *Computer-Aided Civil and Infrastructure Engineering*, 23(8), 625–635.

Duenas-Osorio, L., Craig, J. I., Goodno, B. J., and Bostrom, A. (2007). "Interdependent response of networked systems." *Journal of Infrastructure Systems*, ASCE, 13(3), 185–194.

El-Anwar, O., Ye, J., and Orabi, W. (2016a). "Efficient optimization of post-disaster reconstruction of transportation networks." *Journal of Computing in Civil Engineering*, ASCE, 30(3). https://ascelibrary.org/doi/10.1061/%28ASCE%29CP.1943-5487.0000503

El-Anwar, O., Ye, J., and Orabi, W. (2016b). "Innovative linear formulation for transportation reconstruction planning." *Journal of Computing in Civil Engineering*, ASCE, 30(3). https://ascelibrary.org/doi/10.1061/%28ASCE%29CP.1943-5487.0000504

Elbeltagi, E. and Tantawy, M. (2008). "Asset Management: The Ongoing Crisis." *Proceedings for 2nd Conference on Project Management* (pp. 1–9), Saudi Council of Engineers, Riyadh.

Elhadidy, A., Elbeltagi, E., and Ammar, M. (2015). "Optimum analysis of pavement maintenance using multi-objective genetic algorithms." *Housing and Building National Research Center (HRBC)*, Elsevier, 11(1), 107–113.

Elhakeem, A. and Hegazy, T. (2005). "Graphical approach for manpower planning in infrastructure networks." *Journal of Construction Engineering and Management*, ASCE, 131(2), 168–175.

El Hakea, A. and Fakhr, M. (2023). "Recent computer vision applications for pavement distress and condition assessment." *Automation in Construction*, Elsevier, 146. https://doi.org/10.1016/j.autcon.2022.104664.

Elsawah, H., Bakry, I., and Moselhi, O. (2016). "Decision-support model for integrated risk assessment and prioritization of intervention plans of municipal infrastructure." *Journal of Pipeline Systems Engineering and Practice*, ASCE, 7(4). https://ascelibrary.org/doi/10.1061/%28ASCE%29PS.1949-1204.0000245

Elsawah, H., Guerrero, M., and Moselhi, O. (2014). "Decision-Support Model for Integrated Intervention Plans of Municipal Infrastructure." *International Conference on Sustainable Infrastructure 2014* (pp. 1039–1050), Nov 6–8, ASCE, Long Beach, CA.

Elsayed, M. and Zayed, T. (2013). "Modeling fund allocation to water main rehabilitation projects." *Journal of Performance of Constructed Facilities*, ASCE, 27(5), 646–655.

Fares, H. and Zayed, T. (2009). "Hierarchical fuzzy expert system for risk of failure of water mains." *Journal of Pipeline Systems Engineering and Practice*, ASCE, 1(1), 53–62.

Fares, H., Shahata, K., Elwakil, E., Eweda, A., Zayed, T., Abdelrahman, M., and Basha, I. (2012). "Modeling the performance of pavement marking in cold weather conditions." *Structure and Infrastructure Engineering*, Taylor & Francis, 8(11), 1067–1079.

Farran, M. and Zayed, T. (2012). "New life-cycle costing approach for infrastructure rehabilitation." *Engineering, Construction and Architectural Management*, EmerlandInsight Publishing Group Limited, 19(1), 40–60.

Farran, M. and Zayed, T. (2015). "Fitness-oriented multi-objective optimisation for infrastructures rehabilitations." *Structure and Infrastructure Engineering*, Taylor & Francis, 11(6), 761–775.

Fathy, A., Abu-Samra, S., Elsheikha, M., and Hosny, O. (2015) "Trenchless Technologies Decision-Support System Using Integrated Hierarchical Artificial Neural Networks and Genetic Algorithms." *Pipelines 2015*, Aug 23–26, ASCE, Baltimore, MD.

Faust, K., Abraham, D. M., and DeLaurentis, D. (2013). "Assessment of stakeholder perceptions in water infrastructure projects using system-of-systems and binary probit analyses: A case study." *Journal of Environmental Management*, Elsevier, 128, 866–876.

Faust, K., Abraham, D. M., and McElmurry, S. P. (2015). "Water and wastewater infrastructure management in shrinking cities." *Public Works Management and Policy*, SAGE Journals, 21(2) 128–156.

Francisque, A., Rodriguez, M. J., Sadiq, R., Miranda, L. F., and Proulx, F. (2011). "Reconciling 'actual' risk with 'perceived' risk for distributed water quality: A QFD-based approach." *Journal of Water Supply: Research and Technology-AQUA*, IWA Publishing, 60(6), 321–342.

Frangopol, D. M. and Liu, M. (2007). "Maintenance and management of civil infrastructure based on condition, safety, optimization, and life-cycle cost." *Structure and Infrastructure Engineering*, Taylor & Francis, 3(1), 29–41.

Frangopol, D. M., Dong, Y., and Sabatino, S. (2017). "Bridge life-cycle performance and cost: Analysis, prediction, optimisation and decision-making." *Structure and Infrastructure Engineering*, Taylor & Francis, 13(10), 1239–1257.

Fwa, T. F., Chan, W. T., and Hoque, K. Z. (2000). "Multiobjective optimization for pavement maintenance programming." *Journal of Transportation Engineering*, ASCE, 126(5), 367–374.

Ganjidoost, A., Knight, M. A., Unger, A., and Haas, C. (2015). "Benchmark performance indicators for utility water and wastewater pipelines infrastructure." *Journal of Water Resources Planning and Management*, ASCE, 144(3). https://ascelibrary.org/doi/10.1061/%28ASCE%29WR.1943-5452.0000890

Gharaibeh, N. G., Chiu, Y., and Gurian, P. L. (2006). "Decision methodology for allocating funds across transportation infrastructure assets." *Journal of Infrastructure Systems*, ASCE, 12(1), 1–9.

Goodall, J. L., Castronova, A. M., Huynh, N. N., Caicedo, J. M., and Hall, R. (2015). "Using a service-oriented approach to simulate integrated urban infrastructure systems." *Journal of Computing in Civil Engineering*, ASCE, 29(5). https://ascelibrary.org/doi/abs/10.1061/(ASCE)CP.1943-5487.0000364

Guistolisi, O., Laucelli, D., and Savic, D. (2006). "Development of rehabilitation plans for water mains replacement considering risk and cost-benefit assessment." *Civil Engineering and Environmental Systems*, Taylor & Francis, 23(3), 175–190.

Halfawy, M. R., Pyzoha, D., and El-Hosseiny, T. (2002). "An Integrated Framework for GIS-Based Civil Infrastructure Management Systems." *Proceedings of the Canadian Society for Civil Engineers*, CSCE, Montreal.

Halfawy, M. R., Pyzoha, D., Young, R., Abdel-Latif, M., Miller, R., Windham, L., and Wiegand, R. (2000). "GIS-Based Sanitary Sewer Evaluation Survey." *20th Annual ESRI International User Conference,* Jun 26–30, San Diego, CA.

Hansen, K. L. and Neale, B. S. (2014). "Infrastructure Resilience in the UK: An Overview of Current Approaches." *International Conference on Sustainable Infrastructure 2014*, Nov 6–8, ASCE, Long Beach, CA.

Hawari, A., Alkadour, F., Elmasry, M., and Zayed, T. (2017). "Simulation-based condition assessment model for sewer pipelines." *Journal of Performance of Constructed Facilities*, ASCE, 31(1). https://ascelibrary.org/doi/abs/10.1061/%28ASCE%29CF.1943-5509.0000914

Hegazy, T. (2005) "Computerized system for efficient scheduling of highway construction." *Transportation Research Board (TRB)*, 1907, 8–14.

Hegazy, T. (2006) "A computerized system for efficient delivery of infrastructure MR&R programs." *Journal of Construction Engineering and Management*, ASCE, 132(1), 26–34.

Hegazy, T. and Elhakeem, A. (2011). "Multiple optimization and segmentation technique (MOST) for large-scale bilevel life cycle optimization." *Canadian Journal of Civil Engineering*, 38(3), 263–271.

Hegazy, T., Elhakeem, A., Ahluwalia, S. S., and Attalla, M. (2012). "MOST-FIT: Support techniques for inspection and life cycle optimization in building asset management." *Computer-Aided Civil and Infrastructure Engineering*, 27(2), 130–142.

Hegazy, T., Elhakeem, A., and Elbeltagi, E. (2004). "Distributed scheduling model for infrastructure networks." *Journal of Construction Engineering and Management*, ASCE, 130(2), 160–167.

Hegazy, T., Fazio, P., and Moselhi, O. (1994). "Developing practical neural network applications using back-propagation." *Computer-Aided Civil and Infrastructure Engineering*, 9(2), 145–159.

Huang, I. B., Keisler, J., and Linkov, I. (2011). "Multi-criteria decision analysis in environmental sciences: Ten years of applications and trends." *Science of the Total Environment*, SpringerLink, 409(19), 3578–3594.

Hutto, R. L. (2016). "Should scientists be required to use a model-based solution to adjust for possible distance-based detectability bias?" *Ecological Applications Journal*, Ecological Society of America, 26(5), 1287–1294.

Ismaeel, M. and Zayed, T. (2016). "Performance Assessment Model for Water Networks." *Pipelines 2016*, July 17–20, ASCE, Kansas City, MO.

Ismaeel, M. and Zayed, T. (2018). "Water network performance assessment model." *Journal of Infrastructure Systems*, ASCE, 24(2). https://ascelibrary.org/doi/10.1061/9780784479957.064

Kaddoura, K., Zayed, T., and Hawari, A. H. (2016). "Condition Assessment of Sewer Pipelines Using Multi Attribute Utility Theory (MAUT)." *34th International No-Dig Conference and Exhibition,* Oct 10–12, Beijing.

Kaddoura, K., Zayed, T., and Hawari, A. H. (2018). "Multiattribute utility theory deployment in sewer defects assessment." *Journal of Computing in Civil Engineering*, ASCE, 32(2). https://ascelibrary.org/doi/10.1061/%28ASCE%29CP.1943-5487.0000723

Karvetski, C. W., Lambert, J. H., and Linkov, I. (2009). "Emergent conditions and multiple criteria analysis in infrastructure prioritization for developing countries." *Journal of Multi-Criteria Decision Analysis: Optimization, Learning, and Decision Support*, John Wiley and Sons Ltd, 16(5–6), 125–137.

Khan, Z., Moselhi, O., and Zayed, T. (2015). "Identifying rehabilitation options for optimum improvement in municipal asset condition." *Journal of Infrastructure Systems*, ASCE, 21(2). https://ascelibrary.org/doi/10.1061/%28ASCE%29IS.1943-555X.0000220

Kielhauser, C., Adey, B. T., and Lethanh, N. (2017). "Investigation of a static and a dynamic neighbourhood methodology to develop work programs for multiple close municipal infrastructure networks." *Structure and Infrastructure Engineering*, Taylor & Francis, 13(3), 361–389.

Kimutai, E., Betrie, G., Brander, R., Sadiq, R., and Tesfamariam, S. (2015). "Comparison of statistical models for predicting pipe failures: Illustrative example with the city of Calgary water main failure." *Journal of Pipeline Systems Engineering and Practice*, ASCE, 6(4). https://ascelibrary.org/doi/10.1061/%28ASCE%29PS.1949-1204.0000196

Kleiner, Y., Nafi, A., and Rajani, B. (2010). "Planning renewal of water mains while considering deterioration, economies of scale and adjacent infrastructure." *Water Science and Technology: Water Supply*, 10(6), 897–906.

Lambert, J. H., Karvetski, C. W., Spencer, D. K., Sotirin, B. J., Liberi, D. M., Zaghloul, H. H., Koogler, J. B., Hunter, S. L., Goran, W. D., Ditmer, R. D., and Linkov, I. (2012). "Prioritizing infrastructure investments in Afghanistan with multiagency stakeholders and deep uncertainty of emergent conditions." *Journal of Infrastructure Systems*, ASCE, 18(2), 155–166.

Liu, F. and Wang, K. C. P. (1996). "Pavement performance-oriented network optimization system." *Transportation Research Record*, 1524, 86–93.

Marzouk, M. and Hamid, S. A. (2017). "Budget allocation for water mains rehabilitation projects using Simos' procedure." *Housing and Building National Research Center (HBRC) Journal*, Elsevier, 13(1), 54–60.

Marzouk, M. and Omar, M. (2013). "Multiobjective optimisation algorithm for sewer network rehabilitation." *Structure and Infrastructure Engineering*, Taylor & Francis, 9(11), 1094–1022.

Marzouk, M., Hamid, S. A., and El-Said, M. (2015). "A methodology for prioritizing water mains rehabilitation in Egypt." *Housing and Building National Research Center (HBRC) Journal*, Elsevier, 11(1), 114–128.

Matar, M., Osman, H., Georgy, M., Abou-Zeid, A., and El-Said, M. (2017). "A systems engineering approach for realizing sustainability in infrastructure projects." *Housing and Building National Research Center (HRBC)*, Elsevier, 13(2), 190–201.

Matthews, J., Selvakumar, A., Condit, W., and Sterling, R. (2011). "Gaps of Decision-Support Models for Pipeline Renewal and Recommendations for Improvement." *International Conference on Pipelines and Trenchless Technology*, Oct 26–29, ASCE, Beijing.

Mavrotas, G. (2009). "Effective implementation of the ε-constraint method in multi-objective mathematical programming problems." *Applied Mathematics and Computation*, Elsevier, 213(2), 455–465.

Michele, D. and Daniela, L. (2011). "Decision-Support Tools for Municipal Infrastructure Maintenance Management." *World Conference on Information Technology (WCIT)* (pp. 36–41), vol. 3, Oct 5–7, Elsevier, Istanbul.

Mohamed, E. and Zayed, T. (2013). "Modeling fund allocation to water main rehabilitation projects." *Journal of Performance of Constructed Facilities*, ASCE, 27(5), 646–655.

Mohammed, A., Abu-Samra, S., and Zayed, T. (2017). "Resilience Assessment Framework for Municipal Infrastructure." *MAIREINFRA-The International Conference on Maintenance and Rehabilitation of Constructed Infrastructure Facilities,* July 19–21, Seoul.

Moselhi, O., Hammad, A., Alkass, S., Assi, C., Debbabi, M., and Haider, M. (2005). "Vulnerability Assessment of Civil Infrastructure Systems: A Network Approach." *1st CSCE Specialty Conference on Infrastructure Technologies, Management and Policy,* June 2–4, Toronto.

Moselhi, O., Zayed, T., and Salman, A. (2009). "Selection Method for Rehabilitation of Water Distribution Networks." *International Conference on Pipelines and Trenchless Technology (ICPTT),* Oct 18–21, ASCE, Shanghai.

Moselhi, O., Zayed, T., Khan, Z., and Salman, A. (2010). "Community-driven and reliability-based budget allocation for water networks community-driven and reliability-based budget allocation for water networks." *Construction Research Congress 2010*, ASCE, 1–10, May 8–10, Banff, Alberta, Canada.

Muschalla, D. (2008). "Optimization of integrated urban wastewater systems using multi-objective evolution strategies." *Urban Water Journal*, Taylor & Francis, 5(1), 59–67.

Oh, J. S., Kim, H., and Park, D. (2011). "Bi-objective network optimization for spatial and temporal coordination of multiple highway construction projects." *KSCE Journal of Civil Engineering*, SpringlerLink, 15(8), 1449–1455.

Orabi, W. and El-Rayes, K. (2012). "Optimizing the rehabilitation efforts of aging transportation networks." *Journal of Construction Engineering and Management*, ASCE, 138(4), 529–539.

Orabi, W., El-Rayes, K., Senouci, A., and Al-Derham, H. (2009). "Optimizing postdisaster reconstruction planning for damaged transportation networks." *Journal of Construction Engineering and Management*, ASCE, 135(10), 1039–1048.

Osman, H. (2005). "Risk-based life cycle cost analysis for privatized infrastructure." *Transportation Research Record: Journal of the Transportation Research Board (TRB)*, 1924(1), 192–196.

Osman, H. (2015). "Coordination of urban infrastructure reconstruction projects." *Structure and Infrastructure Engineering*, Taylor & Francis, 12(1), 108–121.

Osman, H. and Ali, M. R. (2012). "Complex Systems Modeling of Infrastructure Assets, Operators, Users, and Politicians Using System Dynamics." *Construction Research Congress 2012*, May 21–23, ASCE, West Lafayette, IN.

Osman, H. and El-Diraby, T. (2006). "Interoperable Decision-support Model for Routing Buried Urban Infrastructure." *Joint International Conference on Computing and Decision Making in Civil and Building Engineering* (pp. 3226–3235), June 14–16, Montreal.

Osman, H., Ammar, M., and El-Said, M. (2017). "Optimal scheduling of water network repair crews considering multiple objectives." *Journal of Civil Engineering and Management*, ASCE, 23(1), 28–36.

Osman, H., Atef, A., and Moselhi, O. (2012). "Optimizing inspection policies for buried municipal pipe infrastructure." *Journal of Performance of Constructed Facilities*, ASCE, 26(3), 345–352.

Rashedi, R. and Hegazy, T. (2014). "Capital renewal optimisation for large-scale infrastructure networks: Genetic algorithms versus advanced mathematical tools." *Structure and Infrastructure Engineering*, Taylor & Francis, 11(3), 253–262.

Rashedi, R. and Hegazy, T. (2016). "Strategic policy analysis for infrastructure rehabilitation using system dynamics." *Structure and Infrastructure Engineering*, Taylor & Francis, 12(6), 667–681.

Roelich, K., Knoeri, C., Steinberger, J. K., Varga, L., Blythee, P. T., Butler, D., Gupta, R., Harrisonh, G. P., Martin, C., and Purnell, P. (2015). "Towards resource-efficient and service-oriented integrated infrastructure operation." *Technological Forecasting and Social Change*, Elsevier, 92, 40–52.

Saad, D. A. and Hegazy, T. (2017). "Economic optimization for the rehabilitation of co-located mixed assets." *Canadian Journal of Civil Engineering*, 44(10), 820–828.

Saad, D. and Hegazy, T. (2015a). "Enhanced benefit–cost analysis for infrastructure fund-allocation." *Canadian Journal of Civil Engineering*, 42(2), 89–97.

Saad, D. and Hegazy, T. (2015b). "Behavioral economic concepts for funding infrastructure rehabilitation." *Journal of Management in Engineering*, ASCE, 31(5). https://ascelibrary.org/doi/10.1061/%28ASCE%29ME.1943-5479.0000332

Saad, D., Mansour, H., and Osman, H. (2016). "Bilevel Optimization of Infrastructure Fund-Allocation Decisions." *Proceedings of Annual General Conference*, June 1–4, CSCE, London.

Saad, D., Mansour, H., and Osman, H. (2017). "Concurrent bilevel multi-objective optimisation of renewal funding decisions for large-scale infrastructure networks." *Structure and Infrastructure Engineering*, Taylor & Francis, 14(5), 594–603.

Sadiq, R., Kleiner, Y., Rajani, B., Tesfamariam, S., Haider, H. (2014). "Water quality–water main renewal planner (Q-WARP): Development and application." *Journal of Water Supply: Research and Technology-AQUA*, IWA Publishing, 63(6), 429–448.

Salman, A., Moselhi, O., and Zayed, T. (2013). "Scheduling model for rehabilitation of distribution networks using MINLP." *Journal of Construction Engineering and Management*, ASCE, 139(5), 498–509.

Savic, D. (2002). "Single-Objective vs. Multiobjective Optimisation for Integrated Decision Support." *Proceedings of the First Biennial Meeting of the International Environmental Modeling and Software Society (IEMSS),* June 24–27, Lugano.

Scaparra, N. and Church, R. (2008). "A bilevel mixed-integer program for critical infrastructure protection planning." *Computers and Operations Research*, 35(6), 1905–1923.

Scheinberg, T. and Anastasopoulos, P. C. (2009). "Pavement Preservation Programming: A Multi-Year Multi-Constraint Optimization Methodology." *Proceedings of 89th Annual Meeting of the Transportation Research Board (TRB),* Washington, DC.

Shahata, K. and Zayed, T. (2009). "Integrated Management Approach for Infrastructure Assets." *INFRA 2009,* Nov 16–18, Mont-Tremblant.

Shahata, K. and Zayed, T. (2010). "Integrated Decision-Support Framework for Municipal Infrastructure Asset." *Conference on Pipelines: Climbing New Peaks to Infrastructure Reliability: Renew, Rehab, and Reinvest* (pp. 1492–1502), ASCE, Keystone, CO.

Shahata, K. and Zayed, T. (2011). "Integrated Decision-Support Framework for Municipal Infrastructure Asset." *Pipeline Division Specialty Conference 2010,* Aug 28–Sep 1, ASCE, Keystone, CO.

Shahata, K. and Zayed, T. (2013). "Simulation-Based Life Cycle Cost (SLCC) modeling and maintenance plan for water mains." *Structure and Infrastructure Engineering*, Taylor & Francis, 9(5), 403–415.

Shahata, K. and Zayed, T. (2016). "Integrated risk-assessment framework for municipal infrastructure." *Journal of Construction Engineering and Management*, ASCE, 142(1), 1–13.

Sharma, V., Al-Hussein, M., Safouhi, H., and Bouferguene, A. (2008). "Municipal infrastructure asset levels of service assessment for investment decisions using analytic hierarchy process." *Journal of Infrastructure Systems*, ASCE, 14(3), 193–200.

Shehab-Eldeen, T., and Moselhi, O. (2011). "A decision-support system for rehabilitation of sewer pipes." *Canadian Journal of Civil Engineering*, 28(3), 394–401.

Sitzabee, W. E. and Harnly, M. T. (2013). "A strategic assessment of infrastructure asset-management modeling." *Air and Space Power Journal*, 27(6), 45–68.

Too, E. G. and Too, L. (2010). "Strategic infrastructure asset management: A conceptual framework to identify capabilities." *Journal of Corporate Real Estate*, EmerlandInsight Publishing Group Limited, 12(3), 196–208.

Tscheikner-Gratl, F., Sitzenfrei, R., Rauch, W., and Kleidorfer, M. (2015). "Integrated rehabilitation planning of urban infrastructure systems using a street section priority model." *Urban Water Journal*, Taylor & Francis, 13(1), 28–40.

Uddin, W., Hudson, W. R., and Haas, R. (2013). *Public Infrastructure Asset Management*, 2nd Edition. McGraw-Hill Education, Columbus, OH.

Ward, B. and Savic, D. (2013). "A multi-objective optimisation model for sewer rehabilitation considering critical risk of failure." *Water Science and Technology*, 66(11), 2410–2417.

Wu, Z. and Flintsch, G. W. (2008). "Pavement preservation optimization considering multiple objectives and budget variability." *Journal of Transportation Engineering*, ASCE, 135(5), 305–315.

Zayed, T. and Mohamed, E. (2013). "Budget allocation and rehabilitation plans for water systems using simulation approach." *Tunnelling and Underground Space Technology*, Elsevier, 36, 34–45.

Zdenko, V. C., Gustaf, O., Barry, L., Michael, S., and Varoui, A. (2015). "Utility analysis and integration model." *American Water Works Association (AWWA) Journal*, 107(8), 64–71.

Zeleny, M. (2011). "Multiple Criteria Decision Making (MCDM): From paradigm lost to paradigm regained?" *Journal of Multi-Criteria Decision Analysis: Optimization, Learning, and Decision Support*, John Wiley and Sons Ltd, 18(1–2), 77–89.

Zhang, H., Keoleian, G., and Lepech, M. (2013). "Network-level pavement asset management system integrated with life-cycle analysis and life- cycle optimization." *Journal of Infrastructure Systems*, ASCE, 19(1), 99–107.

14 Optimization Framework and Steps Toward Coordinated Planning of Municipal Infrastructure

INTRODUCTION

Given the complexity of the problem in hand and the countless number of viable alternatives that asset managers must select from at every point in time throughout the assets' lifecycle, the need for adopting an engine simulates the outcome of those scenarios and selects an optimal solution arose. Several optimization models have been developed to reflect the different phases of the project, PBC structuring and maintenance planning, as shown in Figure 14.1. For the PBC structuring, two scenarios have been built. The first scenario aims at defining the KPIs' thresholds that both aligns with the municipality goals and meets their limited resources (e.g., number of maintenance crews, annual budget, etc.). The second scenario aims at minimizing the overall LCC through obtaining optimal penalties and incentive (P/I) while meeting the pre-defined KPI thresholds. For the maintenance planning optimization, the model aims at selecting an optimized intervention schedule that fits the pre-defined PBC parameters and minimizes the LCC. The optimization models were developed through two software packages: (1) spreadsheet modeling with genetic algorithms (GAs)-based optimization engine; and (2) REMSOFT software equipped with MOSEK linear and mixed-integer programming optimization algorithm. The optimization mathematical formulation is similar for both packages (REMSOFT 2018; MOSEK 2018). The only difference is the expanded capability of MOSEK linear and mixed-integer programming optimization algorithm over the GA such that it was applied to a greater number of corridors as opposed to the GAs spreadsheet model, which was very limited in terms of the number of corridors it could be potentially applied to. Accordingly, the model was applied to two real-life case studies, namely City of Montreal and Town of Kindersley, as will be discussed in chapters 15 and 16.

PBC STRUCTURING OPTIMIZATION

The optimization model for the PBC structuring could be set up into two different modes, as shown in Figures 14.1, 14.2, and 14.3. The first mode is set up with an objective of minimizing the summation of the deviational variables, which are

DOI: 10.1201/9781003327509-18

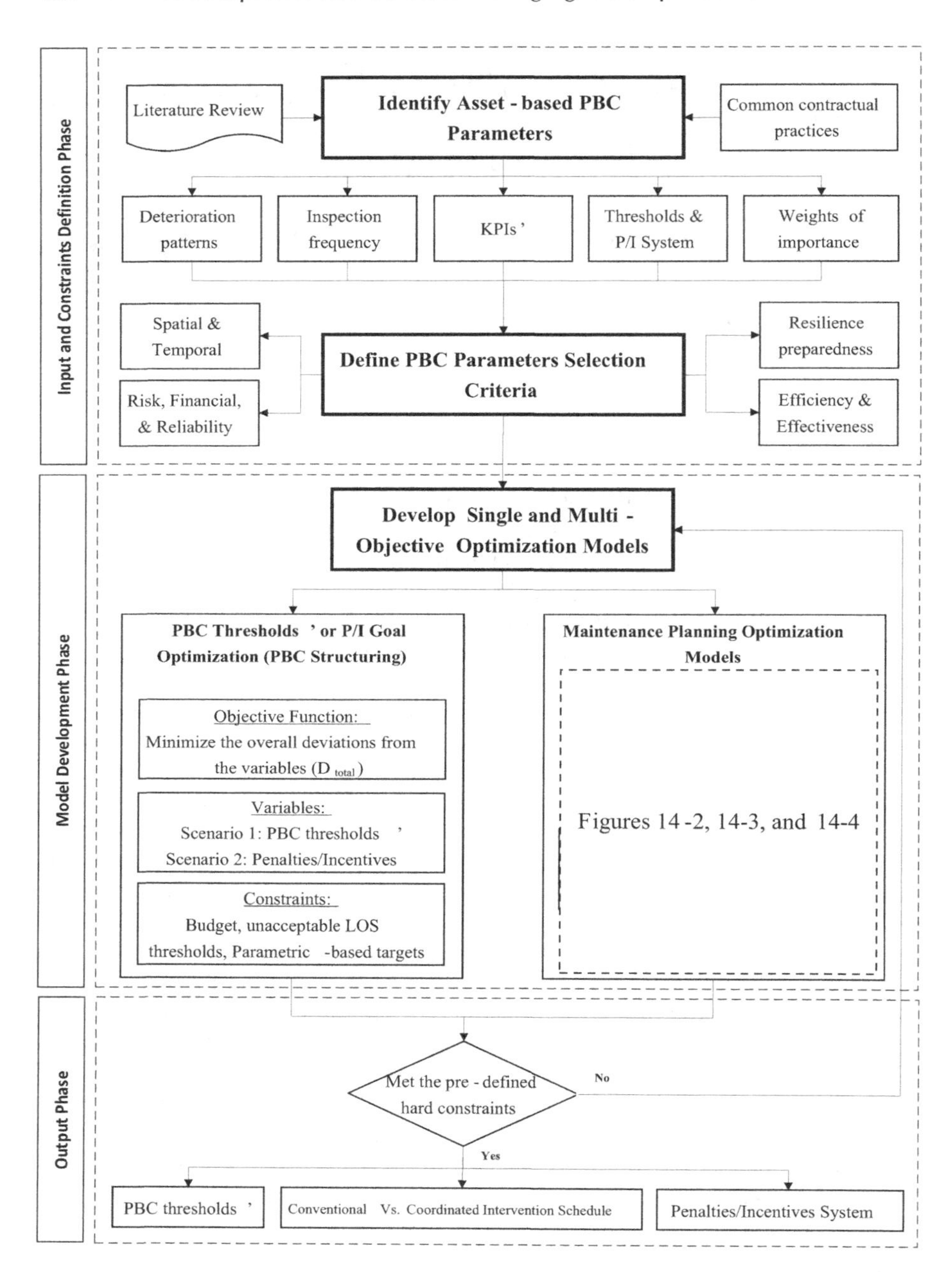

FIGURE 14.1 Multi-objective PBC optimization flowchart.

the PBC KPIs' thresholds, within the available budget and without reaching the unacceptable thresholds. The decision variables are the KPIs' threshold values, as displayed in Equation 14.1. The thresholds of deviational variables play an important role in the PBC as they define the expected KPIs' levels for all the n_s systems.

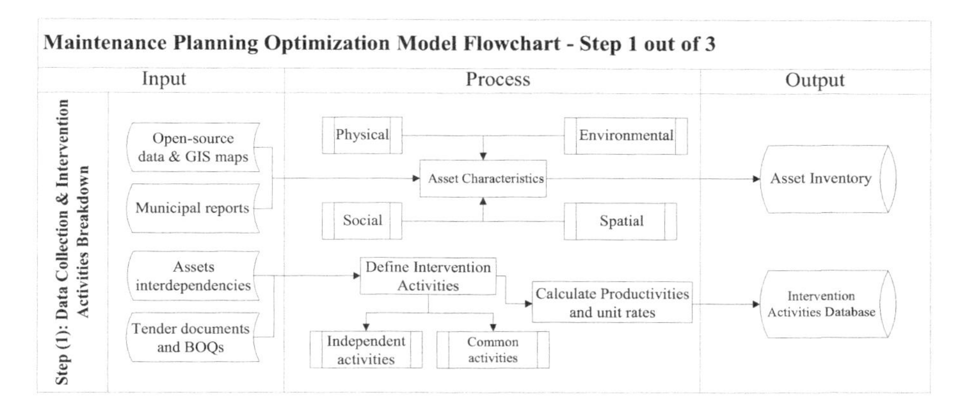

FIGURE 14.2 Data collection and intervention activities breakdown.

Based on those thresholds, the potential private partner will compute the necessary intervention costs and consider unforeseen risks through a certain contingency (%), which varies according to numerous factors, but most importantly, the KPIs' thresholds and the contractual P/I. Accordingly, it is necessary to balance the trade-off between the end users, who are the definite target of the expected KPIs' levels, and the private partner, who are the budget users. In other words, the municipalities shall undertake a trade-off analysis between the KPIs' levels and cost. The key question, in this case, should be "Are the end users willing to pay extra premium/money to elevate the KPIs' levels (e.g., increase level of service/reliability, etc.)?" Non-preemptive goal optimization was used to solve the problem. The optimization mathematical formulation could be summarized in Equations 14.2 through 14.8. The objective function is minimizing the deviational variables from the KPI's thresholds, as shown in Equation 14.2. The constraints are modeled through binary coding rules, such that "0" implies that the constraint threshold has been met and "1" implies that the constraint threshold has not been met. The model constraints are as follows: (1) financial resources (budget); (2) maintenance capacity; (3) number of annual intervention activities; (4) temporal limit; (5) demand/capacity ratio; and (6) spatial disruption (e.g., lane rental approach). Aside from that, the optimization model can be set up in another mode to solve another scenario and obtain an optimal set of P/I that guarantees the successful delivery of the KPIs without an elevated contingency (%) to cover contractual risks and high penalties. Hence, the objective and constraints formulation will be similar to the previous scenario as outlined in Equations 14.2 through 14.8. The deviational variables are computed through Equations 14.9 and 14.10. However, the decision variables will be the P/I for the pre-defined KPIs, as displayed in Equation 14.2. The P/I could be either monetary values that are incorporated in the financial model calculations or contractual period extension, which is incorporated through an extended planning horizon. The mathematical formulation of the financial penalties and incentives for corridor o at a certain point of time (t) could be displayed in Equations 14.11 and 14.12, respectively. The P/I system is integrated with the financial model (Abusamra 2019).

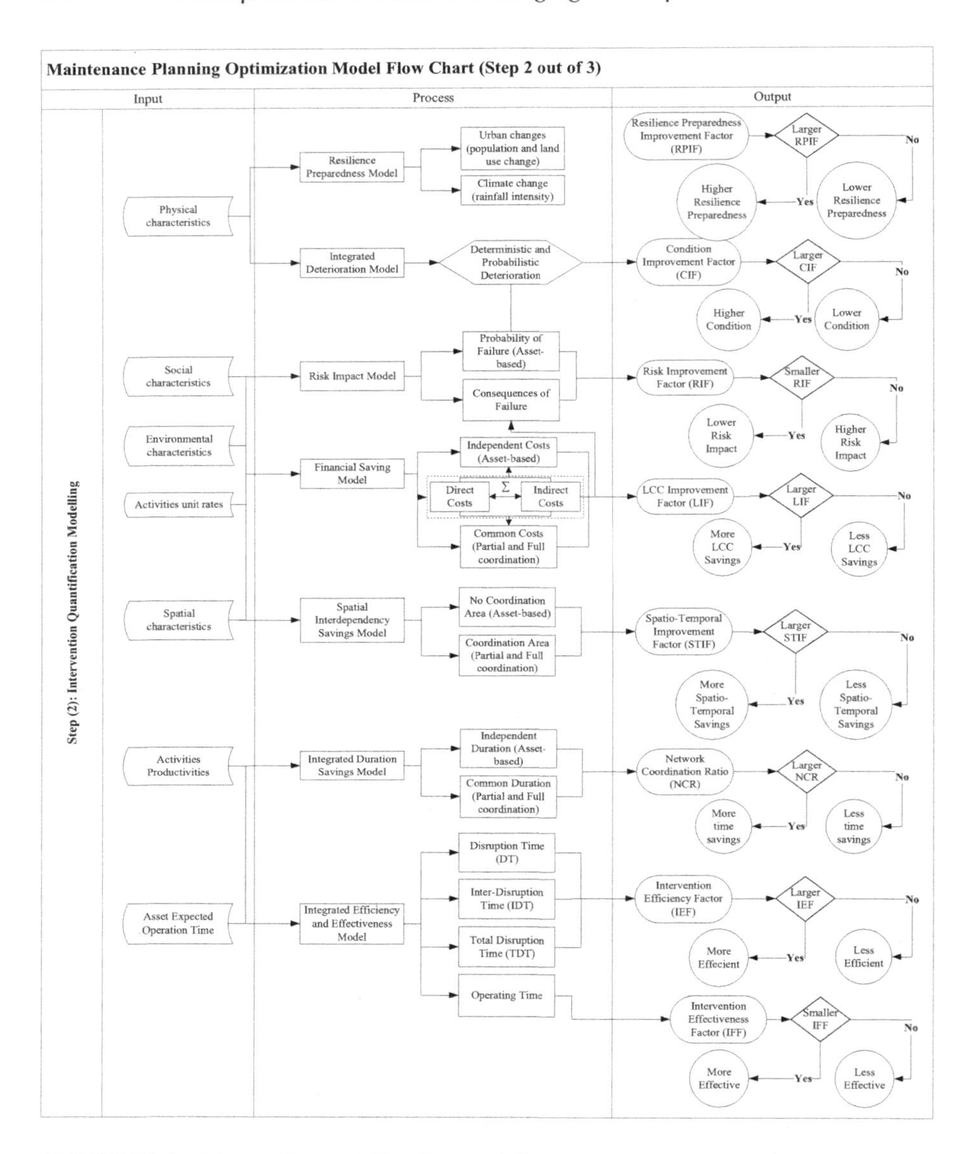

FIGURE 14.3 Intervention quantification modeling.

$$\textbf{Decision variables (Mode 1)} = \begin{bmatrix} TH_{v_i} & \cdots & TH_{v_{n_S}} \\ \vdots & \ddots & \vdots \\ TH_{V_i} & \cdots & TH_{V_{n_S}} \end{bmatrix} \tag{14.1}$$

For $TH_{v_i} = 0, 1, \ldots 100$
$v = 1, 2, \ldots VI_{v_i}$
$i = 1, 2, \ldots n_s$

$$\textbf{Decision variables (Mode 2)} = \left(\begin{bmatrix} P_{u_{v_i}} & \cdots & P_{u_{v_{n_S}}} \\ \vdots & \ddots & \vdots \\ P_{u_{V_i}} & \cdots & P_{u_{V_{n_S}}} \end{bmatrix} \bigcup \begin{bmatrix} I_{u_{v_i}} & \cdots & I_{u_{v_{n_S}}} \\ \vdots & \ddots & \vdots \\ I_{u_{V_i}} & \cdots & I_{u_{V_{n_S}}} \end{bmatrix} \right) \tag{14.2}$$

For $P_{v_i} = 0, 1, \ldots$ (\$)
$I_{v_i} = 0, 1, \ldots$ (\$)
$v = 1, 2, \ldots V$
$i = 1, 2, \ldots n_s$

$$\textbf{Min}(\textbf{Z}) = \sum_{i=1}^{n_S} \sum_{v=1}^{V} \sum_{t=1}^{T} \left[W_i \times W_v \times \left(d_{k_t}^- + d_{m_t}^+ \right) \right] \tag{14.3}$$

Subject to the following constraints:

$$\sum_{i=1}^{n_S} W_i = 1 \tag{14.4}$$

$$\sum_{v=1}^{V} W_v = 1 \tag{14.5}$$

$$LCC \le LCC_B \tag{14.6}$$

$$R_{i_{o_t}} \ge R_{th} \tag{14.7}$$

$$F_{i_{o_t}} \le F_{TH} \tag{14.8}$$

$$d_{k_t}^- = \sum_{i=1}^{n_s} \sum_{h=1}^{H} \frac{KPI_{h_{i_t}} - TH_{h_i}}{TH_{h_i}}; \text{ for all } k \text{ and } t \tag{14.9}$$

$$d_{m_t}^{+} = \sum_{i=1}^{n_s}\sum_{l=1}^{L}\frac{\mathrm{TH}_{l_i} - \mathrm{KPI}_{l_{it}}}{\mathrm{TH}_{l_i}}; \text{ for all } m \text{ and } t \quad (14.10)$$

$$P_{O_t} = \sum_{i=1}^{n_s}\sum_{v=1}^{V}\left(P_{u_{v_i}} \times \mathrm{PA}_{v_{t_i}}\right); \text{ for } \mathrm{KPI}_{h_i} > \mathrm{TH}_{h_i} \text{ or } \mathrm{KPI}_{l_i} < \mathrm{TH}_{l_i} \quad (14.11)$$

$$I_{O_t} = \sum_{i=1}^{n_s}\sum_{v=1}^{V}\left(I_{u_{v_i}} \times \mathrm{IA}_{v_{t_i}}\right); \text{ for } \mathrm{KPI}_{h_i} \le \mathrm{TH}_{h_i} \text{ or } \mathrm{KPI}_{l_i} \ge \mathrm{TH}_{l_i} \quad (14.12)$$

where TH_{v_i} is the threshold values (decision variables for the first optimization mode) defined for each KPI (v) and system (i) (varies according to the KPI); $P_{u_{v_i}}$ is the financial penalty unit cost (decision variables for the second optimization mode) defined for each KPI (v) and system (i) ($); $I_{u_{v_i}}$ is the incentive unit cost (decision variables for the second optimization mode) defined for each KPI (v) and system (i) ($); Z is the summation of the deviational variables of n_s system throughout the planning horizon T (%); W_v represents the deferential weights among the conflicting goals (%); v is the KPIs' counter (number); V is the total number of KPIs (number); LCC_B is the available intervention budget across the planning horizon ($); C_{th} is the condition state threshold for all n_s systems (varies from one system to another) (%); $d_{k_t}^{-}$ is the summation of all the negative deviational variables at point of time (t) (%); $d_{m_t}^{+}$ is the summation of all the positive deviational variables at point of time (t) (%); h and l are the counters of the positive and negative deviational variables, respectively (number); H and L are the total number of positive and negative deviational variables, respectively (number); KPI_{h_i} and KPI_{l_i} are the values of positive KPI (h) and negative KPI (l), respectively, for system (i) at the time of assessment (t) (varies according to the KPI); TH_{h_i} and TH_{l_i} are the KPIs' threshold values of the positive KPI (h) and negative KPI (l), respectively, for system i (varies according to the KPI); $\mathrm{PA}_{v_{t_i}}$ is a binary financial penalty applicability index for KPI (v) at time (t) for system (i) (0 or 1); and $\mathrm{IA}_{v_{t_i}}$ is a binary financial incentive applicability index for KPI (v) at time (t) for system (i) (0 or 1).

MAINTENANCE PLANNING OPTIMIZATION

During the operations and maintenance phase, the need for optimization is paramount for both municipalities and private partner, who takes the responsibility of maintaining the assets, as there will be countless possible and valid alternatives for each asset in the corridor at each point of time. For instance, the number of solutions for 20 corridors for one asset class, throughout 25 years planning horizon, incorporating five types of interventions is $5^{20\times 25}$. Imagine the number of assets for any medium- or large-sized municipalities along with limited budget and increasing demand for higher level of service. Municipalities are faced with extra pressure on to optimize the utilization of their expenditures and fulfill the end users' expectations within the tight available budgets. Another challenge is the lengthy planning horizon, driven by the assets' service life, that makes it more computationally complicated and challenging, setting extra uncertainty for asset managers while taking intervention decisions. To overcome those challenges and provide flexibility to asset managers to select their preferred option that fits into their objective setting, three optimization models were set up for the maintenance planning phase, as shown in Figure 14.2: (1) single-objective optimization model; (2) non-preemptive goal optimization; and (3) multi-objective hierarchical goal optimization.

Single Objective

The single-objective formulation aims at maximizing the NPI while meeting the contractual KPIs' thresholds. GA engine combined with pre-applied metaheuristic rules, which minimize the optimization engine search space, were applied. GA is derived from the biological systems. It relies on simulating the natural survival of the fittest where the solution is represented as a string of chromosomes, which consist of several genes. The genes' exchanging process within the chromosomes is carried out through mutation and crossover operations, and new solutions are evaluated to replace the weaker member in the population and produce better solutions. This process continues until a near-optimum solution is generated. The performance of GAs is affected by four main parameters: (1) number of generations; (2) population size; (3) mutation rate; and (4) crossover rate (Elbeltagi and Tantawy 2008). The system utilized advanced spreadsheet modeling and Evolver™ Version 7.0 as an optimization engine (Evolver 2016). It functions through a powerful engine that is designed to fit the municipalities' needs and meet the KPIs' defined thresholds. To reduce the search space, several pre-determined metaheuristic rules are applied as

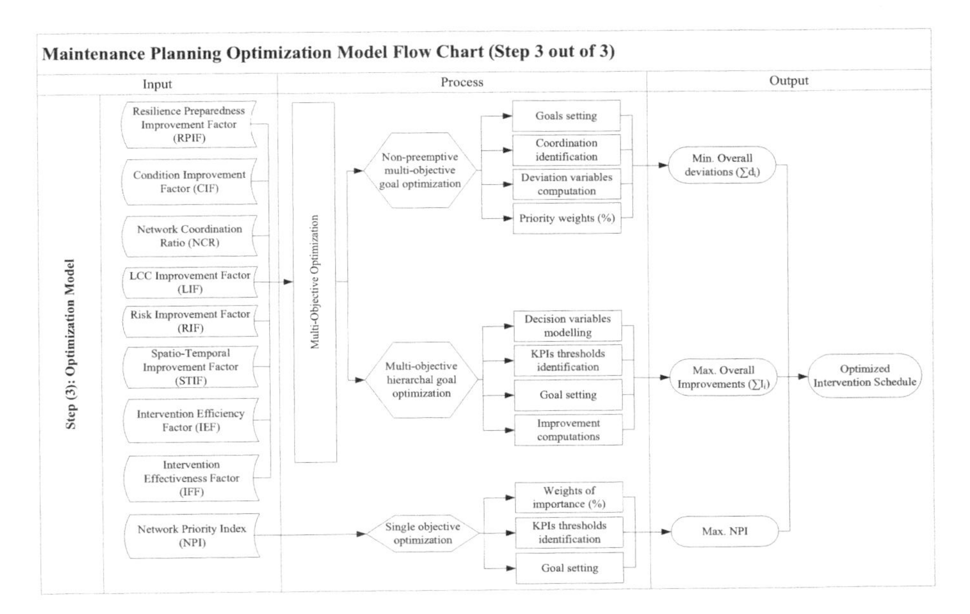

FIGURE 14.4 Maintenance planning optimization models.

follows: (1) number of annual intervention actions; (2) corridor-based inter-disruption time between carrying out one intervention action and another to minimize the service disruptions; and (3) number of corridor interventions across the planning horizon. The decision variables are formulated through integer programming, where the ranges of decision variables are defined according to the number of intervention activities that need to be addressed in the model. The wider the range of the decision variables, the exponentially more computationally complicated the optimization problem is, as it generates a greater spectrum of possible combinations and requires an evolutionary optimization algorithm to reach a near-optimum solution for this combinatorial-in-nature problem. In this problem, the decision variables vary between 0 and 10, as shown in the action-level decision variables of Table 14.1. Thereafter, the constraint check is undertaken annually and uses binary coding rules, where "0" represents meeting the constraint and "1" represents the failure to meet the constraint, to guarantee that the KPIs are met on an annual basis. The mathematical formulation of the single-objective optimization problem is shown in Equations 14.13 through 14.20.

TABLE 14.1
Optimization Decision-Making Levels and Decision Variables

Coordination Level (1)		System Level (2)		Action Level (3)	
Decision Variable	Description	Decision Variable	Description	Decision Variable	Description
0	Do Nothing	0	Do Nothing	0	Do Nothing
1	Conventional (1 system)	1	Roads	1	Surface overlay
				2	Resurfacing
		2	Water	3	Leaks repair
				4	Pipe replacement
		3	Sewer	5	Leaks repair
				6	Pipe replacement
2	Partially coordinated (2 systems)	4	Roads and Water	7	Pipe replacement and road resurfacing
		5	Roads and Sewer	8	Pipe replacement and road resurfacing
		6	Water and Sewer	9	Pipe replacement
3	Full coordination (3 systems)	7	Full Coordination	10	Pipe replacement and road resurfacing

Outer optimization layer (Decision-making layer) — *Inner optimization layers*

$$\textbf{Decision variables} = \begin{bmatrix} I_{t_o} & \cdots & I_{T_o} \\ \vdots & \ddots & \vdots \\ I_{tO} & \cdots & I_{TO} \end{bmatrix} \tag{14.13}$$

For $I_{t_o} = 0, 1, \ldots 10$
$t = 1, 2, \ldots T$
$o = 1, 2, \ldots O$

$$\textbf{Max}(\textbf{NPI}) = \sum_{o=1}^{O} \left(\frac{L_o}{\sum_{o=1}^{O} L_o} \times \text{CPI}_{o_t} \right) \tag{14.14}$$

Subject to the following constraints:

$$\text{LCC} \le \text{LCC}_B \tag{14.15}$$

$$\text{CCD}_t \le D_a \tag{14.16}$$

$$\text{IN}_t \le \text{IN}_a \tag{14.17}$$

$$R_{i_{o_t}} \ge R_{\text{th}} \tag{14.18}$$

$$F_{i_{o_t}} \le F_{\text{TH}} \tag{14.19}$$

$$\text{RI}_{o_t} \le \text{RI}_{\text{th}} \tag{14.20}$$

where D_a is the available duration based on the maintenance capacity (hours); IN_t is the number of annual interventions at a certain point of time (t) (number); IN_a is the allowable number of annual interventions to avoid extra disruption for the surrounding community (number); and RI_{th} is the RI threshold (0–5).

Non-preemptive Goal Optimization

The non-preemptive goal optimization model differs from the single objective as it accounts for all the KPIs while undertaking intervention decisions. Similar to the single objective, integrated goal optimization and GAs combined with pre-applied metaheuristic rules, which minimize the optimization engine search space, were applied. The formulation of decision variables takes place through integer programming where the variables formulation could be summarized as follows: (1) "0" represents the "Do nothing" and the "road maintenance" scenarios where either no maintenance took place at this point of time or only road maintenance took place. This road maintenance was combined with the "Do nothing" scenario as there is only a one-way functional interdependency where a disruption caused by the road will affect neither the water nor the sewer networks. (2) "1 and 2" represent the "only water or sewer rehabilitation" scenario where either the water or the sewer network is undergoing an intervention, and the road is not. In this case, the geographical interdependency takes place as the two assets are geographically located at the same space and the disruption of the water network will partially/fully affect the road service. In addition, the maintenance of the water network requires excavating the corridor, which implies applying a "reconstruction" for the road corridor and returning it to a pristine condition state. (3) "3, 4, and 5" represent the "partially coordinated" scenario where the intervention actions are partially coordinated for the roads and water networks; roads and sewer networks; or water and sewer networks, respectively. In this case, the road network will undertake a "reconstruction," given the fact that it will be impossible to carry out "slurry seal" or "crack filling" when you are rehabilitating/replacing the water network, and (4) "6" represents the "full coordination" scenario where the roads, water, and sewer networks' interventions are fully coordinated. The wider the range of the defined decision variables, the exponentially more complicated the optimization problem is, generating a greater spectrum of possible combinations and requiring an evolutionary optimization algorithm to reach a near-optimum solution for the combinatorial-in-nature problem. Dynamic programming has been utilized due to the complexity of the optimization problem and the lengthy planning horizon. A 5-year segmentation analysis was chosen for applying dynamic programming and the optimization has been subsequently done across the whole planning horizon in a chronological sequence. Goal programming or goal optimization has been chosen for the problem in hand, given the fact that the problem features conflicting goals and multiple assets. The objective is linked to the variables through "Goal Constraints." However, the objective is clearly formulated to minimize the sum of deviations for the prescribed goal values defined by the user. To combine the objectives, a percentile ranking approach was utilized by calculating the percentage deviation from a goal rather than the absolute deviation (Schniederjans 1995). Finally, the deviational variables are formulated to fit the pre-defined set of KPIs, as shown in Equations 14.21 through 14.29.

$$\mathbf{Min}(\mathbf{Z}) = \sum_{i=1}^{n_s}\sum_{v=1}^{V}\sum_{t=1}^{T}\left[W_i \times W_v \times \left(d_{k_t}^- + d_{m_t}^+\right)\right] \tag{14.21}$$

Subject to the following constraints:

$$\sum_{i=1}^{n_s} W_i = 1 \tag{14.22}$$

$$\sum_{v=1}^{V} W_v = 1 \tag{14.23}$$

$$\text{LCC} \le \text{LCC}_B \tag{14.24}$$

$$\text{CCD}_t \le D_a \tag{14.25}$$

$$\text{IN}_t \le \text{IN}_a \tag{14.26}$$

$$R_{i_{ot}} \ge R_{\text{th}} \tag{14.27}$$

$$F_{i_{ot}} \le F_{\text{TH}} \tag{14.28}$$

$$\text{RI}_{ot} \le \text{RI}_{\text{th}} \tag{14.29}$$

Multi-objective Hierarchical Tri-Level Goal Optimization

The complexity of the problem on hand arises from the spatial interdependency among the assets under study and the varying intervention scenarios. Thus, it would be computationally impossible to manually reach an optimal solution due to the outsized search space. The scenario of "n_s" systems, "o" corridors, "t" planning horizon, and "c" coordination scenarios will yield a total of $c^{n_s \times o \times t}$ possible solutions. Even though previous scholars utilized dynamic programming and phased optimization for fund allocation problems (Atef et al. 2012; Atef and Moselhi 2013a, b; Hegazy and Elhakeem 2011; Scheinberg and Anastasopoulos 2009), they result in near-optimal solutions, based on a micro level, which are not necessarily optimal for the macro-level problem on hand. The fact that it disables the decision-makers to "*see the forest for the trees*" might impact their outcome and result in suboptimal solutions (Colson et al. 2007). Thus, the proposed integrated non-preemptive, multi-objective hierarchical goal optimization and GAs or MOSEK approach opts at attaining an optimum or near-optimum solution for n_s systems in "o" corridors to reach a globally optimal solution for the overall network. It functions through eight integrated models that compute the duration, cost, space utilized, efficiency, effectiveness, condition, resilience preparedness, and risk of each corridor, accounting for all the possible combinations. Thenceforth, the decision-making is undertaken through three layers where two of them are inner layers that act as an output of the outer optimization problem, as shown in Figure 14.5. Those layers represent the hierarchical levels where the outer layer represents the coordination-level decisions;

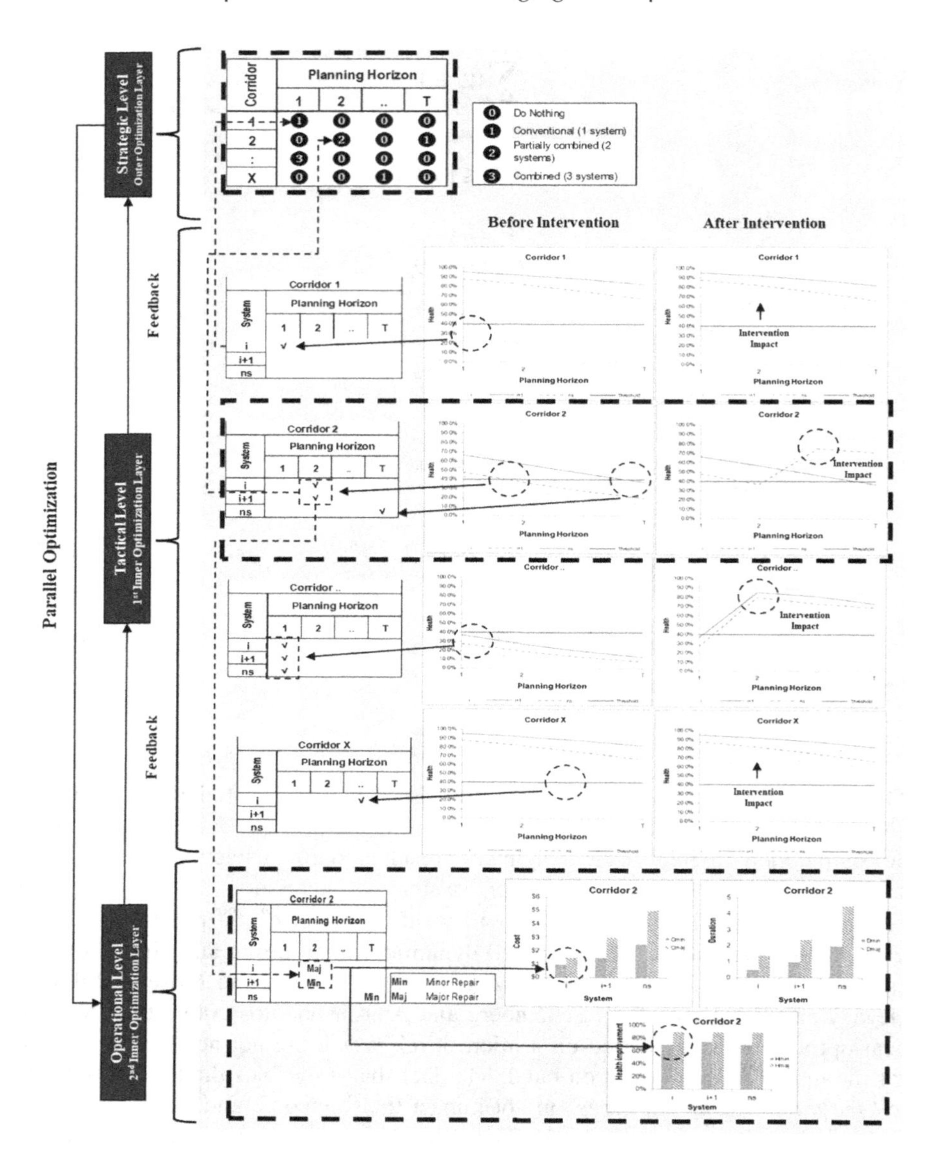

FIGURE 14.5 Multi-objective hierarchical goal optimization model –visual example.

the inner layers represent the systems and actions levels' decisions, respectively. The outer optimization layer, decision-making layer, aims at taking coordination decisions on a network level. In this level, the model answers two questions. The first is "should we undertake an intervention for this corridor?" and if the answer is yes, the second questions is "how many systems should undertake interventions at this point in time?" The answers to those questions are then processed within the two inner layers. The first inner layer, systems layer, deals with each system within each corridor separately. Based on the health of each system within the corridor, the model aims at answering one question, "Which system(s) need intervention(s)?" For instance, if the answers of the outer layer were "yes for corridor 2" and "two in the second year," the model will select the two systems with the least condition out of the n_s systems for interventions. Thus, in the case shown in Figure 14.5, the model selected system i and $i+1$ for intervention, given that they have the least condition and the outer optimization layer selected 2 systems "partial coordination". Finally, the second inner layer, actions layer, deals with the intervention actions along with their associated costs, duration, risk, resilience and condition improvement. Based on the answer to the predecessor layer's question, the model will answer one question, "What type of intervention is required to enhance the system condition/resilience state within the least cost, duration?" For instance, let's continue the previous case of corridor 2 where the model selected system i and $i+1$ for interventions. In this case, the second inner layer selects the suitable intervention type (e.g., minor or major) for this corridor, based on the weights of importance associated with the conflicting objectives. If the municipality has a limited budget, the model will select the alternative with minimum cost to meet their tight budget. However, if the municipality is looking for a better level of service, the model will select the alternative that best enhances the system's condition state. The outcome of the second inner layer directly feeds the objective with the financial, temporal, condition, spatial, efficiency, risk, resilience preparedness, and effectiveness improvement information of the intervention scenario for all the corridors in the network. This newly developed multi-objective hierarchical optimization approach substantially reduced the search space through removing the unviable solutions (e.g., undertake a pipe replacement for a newly installed pipe, do resurfacing for a newly constructed road, etc.). To better imagine the huge savings in the computational time, let's assume a case of 20 corridors, with 3 systems in each corridor, 2 intervention types for each system (minor and major), and 25 years planning horizon. In typical one-level decision-making, the number of decision variables will range from 0 (Do nothing) to 10 to account for all the coordination scenarios, systems, and their corresponding types of interventions. Thus, the number of possible solutions will be $11^{20\times25}$. In the multi-objective hierarchical decision-making, the decision variables will range from 0 (Do nothing) to 3 (number of systems per intervention). In that case, the number of possible solutions will be $4^{20\times25}$. The reduction of the search space, represented through savings in the number possible solutions, for the hierarchical approach would be the difference between both approaches, which is $7^{20\times25}$, almost three times smaller number of possible solutions, compared to the one-level decision-making. Given the complexity of the problem in hand, integrated goal optimization, integer programming, and GAs or MOSEK were utilized to enable decision-makers trade-off their interventions based on conflicting goals as displayed in the equations below. The $s_{o_{tcir}}$ integer programming-based decision variable is used to represent the three-level dimensional space of "o" corridors, "t" planning horizon, "c" coordination scenario (e.g., conventional, partially coordinated, or full coordination), "i" system(s) selected for intervention, and "r" intervention type. The list of decision variables at each optimization layer is displayed in Table 14.1. For instance, if $s_{3_{52138}}$ is equal to 2, then corridor 3 at year 5 will experience a partial coordination for systems 1 and 3 using intervention 8. The model decision variables could be mathematically formulated as displayed in Equation 14.30. The model functions through constraints to ensure proper condition state for all the n_s systems as displayed in Equations 14.32 through 14.34. The multi-objective optimization is beneficial for asset managers facing conflicting goal preferences. Those preferences might vary among different decision-making bodies such as asset managers, maintenance contractors, citizens/users, politicians, etc. The multi-objective formulation accounts for all the decision-makers preferences through setting weights of importance for each one. It computes the improvement factor of each goal for each optimization scenario and compares it with the conventional scenario. Thereafter, weighted-sum mean is used to combine multiple goals into a single function after assigning different weights of importance for each goal, which should sum up to 1 for all the goals/preferences. Each goal is formulated in the form of an improvement deviational variable (I_j) from the conventional scenario. Finally, the objective function is formulated to maximize the overall improvements, resulting from the weighted sum of the considered goals, as outlined in Equations 14.31, 14.35, and 14.36.

$$\textbf{Decision variables} = \begin{bmatrix} s_{1_{1cir}} & \cdots & s_{O_{1cir}} \\ \vdots & \ddots & \vdots \\ s_{1_{Tcir}} & \cdots & s_{O_{Tcir}} \end{bmatrix} \tag{14.30}$$

For $o = 1, 2, \ldots O$
$t = 1, 2, \ldots T$
$c = 1, 2, \ldots n_{s+}1$
$i = 1, 2, \ldots n_s$
$r = 1, 2, \ldots R$

$$\mathbf{Max}(\mathbf{G}) = \sum_{u=1}^{U}\sum_{k=1}^{K}\sum_{m=1}^{M}\sum_{t=1}^{T}\left[W_u \times \left(I^-_{k_t} + I^+_{m_t}\right)\right] \tag{14.31}$$

Subject to the following constraints:

$$\sum_{u=1}^{U} W_u = 1 \tag{14.32}$$

$$\text{CCS}_{i_{\text{SC}_{op}}} \geq \text{CCS}_{i_{\text{TH}}} \tag{14.33}$$

$$F_{i_{ot}} \leq F_{\text{TH}} \tag{14.34}$$

$$I^-_{k_t} = \sum_{i=1}^{n_s}\sum_{h=1}^{H} \frac{\text{TH}_{l_i} - \text{KPI}_{l_{it}}}{\text{TH}_{l_i}}; \text{ for all } k \text{ and } t \tag{14.35}$$

$$I^+_{m_t} = \sum_{i=1}^{n_s}\sum_{l=1}^{L} \frac{\text{KPI}_{h_{it}} - \text{TH}_{h_i}}{\text{TH}_{h_i}}; \text{ for all } m \text{ and } t \tag{14.36}$$

where $s_{o_{tcir}}$ is an integer-programming-based decision variable that represents the three-dimensional space of "*o*" corridors, "*t*" planning horizon, "*c*" coordination scenario, "*i*" system(s) selected for intervention, and "*r*" intervention type (number); R is the total number of intervention types available for each system i (number); G represents the maximized value for all the negative (I_k^-) and positive improvements (I_m^+) for U goals (%); u is the improvement deviational variables counter (number); U is the total number of improvement deviational variables (number); W_u represents the deferential weights among the conflicting goals (%); $I_{k_t}^-$ is the summation of all the negative improvement deviational variables at point of time (t) (%); $I_{m_t}^+$ is the summation of all the positive improvement deviational variables at point of time (t) (%); h and l are the counters of the positive and negative improvement deviational variables, respectively (number); H and L are the total number of positive and negative deviational variables, respectively (number); $\mathrm{CCS}_{i\mathrm{SC}_{o_p}}$ is the corridor condition state (CCS) of the resulting intervention scenario of system (i) within corridor o (%); and $\mathrm{CCS}_{i_{\mathrm{TH}}}$ is the CCS threshold for all n_s systems (varies from one system to another) (%).

REFERENCES

Abusamra, S. (2019). "Coordination and Multi-Objective Optimization Framework for Managing Municipal Infrastructure under Performance-Based Contracts." Doctoral of Philosophy Thesis, Concordia University, Montreal.

Atef, A. and Moselhi, O. (2013a). "Budget Allocation for Vulnerable and Interdependent Civil Infrastructure Networks." *9th Construction Specialty Conference,* May 31–June 2, CSCE, Montreal.

Atef, A. and Moselhi, O. (2013b). "Understanding the Effect of Interdependency and Vulnerability on the Performance of Civil Infrastructure." *International Symposium on Automation and Robotics in Construction (ISARC),* Aug 12–14, Montreal.

Atef, A., Osman, H., and Moselhi, O. (2012). "Multi-objective genetic algorithm to allocate budgetary resources for condition assessment of water and sewer networks." *Canadian Journal of Civil Engineering*, 39(9), 978–992.

Colson, B., Marcotte, P., and Savard, G. (2007). "An overview of bilevel optimization." *Annals of Operations Research*, 153(1), 235–256.

Elbeltagi, E. and Tantawy, M. (2008). "Asset Management: The Ongoing Crisis." *Proceedings for 2nd Conference on Project Management* (pp. 1–9), Saudi Council of Engineers, Riyadh.

Evolver Version 7.0 [Computer Software] (2016). Palisade, Ithaca.

Hegazy, T. and Elhakeem, A. (2011). "Multiple optimization and segmentation technique (MOST) for large-scale bilevel life cycle optimization." *Canadian Journal of Civil Engineering*, 38(3), 263–271.

MOSEK Optimization Version 8 [Computer Software] (2018). MOSEK ApS, Copenhagen.

REMSOFT Asset Management Modelling Software [Computer Software] (2018). REMSOFT, Fredericton.

Scheinberg, T. and Anastasopoulos, P. C. (2009). "Pavement Preservation Programming: A Multi-Year Multi-Constraint Optimization Methodology." *Proceedings of 89th Annual Meeting of the Transportation Research Board (TRB),* Washington, DC.

Schniederjans, M. (1995). *"Goal Programming: Methodology and Applications."* Kluwer Academic, Boston, MA.

Part 5

Case Studies

15 City of Montreal Case Study

INTRODUCTION

A 9 km stretch from the city of Montreal roads, water, and combined sewer networks was selected for the analysis. The network comprises 20 corridors and was equally divided into four areas with five corridors for each area. The dataset scale/size in terms of the number of corridors is scaled down several times to enable the use of the available optimization engines. The condition states of the 20 corridors were assumed to represent the overall network condition states of each system. Time value of money was considered with an interest rate of 2% (Trading Economics 2023; Bank of Canada 2023). Furthermore, the study planning horizon was 25 years. The weights of the systems were assumed according to the overall life-cycle costs (LCC) of each system across 100 years, using the longest life method. The results displayed 45%, 25%, and 30% for the roads, water, and sewer systems, respectively. However, those weights are subject to change according to the stakeholders' preferences (e.g., condition, replacement cost, and crews' availability). The physical and financial-related data (e.g., physical state, operation and maintenance costs, and water and sewer pipe breaks) was extracted from two sources: (1) interviews with city officials (Hachey 2018; Sabourin and Abusamra 2018) and (2) city of Montreal official website (Ville de Montreal 2022a, b).

ASSET INVENTORY

The asset inventory contains all the assets that are spatially located in the same corridor. In this case study, 20 corridors were considered for the analysis as displayed in Table 15.1. This information includes the corridor length, road width, number of lanes, area, average annual daily traffic, water or sewer pipe material and diameter, excavation depth, soil type, demand category, age/year or installation, etc. The asset inventory acts as a central database for the computational models such that all the necessary information about the corridor can be extracted directly from the inventory and used for further analysis.

TEMPORAL DATASET

The temporal dataset includes the unit rates of different intervention actions and unit rate breakdown for each intervention activity. The unit rates have been adopted from the city of Montreal and the literature as displayed earlier in Table 9.3. It is worth noting that the temporal dataset is similar for both case studies. However, they differ in the number of corridors, current asset age, asset materials, pipe diameters, etc.

DOI: 10.1201/9781003327509-20

TABLE 15.1
City of Montreal Dataset

General		Road Network						Water Network		Sewer Network	
Corridor ID #	Corridor Length (m)	Number of Lanes	Lane Width (m)	Section Area (m²)	AADT	Traffic Growth Rate (%)	Current Condition (%)	Year of Installation	Pipe Diameter (cm)	Year of Installation	Pipe Diameter (cm)
1	370	3	3	3,330	12,000	5	90	1953	450	1920	600
2	370	4	3	4,440	8,000	5	70	1982	150	1900	525
3	452	4	3	5,424	10,000	5	85	1976	250	1893	375
4	393	2	3	2,358	11,000	5	65	1958	200	1950	300
5	419	3	3	3,771	7,000	5	70	1965	150	1960	250
6	766	4	3	9,192	9,500	5	90	1991	100	1970	150
7	451	4	3	5,412	10,500	5	70	1992	500	1980	200
8	311	2	3	1,866	8,500	5	85	1977	500	1990	450
9	425	4	3	5,100	6,800	5	65	1982	350	2000	200
10	783	4	3	9,396	7,500	5	70	1991	500	1975	525
11	318	3	3	2,862	9,000	5%	90%	1972	500	1943	375
12	162	4	3	1,944	6,000	5%	70%	1960	250	1955	300
13	498	4	3	5,976	5,000	5%	85%	1979	250	1965	250
14	686	4	3	8,232	11,000	5%	65%	1953	100	1975	150
15	207	2	3	1,242	10,000	5%	70%	1960	450	1985	200
16	715	3	3	6,435	6,000	5%	90%	1977	200	1905	450
17	270	2	3	1,620	9,000	5%	70%	1986	150	1965	200
18	217	2	3	1,302	12,000	5%	85%	1992	350	1968	600
19	519	2	3	3,114	9,000	5%	65%	1975	300	1978	525
20	560	4	3	6,720	8,000	5%	70%	1987	150	1982	600

TABLE 15.2
Area Savings for Different Intervention Coordination Scenarios

Corridor/ Intervention Scenario	Conventional Intervention Scenario (%)	Partial Coordination Intervention Scenario (%)	Full Coordination Intervention Scenario (%)
Corridor 1	0	20	32
Corridor 2	0	23	38
Corridor 3	0	14	26

Spatial Dataset

Given the lack of available GIS combined dataset for the city of Montreal's roads, water, and combined sewer networks, the spatial dataset divided the corridors into four areas. Each area includes five corridors such that the selected corridors are assumed to include the three assets spatially located. Furthermore, the area savings for the partial and full coordination intervention scenarios as opposed to the conventional intervention scenario were computed for each corridor, depending on several factors such as road width, pipe diameter, excavation depth, intervention type, etc. Table 15.2 displays the % of area savings arising from partial or full coordination.

Financial Dataset

The financial dataset includes the costs of different intervention actions, cost breakdown for each intervention activity, indirect/user costs associated with disrupting the public, and intervention cost in partial and full coordination scenarios. The costs have been adopted from the city of Montreal and recent Canadian bill of quantities' (BOQs) and tender documents, as displayed earlier in Tables 9.4, 9.5, 15.3, and 15.4 (Ville de Montreal 2022a, b; Abusamra 2019; Qin and Cutler 2014; TDOT 2015). It is worth noting that the financial dataset is similar for both case studies.

Physical Dataset

The physical dataset includes all the information with respect to the condition, age, physical characteristics, etc. The physical characteristics vary from one asset to another and even within the same asset. For instance, the roads' age could vary from 15 to 25 years depending on the operational and climatic conditions, road design, and sub-surface condition. Furthermore, the roads' deterioration pattern varies according to the expected service life of the asset, and physical, operational, and climatic conditions (Abu-Samra 2017). Thus, the models were classified into three categories based on the road structural design as well as the traffic: (1) low traffic, (2) medium traffic, and (3) high traffic (Amador and Magnuson 2011). For the water pipes, the age of the pipes varies from 60 to 80 years according to the pipe material and diameter. Accordingly, the Weibull distribution will vary from one pipe to another according to its age as well as its associated shape (beta) and scale (alpha) parameters as highlighted

TABLE 15.3
Replacement Rules and Pipes' Replacement Costs (Abusamra 2019)

Original Pipe Diameter (mm)	Replacement Pipe Diameter (mm)			Unit Cost for Same Pipe Size ($/m)				
	Low Demand	Medium Demand	High Demand	Pipe	E/B[a]	PR[b]	Installation	Total
300	375	450	600	$ 80.70	$ 65.00	$ 37.00	$ 225.00	$ 407.70
375	450	600	750	$ 99.60	$ 65.00	$ 37.00	$ 275.00	$ 476.60
450	600	675	900	$ 105.70	$ 65.00	$ 37.00	$ 300.00	$ 507.70
525	675	825	1,050	$ 111.80	$ 65.00	$ 37.00	$ 325.00	$ 538.80
600	750	900	1,200	$ 186.10	$ 65.00	$ 37.00	$ 366.70	$ 654.80
675	825	900	1,200	$ 260.30	$ 65.00	$ 37.00	$ 408.30	$ 770.60
750	900	1,200	1,500	$ 334.60	$ 65.00	$ 37.00	$ 450.00	$ 886.60
825	1,050	1,350	1,500	$ 378.90	$ 65.00	$ 37.00	$ 575.00	$ 1,055.90
900	1,200	1,350	1,800	$ 423.30	$ 65.00	$ 37.00	$ 700.00	$ 1,225.30
1,050	1,350	1,500	1,800	$ 544.80	$ 65.00	$ 37.00	$ 862.50	$ 1,509.30
1,200	1,500	1,800	2,400	$ 666.20	$ 65.00	$ 37.00	$ 1,025.00	$ 1,793.20
1,350	1,800	2,400	2,400	$ 857.40	$ 65.00	$ 37.00	$ 1,237.50	$ 2,196.90
1,500	1,800	2,400	2,400	$1,048.50	$ 65.00	$ 37.00	$ 1,450.00	$ 2,600.50
1,800	2,400	2,400	2,400	$1,239.70	$ 65.00	$ 37.00	$16,662.50	$18,004.20
2,400	2,400	2,400	2,400	$1,430.90	$ 65.00	$ 37.00	$ 1,875.00	$ 3,407.90

[a] E/B refers to the excavation and backfilling activities ($/m^3)
[b] PR refers to pavement restoration activities ($/m^2)

TABLE 15.4
User Costs (Qin and Cutler 2014; TDOT 2015)

Activities	Unit Cost ($/unit)
Passenger cars	22.09
Trucks	32.26

earlier. The pipe materials were classified into the following categories: (1) iron pipes that include cast iron and steel pipes, (2) plastic pipes that include PVC pipes, and (3) concrete pipes that include asbestos cement pipes and prestressed concrete pipes. The pipe diameters were classified into the following: (1) small pipes that include all the pipes with diameters less than 10 inches, (2) medium pipes that include all the pipes with diameters between 10 and 18 inches, and (3) large pipes that include all the pipes with diameters larger than 18 inches. According to those categories, different deterioration curves were developed with different ages, shape, and scale parameters. For the combined sewer and stormwater pipes, the age of the pipes varies from 80 to 100 years according to the pipe material and diameter. Accordingly, the Weibull distribution will vary from one pipe to another according to its age as well as its associated shape (beta) and scale (alpha) parameters as highlighted earlier in the previous chapter. The pipe materials were classified into the following categories: (1) iron pipes that include corrugated steel pipes, (2) plastic pipes that include PVC pipes and vinyl pipes, and (3) concrete pipes that include asbestos cement pipes, prestressed concrete pipes, and vitrified clay pipes. The pipe diameters were classified into the following: (1) small pipes that include all the pipes with diameters less than 10 inches, (2) medium pipes that include all the pipes with diameters between 10 and 18 inches, and (3) large pipes that include all the pipes with diameters larger than 18 inches. According to those categories, different deterioration curves were developed with different ages, shape, and scale parameters. The deterioration curves of the roads, water, and combined sewer pipes are displayed in Figures 15.1, 15.2, and 15.3. It is worth noting that the deterioration curves for both the cases are similar. However, they differ in the number of corridors, current asset age, asset materials, pipe diameters, etc.

Resilience Preparedness Dataset

Given the fact that demand curves were not available for the city of Montreal, the resilience preparedness model was excluded from the city of Montreal case study and was only applied to the town of Kindersley case study. Thus, details about the demand data for water and combined sewer and stormwater networks will be further discussed in the town of Kindersley case study.

Risk Dataset

The risk dataset includes the probability and consequences of failure. For the probability of failure (POF), Table 11.7 was used to classify the POF for the roads, water, and combined sewer and stormwater networks. As highlighted earlier, the POF

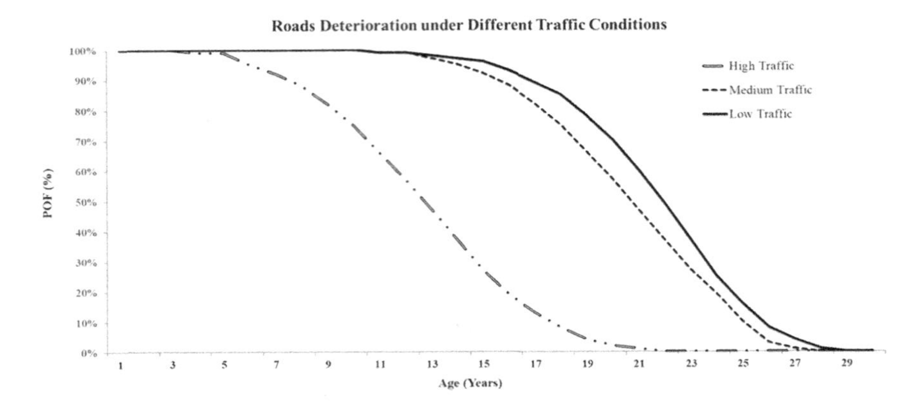

FIGURE 15.1 Roads' deterioration under different traffic operational conditions.

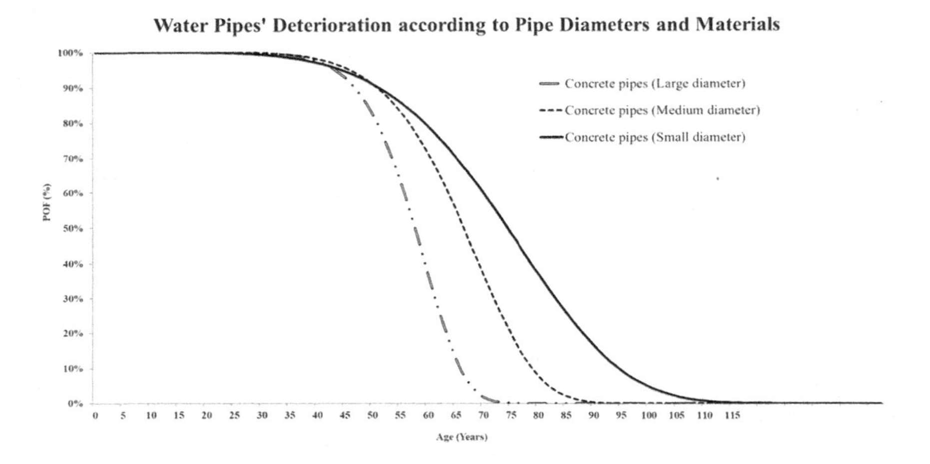

FIGURE 15.2 Water pipes' deterioration according to different pipe diameters and materials.

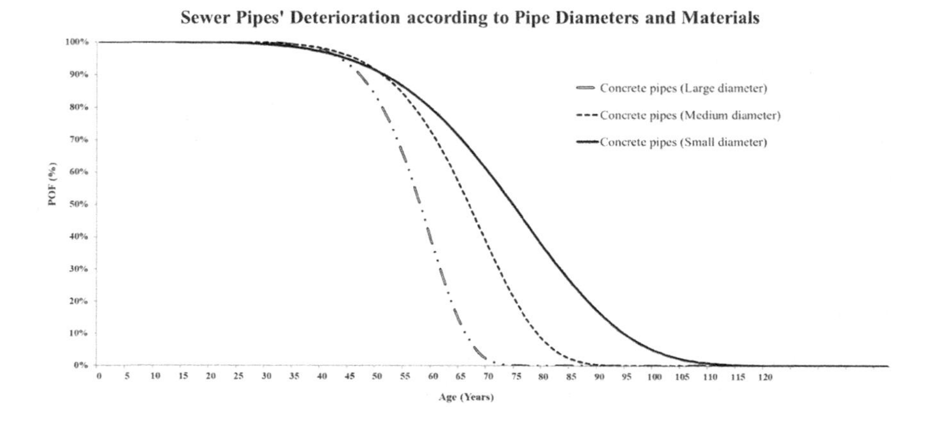

FIGURE 15.3 Sewer pipes' deterioration according to different pipe diameters and materials.

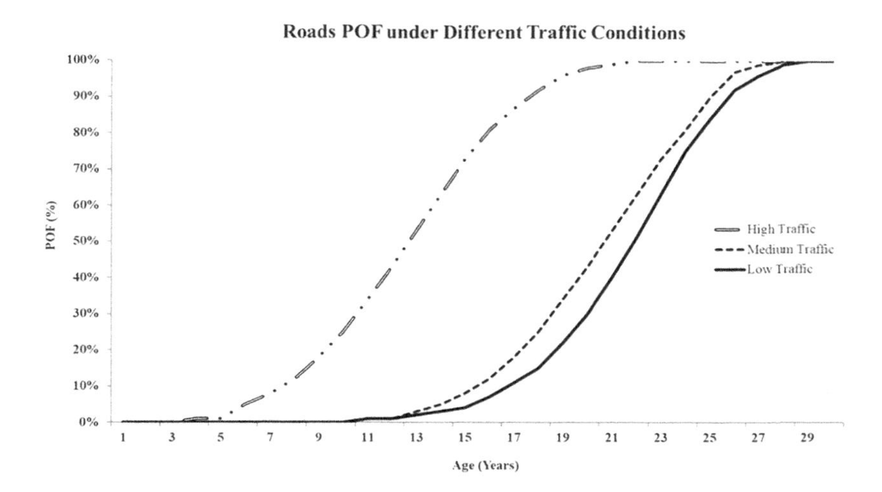

FIGURE 15.4 Roads' POF under different traffic operational conditions.

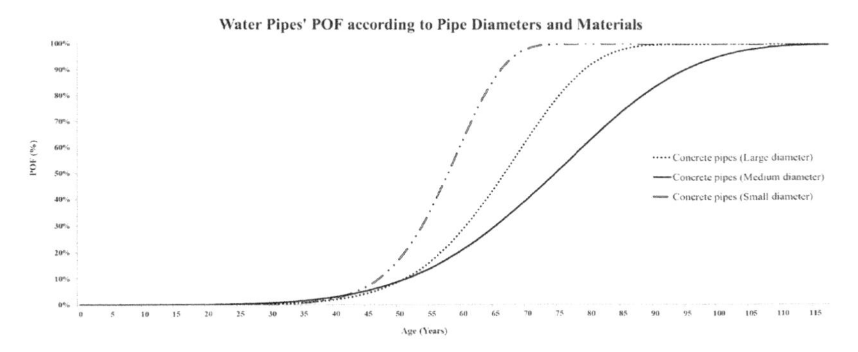

FIGURE 15.5 Water pipes' POF according to different pipe diameters and materials.

of each asset was computed from the reliability using Equation 11.3. Thus, there are different POF curves for different assets and asset categories as displayed in Figures 15.4, 15.5, and 15.6 for roads, water, and combined sewer and stormwater assets, respectively. With respect to the consequences of failure (COF), the scoring ranges between 1 and 5 from insignificant to catastrophic, respectively, as displayed in Table 11.2. Thus, according to the asset characteristics, the COF could be computed. It is worth noting that the risk dataset is similar for both case studies. However, they differ in the number of corridors, current asset age, asset materials, pipe diameters, etc.

RESULTS OF MULTI-DIMENSIONAL PERFORMANCE ASSESSMENT MODELS (PRE-OPTIMIZATION)

In this section, we will address the multi-dimensional performance assessment results for the temporal, financial, and condition indicators only, given the fact that the reliability, risk (POF), spatial, efficiency, and effectiveness are dynamic and

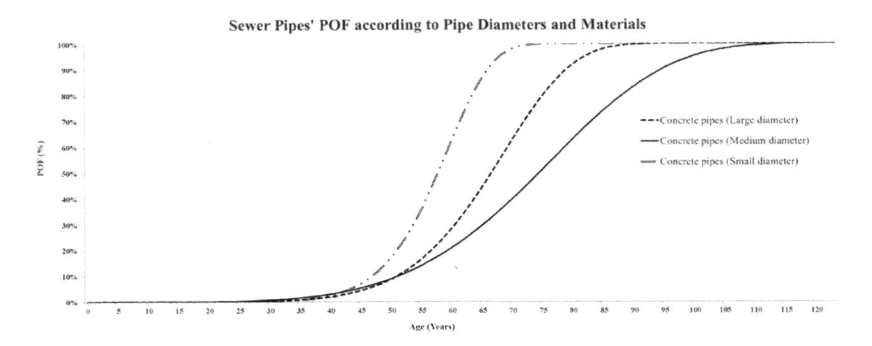

FIGURE 15.6 Sewer pipes' POF according to different pipe diameters and materials.

rely on the existence of a full intervention plan to be computed. Thus, the full-scale implementation of the multi-dimensional performance assessment indicators will be further discussed in the following section after running the optimization.

Duration Savings

The duration savings model aims at computing the temporal savings resulting from coordinating the interventions of the co-located systems. The results of this model represent a static standalone outcome of the time savings such that corridor 8 was randomly selected to visualize the potential temporal savings of the three intervention scenarios. However, the results might vary according to several aspects such as corridor length, system's condition, intervention type, pipe material, and pipe diameter, etc. The dynamic results will be analyzed after running the optimization scenarios to display the network indicators and the overall savings across the planning horizon. The detailed computations of the corridor intervention duration for the different intervention types (e.g., minor or major rehabilitation) and scenarios (e.g., conventional, partially coordinated, and full coordination) were carried out and the outcome is a list of intervention scenarios with their associated durations as well as the maximum temporal coordination savings that can be attained as opposed to the conventional intervention scenario. As shown in Tables 15.5 and 15.6, the results have shown substantial temporal savings with a corridor coordination ratio ranging between 16% and 38% for the partially coordinated and full coordination intervention scenarios, respectively, as opposed to the conventional one. The temporal savings reflect the coordination of the intervention activities where the common activities

TABLE 15.5
Temporal Results for the Intervention Scenario

Asset Cost (Water)		Asset Cost (Sewer)					
Leaks Case	**Replacement Case**	**Leaks Case**	**Replacement Case**	**Roads and Water**	**Roads and Sewer**	**Water and Sewer**	**Full Coordination**
60.46	69.56	59.88	69.76	69.56	69.76	84.66	94.66

TABLE 15.6
Corridor Coordination Ratio for the Intervention Scenarios

Corridor ID#	Corridor Coordination Ratio			
	Case 1—Roads and Water	Case 2—Roads and Sewer	Case 3—Water and Sewer	Case 4—Full Coordination
Corridor 8	31%	29%	16%	38%

TABLE 15.7
LCC for the Conventional Intervention Scenario

Corridor ID#	Asset Cost (Roads)		Asset Cost (Water)		Asset Cost (Sewer)	
	Cracks Case	Resurfacing Case	Leaks Case	Replacement Case	Leaks Case	Replacement Case
Corridor 8	$47,968.80	$3,57,735.59	$46,802.29	$8,38,568.61	$50,802.29	$8,12,168.81

have been carried out once, instead of n_a or n_s times for the partially coordinated and conventional intervention scenarios. Furthermore, undertaking the full coordination intervention increased the number of parallel activities, which increased the temporal savings as opposed to the conventional approach by which n_s interventions are separately undertaken for each system.

Financial Savings

The financial savings model aims at computing the financial savings resulting from coordinating the interventions of the co-located systems. Unlike the temporal model, the financial savings model functions through two modules, namely direct costs and indirect costs calculation modules. The output of the duration savings model is inputted to the indirect costs calculation module to precisely estimate the service disruption duration and compute the indirect costs accordingly. The results of this model represent a static standalone outcome of the cost savings such that corridor 8 was randomly selected to visualize the potential financial savings of the three intervention scenarios. However, the results vary according to several aspects such as corridor length, system's condition, intervention type, pipe material, and pipe diameter, etc. The dynamic results will be shown after running the optimization scenarios to display the network indicators and the overall savings across the planning horizon. The detailed computations of the corridor intervention direct and indirect costs for the different intervention types (e.g., minor or major rehabilitation) and scenarios (e.g., conventional, partially coordinated, and full coordination) were carried out and the outcome is a list of intervention scenarios with their associated direct and indirect as well as the maximum financial coordination savings that can be attained as opposed to the conventional intervention scenario. The direct and indirect costs savings results are displayed in Tables 15.7 and 15.8. The life-cycle costs (LCC) savings, represented through the life-cycle costs improvement factor (LIF), are displayed in Tables 15.9, 15.10, and 15.11.

TABLE 15.8
LCC for the Partially Coordinated and Full Coordination Intervention Scenarios

	Corridor Cost (Case 1—Roads and Water)			Corridor Cost (Case 2—Roads and Sewer)			Corridor Cost (Case 3—Water and Sewer)			Corridor Cost (Case 4—Full Coordination)
Corridor ID#	**Roads and Water Case**	**Sewer Case**	**Overall (Case 1)**	**Roads and Sewer Case**	**Water Case**	**Overall (Case 2)**	**Water and Sewer Case**	**Roads Case**	**Overall (Case 3)**	**Roads, Water and Sewer**
Corridor 8	$8,38,568.61	$8,12,168.81	$16,50,737.42	$8,12,168.81	$8,38,568.61	$16,50,737.42	$12,48,703.05	$3,57,735.59	$16,06,438.64	$12,48,703.05

TABLE 15.9
Direct Costs Savings for the Intervention Scenarios

	LCC Improvement Factor—Direct Costs			
Corridor ID#	**Case 1—Roads and Water**	**Case 2—Roads and Sewer**	**Case 3—Water and Sewer**	**Case 4—Full Coordination**
Corridor 8	17.81%	19.81%	22.02%	37.83%

TABLE 15.10
Indirect Costs Savings for the Intervention Scenarios

	LCC Improvement Factor—Indirect Costs			
Corridor ID#	**Case 1—Roads and Water**	**Case 2—Roads and Sewer**	**Case 3—Water and Sewer**	**Case 4—Full Coordination**
Corridor 8	36.69%	32.69%	16.13%	41.82%

TABLE 15.11
LCC Savings for the Intervention Scenarios

	LCC Improvement Factor—Direct and Indirect Costs			
Corridor ID#	**Case 1—Roads and Water**	**Case 2—Roads and Sewer**	**Case 3—Water and Sewer**	**Case 4—Full Coordination**
Corridor 8	19.62%	21.35%	20.52%	39.49%

The direct and indirect intervention costs for maintaining all the systems in the corridor are summed up for the three intervention coordination scenarios. The results displayed substantial financial savings in the direct and indirect costs ranging between 16% and 41% for the partially coordinated and full coordination intervention scenarios as opposed to the conventional one. Accordingly, the results displayed huge financial savings with a LIF ranging between 19% and 39% for the partially coordinated and full coordination intervention scenarios as opposed to the conventional one. The financial savings reflect the coordination of the intervention activities where the common activities have been carried out once, instead of n_a or n_s times for the partially coordinated and conventional intervention scenarios. Furthermore, the fact that the full coordination intervention results in a smaller number of service disruptions as well as less disruption duration, as highlighted in the duration savings model, reduces the indirect costs and accordingly increases the financial savings.

Condition/Reliability Improvement

The integrated deterioration model computes the condition/reliability of each system and combines them into an overall corridor condition. A regression deterioration model was developed for the roads, given the availability of historical road condition data while a Weibull deterioration model was developed for the water and sewer networks. To display the corridor condition computation, a randomly selected sample of the reliability curves for the roads, water, and sewer networks was selected as displayed in Figures 15.7 through 15.10. The expected service lives of the systems vary according to several factors (e.g., pipe material, pipe diameter, and road structural category). The average service lives of the roads, water, and sewer networks were estimated at 15–30, 60–80, and 80–100 years, respectively. As shown in Figures 15.7 through 15.9, there are two reliability curves to represent each corridor where the first reliability curve displays the impact of the typical aging-based deterioration, and the second reliability curve displays the impact of undertaking an intervention at any point of time. The corridor reliability is computed based on the weights of importance among the systems, as displayed in Figure 15.10. To visualize the interventions' reliability improvement throughout the lifecycle, two intervention programs need to be compared together.

CITY OF MONTREAL POST-OPTIMIZATION RESULTS

The optimization model was run on both the performance-based contracts (PBC) structuring and maintenance planning optimization modes across 25 years planning horizon using Genetic Algorithms' (GA's) optimization engine. For the PBC structuring optimization, the results are the key performance indicators' (KPIs) thresholds as well as the financial penalties and incentives for each KPI. However, the maintenance planning optimization aims at reaching a near-optimal intervention plan for several scenarios as follows: (1) conventional optimization that reaches a

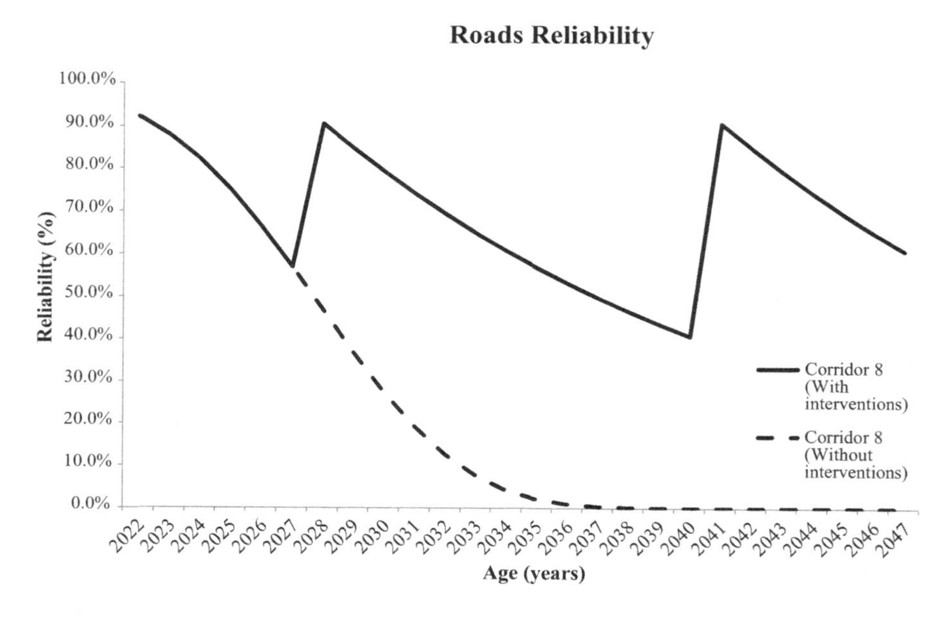

FIGURE 15.7 Sample from the roads' reliability model.

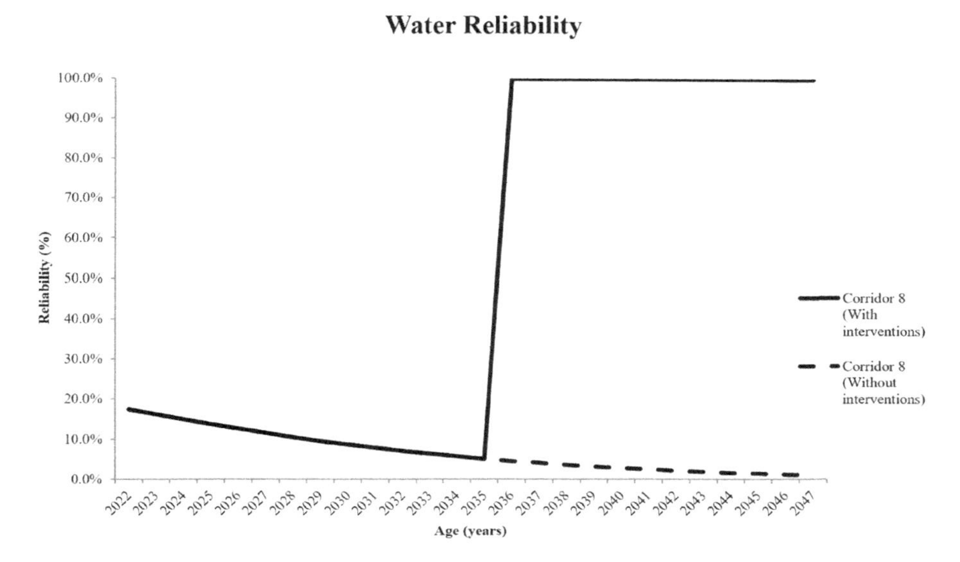

FIGURE 15.8 Sample from the water reliability model.

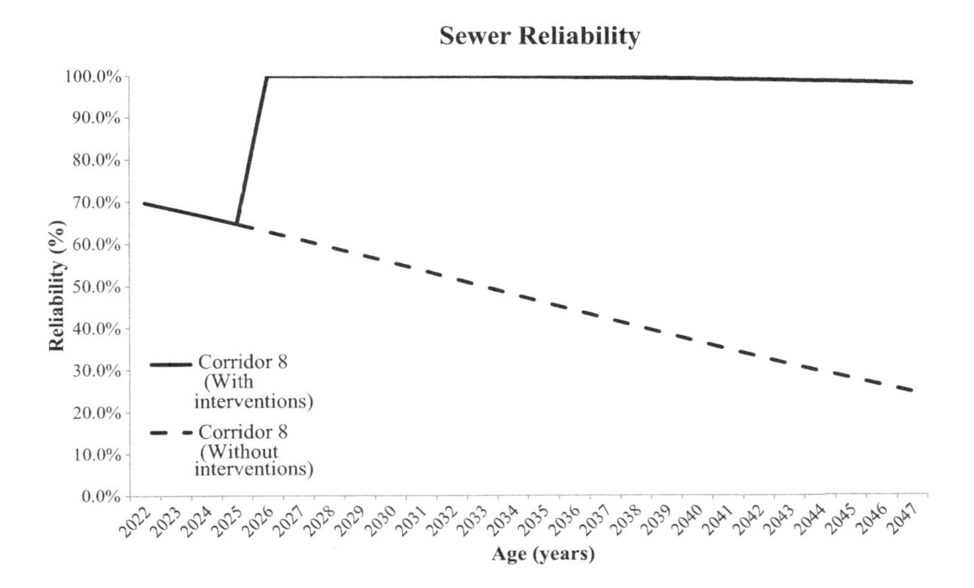

FIGURE 15.9 Sample from the sewer reliability model.

near-optimal solution for each asset under study and (2) coordination optimization that accounts for conventional, partial, and full coordination scenarios while reaching a near-optimal intervention plan. Furthermore, the sensitivity analysis analyzes the impact of changing the reliability KPI threshold on the other KPIs, and comparing it with the baseline scenario (Abusamra 2019).

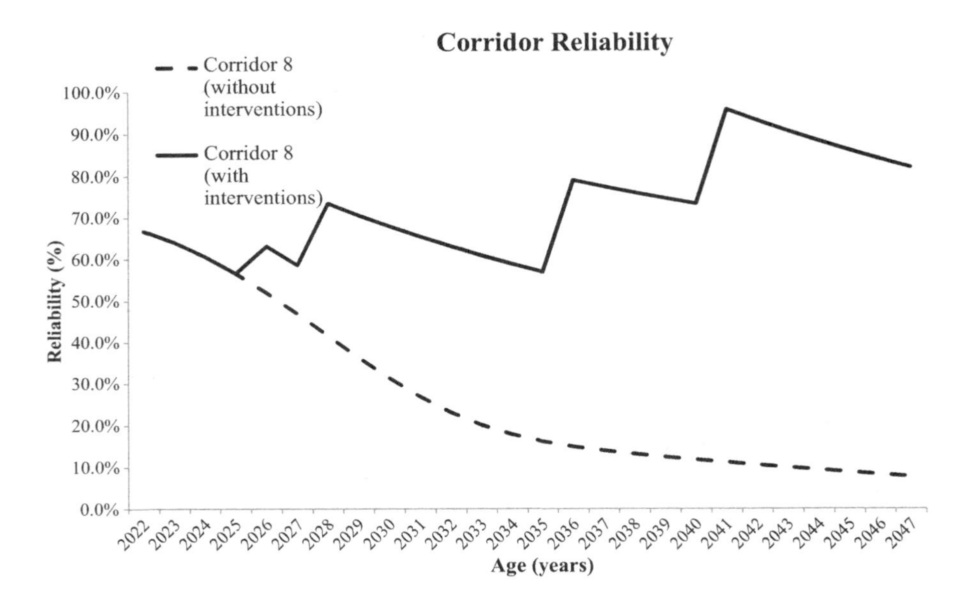

FIGURE 15.10 Sample from the integrated corridor reliability model.

TABLE 15.12
Optimization Parameters

Optimization Parameters	Value
Engine	GA (Evolver 7.5)
Crossover	80%
Mutation	20%
Population	200
Stopping criteria	Progress-based – No improvement in the objective function for 100,000 trials

Optimization Results

PBC Structuring Optimization

The PBC structuring optimization model was run in two modes: (1) KPIs' thresholds and (2) P/I values. The first mode, KPIs' thresholds, aims at obtaining the optimum KPIs' threshold limits that ensure the delivery of safe, sustainable, and financially feasible services to the public with minimal risks of failure and tolerable temporal and spatial public disruption. The optimization model was applied to the city of Montreal dataset with a planning horizon of 25 years, minimum and maximum values for the thresholds, as well as annual budget and unacceptable physical reliability set as constraints while running the model. The optimization model featured a GA-based optimization engine with the attributes outlined in Table 15.12. The results were divided into single asset-based thresholds and corridor-based thresholds, given the fact that the model accounts for both conventional and full coordination scenarios. The optimization model was run three separate times to reach the single asset-based KPIs' thresholds and one time to reach the corridor-based KPIs' thresholds (Table 15.13). It is worth noting that both the safety performance measures along with their penalties and incentives (P/I) and the roads' operational KPIs along with their associated thresholds were adopted (Abu-Samra 2015). The roads' operational KPIs along with their associated thresholds are displayed in Table 15.14. The second mode, P/I values, aims at obtaining the optimum set of financial penalties and incentives, corresponding to the KPIs' threshold defined earlier in the first mode. The model was applied to the city of Montreal dataset with minimum and maximum values for the thresholds as well as annual budget and unacceptable physical reliability set as constraints while running the model. Furthermore, the penalties and incentives' applicability criteria were mathematically defined to be included in the financial computations, as highlighted in Equations 9.25, 9.26, and 9.27. Similar to the KPIs' threshold optimization mode, the optimization model featured a GA-based optimization engine with similar attributes defined in Table 15.12. The results were divided into single asset-based thresholds and corridor-based thresholds, given the fact that the model accounts for both conventional and full coordination scenarios, as displayed in Table 15.15. The optimization model was run three separate times to reach the single conventional asset-based P/I and one time to reach the corridor-based P/I. It is worth noting that penalties and incentives are applied to an asset-basis or corridor-basis, depending on which threshold was not met. In case both thresholds were not met, the corridor-based penalties or incentives value precedes and is accordingly applied to the financial model. Furthermore, in case the KPIs' thresholds are not met or are met for a certain corridor at a certain point of time, the highest penalty or incentive value is applied. The model did not consider other forms of penalties or incentives such as reduction or expansion of the contractual period, etc.

Maintenance Planning Optimization

The maintenance planning optimization featured a multi-objective hierarchical optimization with an objective of maximizing the improvements to the KPIs over the 25 years planning horizon. The weights of importance for the multi-dimensional assessment indicators are defined in Table 15.16. The multi-objective hierarchical optimization featured a GA-based optimization engine with the attributes outlined in Table 15.12. The coordination improvements' optimization results were summarized for all the multi-dimensional assessment indicators as opposed to the conventional scenario as outlined in Table 15.17. The results favored the full coordination intervention scenario as opposed to the conventional one in terms of (1) number of interventions, (2) disruption time, (3) LCC including direct and indirect costs resulting from the intervention actions, (4) consumed space for undertaking the interventions, (5) assets' reliability, (6) assets' risk of failure, and (7) operating time (free of intervention) time. To compute the improvements, the conventional system was modeled using meta-heuristic rules to ensure that the minimum acceptable reliability threshold for each asset is met. As shown in Figures 15.11 and 15.12, the reliability and risk exposure of the full coordination intervention program were better with a CIF and RIF of 10% compared to the conventional intervention program. As shown in Figures 15.13

TABLE 15.13
PBC Structuring Optimization Results – KPI Thresholds

Description	KPIs' Thresholds (per Corridor)				Unit (per Corridor)
	Roads	Water	Sewer	Corridor	
Available Budget	Varies*	Varies*	Varies*	Varies*	N/A
Amount of time allotted for disruption every 5 years	5	30	30	35	Business days every 5 years
Maximum amount of time allotted for a single disruption	3	30	30	35	Business days per intervention
Spatial extent of a single disruption	500	500	500	500	Linear meter per intervention
Maximum number of disruptions every 5 years	1	1	1	1	Number of disruptions per year
Maximum number of revisits every 5 years (excluding road preventive maintenance, e.g., crack sealing and potholes repair)	2	1	1	2	Number of visits every 5 years
Minimum acceptable reliability at any point of time (Detailed KPIs for the roads are discussed in Table 15.13)	65%	50%	50%	60%	%
Maximum acceptable risk threshold at any point of time	35%	50%	50%	40%	%
Maximum demand/capacity ratio at any point of time to avoid combined sewer and stormwater overflooding or unmet water demand	N/A	80%	75%	75%	%
Maximum number of disruptions to undertake interventions for all the corridor assets	N/A	N/A	N/A	3	Number of interventions
Maximum number of years to undertake interventions for all the corridor assets	N/A	N/A	N/A	12	years
Minimum spacing between interventions within the same corridor	N/A	N/A	N/A	4	years
Minimum amount of operating time (e.g., free of intervention) excluding the preventive maintenance actions (e.g., crack sealing and potholes repairs)	N/A	N/A	N/A	8	years
Maximum number of accidents per year due to poor asset condition (e.g., severity level is accounted for in the accidents' police report – fatalities are treated separately)	N/A	N/A	N/A	12	Number of accidents per year
Maximum number of accidents per year due to poor maintenance management (e.g., signs and flagman)	N/A	N/A	N/A	2	Number of accidents per year
Maximum response time for accidents' removal	N/A	N/A	N/A	4	hours
Maximum response time for repairing potholes	N/A	N/A	N/A	5	Business days
Maximum response time for barriers' removal	N/A	N/A	N/A	2	Business days
Maximum response time for repairing defective guardrails	N/A	N/A	N/A	2	Business days

*The available budget varies depending on many factors including size of the asset portfolio, government priorities and budget allocation, etc.

TABLE 15.14
Road Operational KPIs and Their Associated Thresholds

KPI	Description	KPIs' Thresholds (Roads)	Unit
Surface rating	Minimum acceptable surface rating for the corridor (e.g., surface distresses-based deductions are used to obtain the surface rating)	8	Scale from 0 to 10 (0: failing road surface; 10: excellent/ pristine road surface)
Rutting	Maximum acceptable rutting depth	9	Millimeters (mm)
Alligator cracking	Maximum acceptable extent (%) of the alligator cracking within the corridor	30%	%
International roughness index	Minimum acceptable roughness index for the corridor (e.g., measures the characteristic of the longitudinal profile of a traveled wheel track)	2.6	Meter per kilometer (m/km) - scale from 0 to 5 (0: unacceptable; 5: excellent)

and 15.14, the financial savings, representing the direct and indirect costs showed that the full coordination intervention program saves 50% indirect costs compared to the conventional intervention program, implying a fewer number of interventions and less delay time for service disruptions due to combining the interventions of the co-located systems sharing the same spatial location. Furthermore, it resulted in an overall 18% LCC savings compared to the conventional intervention program because of combining the intervention actions of the three systems and given the existence of common and joint activities. Similarly, as displayed in Figure 15.15, the full coordination intervention program displayed 12% less disruption time and 16% less amount of consumed space as opposed to the conventional intervention program. Those savings reflect the coordination of the intervention activities such that the common activities have been carried out once, instead of n_a or n_s times for the partially coordinated and conventional intervention scenarios, respectively. Furthermore, undertaking the full coordination intervention increased the number of parallel activities, which increased the temporal savings as opposed to the conventional approach in which n_s interventions are separately undertaken for each system. In terms of efficiency and effectiveness, specific conclusions were drawn by analyzing the corridor-based results of the full coordination intervention program as opposed to the conventional intervention program. The full coordination intervention program resulted in a more effective intervention plan in terms of the number of service disruptions carried out for the same corridor. The revisiting schedule represents the frequency of carrying out interventions in the same corridor across the planning horizon, as displayed in Table 15.18. The full coordination intervention program was shown to be more efficient in more than 70% of the corridors with a fewer number of interventions. Those savings resulted in temporal and financial savings as well as less end-user service disruption. The coordinated intervention program was shown to be 30% more efficient with a fewer number of interventions resulting from the coordination. Furthermore, it displayed a 26% intervention effectiveness factor (IFF)

TABLE 15.15
PBC Structuring Optimization Results – Penalties and Incentives

Penalties Application Criteria	Penalty Application Frequency	Incentives Application Criteria	Incentive Application Frequency
Applied whenever the maintenance contractor fails to comply with the pre-defined contractual threshold	Penalty value per year (lump sum)	Applied after meeting the pre-defined contractual threshold for four consecutive years	Incentive value per 4 years (lump sum)
Applied whenever the maintenance contractor fails to comply with the pre-defined contractual threshold	Penalty value per day per time	Applied after meeting the pre-defined contractual threshold for two consecutive times (10 years)	Incentive value per 10 years (lump sum)
Applied whenever the maintenance contractor fails to comply with the pre-defined contractual threshold	Penalty value per day per time	Applied after meeting the pre-defined contractual threshold for two consecutive times	Incentive value per time (lump sum)
Applied whenever the maintenance contractor fails to comply with the pre-defined contractual threshold	Penalty value per extra meter per intervention	Applied after meeting the pre-defined contractual threshold for three consecutive times	Incentive value per time (lump sum)
Applied whenever the maintenance contractor fails to comply with the pre-defined contractual threshold	Penalty value per extra day per intervention	Applied after meeting the pre-defined contractual threshold for three consecutive times	Incentive value per time (lump sum)
Applied whenever the maintenance contractor fails to comply with the pre-defined contractual threshold	Penalty value per extra day per intervention	Applied after meeting the pre-defined contractual threshold for three consecutive times	Incentive value per time (lump sum)
Applied whenever the maintenance contractor fails to comply with the pre-defined contractual threshold	Penalty value per year (lump sum)	Applied after meeting the pre-defined contractual threshold for four consecutive years	Incentive value per time (lump sum)
Applied whenever the maintenance contractor fails to comply with the pre-defined contractual threshold	Penalty value per year (lump sum)	Applied after meeting the pre-defined contractual threshold for four consecutive years	Incentive value per time (lump sum)
Applied whenever the maintenance contractor fails to comply with the pre-defined contractual threshold	Penalty value per year (lump sum)	Applied after meeting the pre-defined contractual threshold for four consecutive years	Incentive value per time (lump sum)
Applied whenever the maintenance contractor fails to comply with the pre-defined contractual threshold	Penalty value per additional intervention	Applied after meeting the pre-defined contractual threshold	Incentive value per time per reduced intervention (lump sum)

(*Continued*)

TABLE 15.15 (*Continued*)
PBC Structuring Optimization Results – Penalties and Incentives

Penalties Application Criteria	Penalty Application Frequency	Incentives Application Criteria	Incentive Application Frequency
Applied whenever the maintenance contractor fails to comply with the pre-defined contractual threshold	Penalty value per year per time	Applied after meeting the pre-defined contractual threshold	Incentive value per time (lump sum)
Applied whenever the maintenance contractor fails to comply with the pre-defined contractual threshold	Penalty value per year per time	Applied after meeting the pre-defined contractual threshold	Incentive value per time (lump sum)
Applied whenever the maintenance contractor fails to comply with the pre-defined contractual threshold	Penalty value per year per time	Applied after meeting the pre-defined contractual threshold	Incentive value per time (lump sum)
Applied whenever the maintenance contractor fails to comply with the pre-defined contractual threshold	Penalty value per extra accident	Applied whenever the actual number of accidents per year becomes less than the pre-defined contractual threshold for two consecutive years	Incentive value per reduced accident
Applied whenever the maintenance contractor fails to comply with the pre-defined contractual threshold	Penalty value per extra accident	Applied whenever the actual number of accidents per year becomes less than the pre-defined contractual threshold for two consecutive years	Incentive value per reduced accident
Applied whenever the maintenance contractor fails to respond within the pre-defined response time	Penalty value per additional hour per accident	Applied whenever the maintenance contractor responds within the pre-defined response time for three consecutive times	Incentive value per reduced hour per accident
Applied whenever the maintenance contractor fails to respond within the pre-defined response time	Penalty value per additional day per pothole	Applied whenever the maintenance contractor responds within the pre-defined response time for three consecutive times	Incentive value per reduced day per pothole
Applied whenever the maintenance contractor fails to respond within the pre-defined response time	Penalty value per additional day per barrier	Applied whenever the maintenance contractor responds within the pre-defined response time for three consecutive times	Incentive value per reduced day per barrier
Applied whenever the maintenance contractor fails to respond within the pre-defined response time	Penalty value per additional day per defective guardrail	Applied whenever the maintenance contractor responds within the pre-defined response time for three consecutive times	Incentive value per reduced day per defective guardrail

as opposed to the conventional intervention program with longer operating times for the systems under study. Based on the pre-defined KPIs' weights of importance, the full coordination intervention program has revealed an overall improvement of 15% compared to the conventional intervention program because of all the savings realized in the KPIs (Abusamra 2019).

Sensitivity Analysis

Sensitivity analysis is performed to study and verify the sensitivity of the increasing or decreasing the minimally acceptable reliability thresholds on the other KPIs. It can answer many what-if questions such as: "Should we pay more for enhancing the network reliability?" And if the answer is yes, "what is the cost premium between the proposed intervention program and the optimal intervention program?" The sensitivity analysis was undertaken and four new optimization cases ranging between −20% and +20% with 10% increments were run. After running the optimization on the four scenarios, the improvement deviational

TABLE 15.16
Assessment Indicators' Weights of Importance

Performance Indicator	Basis	Weights of Importance (%)
Time (*I1*)	Intervention duration	10
Space (*I2*)	Intervention spatial and interdependency	10
Cost (*I3*)	Lifecycle costs	25
Efficiency (*I4*)	Intervention crew	5
Effectiveness (*I5*)	Intervention quality	5
Condition (*I6*)	Physical state, reliability, and LOS	25
Risk (*I7*)	Probability and consequences of failure	20

TABLE 15.17
Optimization Improvement Results

Assessment Index/Coordination Scenario	Index	Improvement (%)
Time (I_1)	NCR	12
Space (I_2)	STIF	16
Cost (I_3)	LIF	18
Efficiency (I_4)	IEF	30
Effectiveness (I_5)	IFF	26
Condition (I_6)	CIF	10
Risk (I_7)	RIF	10
Overall improvement (Z)		**15**

TABLE 15.18
Corridor Revisiting Schedule

Corridor ID #	System	Integration Status	Frequency Difference	2016	2017	2018	2019	2020	2021	2022	2023	2024	2025	2026	2027	2028	2029	2030	2031	2032	2033	2034	2035	2036	2037	2038	2039	2040	2041
8	R	CONV	⇧3																										
		COOR																											
	S	CONV	⇧3																										
		COOR																											
	W	CONV	⇧3																										
		COOR																											

CONV: Conventional Intervention Program
COOR: Full coordination Intervention Program

White represents a year with no intervention action taking place for this asset. Grey represents a year when an intervention took place for that asset.

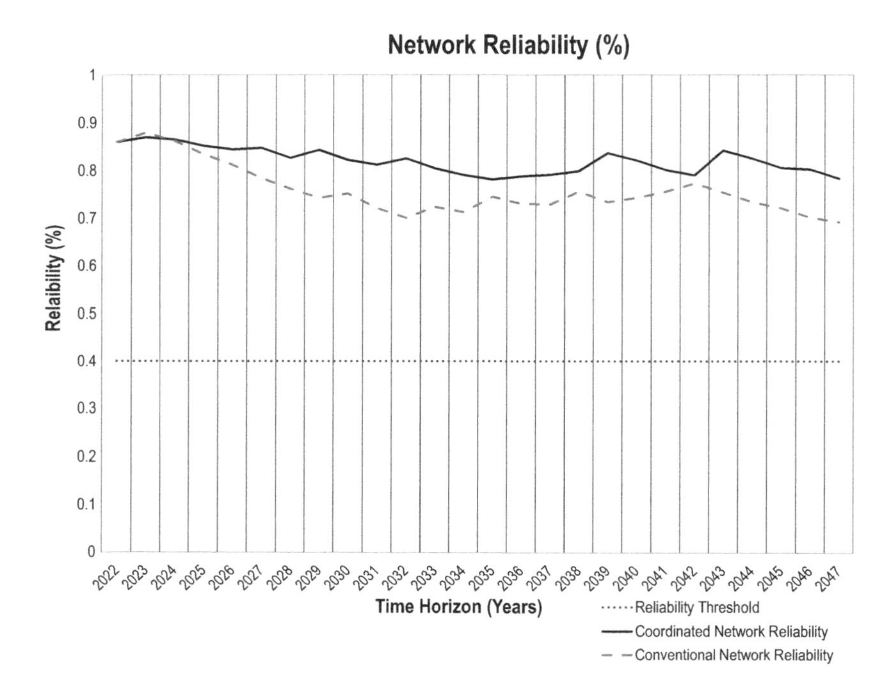

FIGURE 15.11 Network reliability.

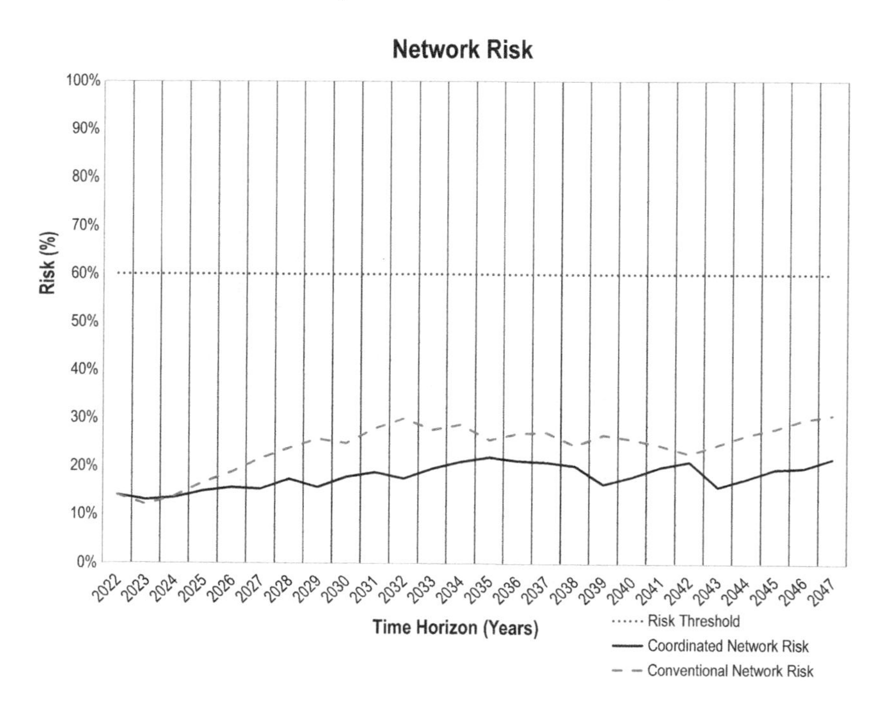

FIGURE 15.12 Network risk.

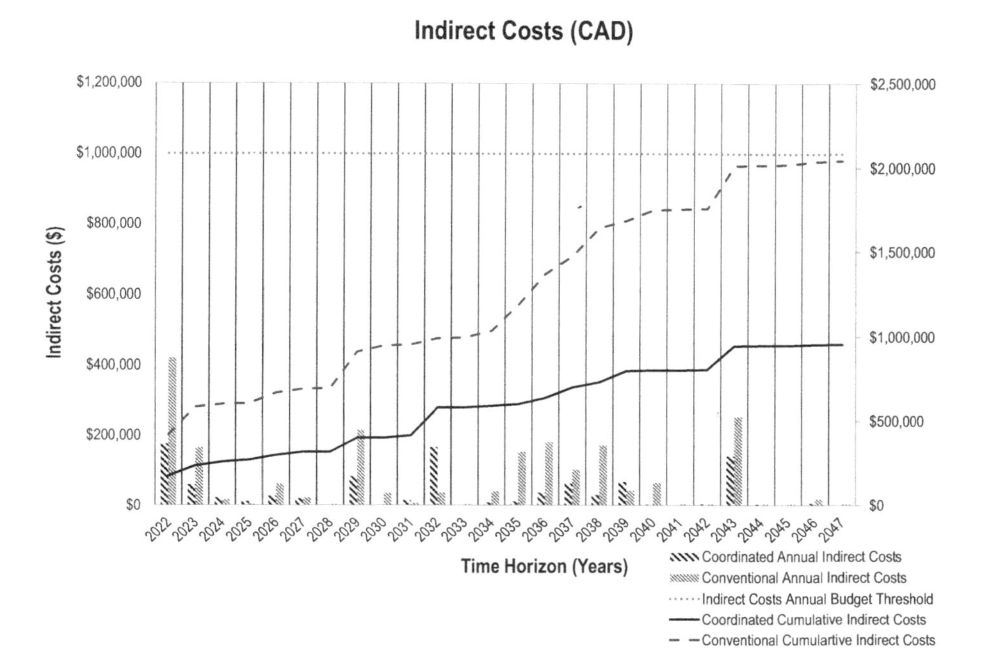

FIGURE 15.13 Indirect annual and cumulative costs.

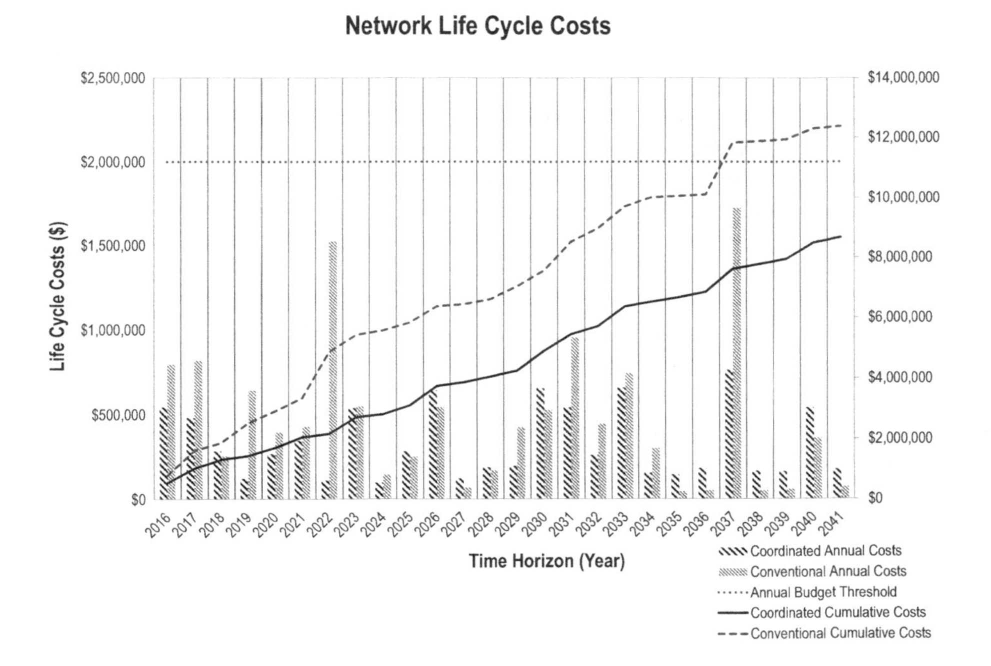

FIGURE 15.14 LCC annual and cumulative costs.

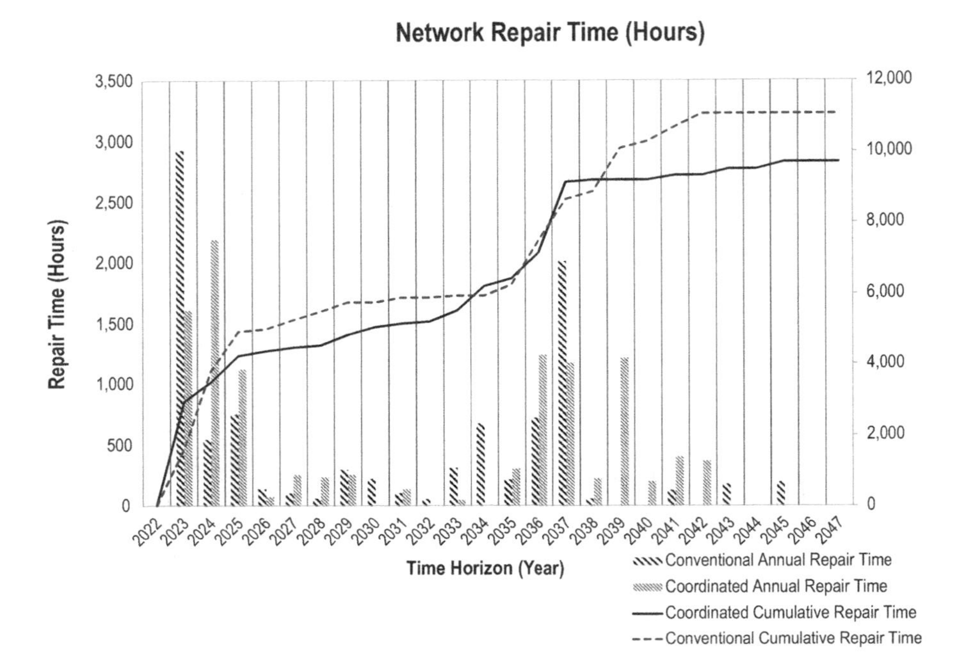

FIGURE 15.15 Network annual and cumulative repair time.

variables were computed as outlined in Table 15.19. Afterwards, the sensitivity analysis was undertaken to compare the cases' improvement deviations variables' outcomes with the baseline case and accordingly plot the difference. As outlined in Table 15.20, the system was shown to be very sensitive to changes in the reliability threshold. For instance, if the municipality increased their acceptable reliability threshold by 10%, it would potentially result in 42% additional disruption time and 31% additional space, 33% extra repair costs for undertaking additional interventions, 13% reduced efficiency given the extra interventions that were undertaken across the planning horizon, 31% less effectiveness implying less operating time, 30% increase in the average network reliability and risk, and 12% decrease in the overall improvement as opposed to the baseline scenario. Similarly, other scenarios were undertaken, and the results are displayed in Figure 15.16. In summary, slight changes in reliability significantly affect the other KPIs. The repair time and cost were shown to be the most sensitive items to the changes in the reliability thresholds. However, effectiveness was shown to be the least sensitive item to the changes in the reliability thresholds.

TABLE 15.19
Sensitivity Analysis Improvement Deviational Variables Results

Assessment Index/ Scenario	Index	Baseline Improvement (%)	Case 1 (−20%)	Case 2 (−10%)	Case 3 (10%)	Case 4 (20%)
Time (I_1)	NCR	12	15%	19%	7%	10%
Space (I_2)	STIF	16	20%	22%	11%	15%
Cost (I_3)	LIF	18	21%	24%	12%	17%
Efficiency (I_4)	IEF	30	39%	33%	26%	21%
Effectiveness (I_5)	IFF	26	21%	23%	18%	25%
Condition (I_6)	CIF	10	6%	9%	13%	15%
Risk (I_7)	RIF	10	6%	9%	13%	15%
Overall improvement (Z)		*15%*	14%	17%	13%	16%

TABLE 15.20
KPIs' Summary Results

Assessment Index/ Scenario	Index	Case 1 (−20%)	Case 2 (−10%)	Baseline Improvement (%)	Case 3 (10%)	Case 4 (20%)
Time (I_1)	NCR	25%	58%	0	−42%	−17%
Space (I_2)	STIF	25%	38%	0	−31%	−6%
Cost (I_3)	LIF	17%	33%	0	−33%	−6%
Efficiency (I_4)	IEF	30%	10%	0	−13%	−30%
Effectiveness (I_5)	IFF	−19%	−12%	0	−31%	−4%
Condition (I_6)	CIF	−40%	−10%	0	30%	50%
Risk (I_7)	RIF	−40%	−10%	0	30%	50%
Overall improvement (Z)		−1%	16%	0	−12%	8%

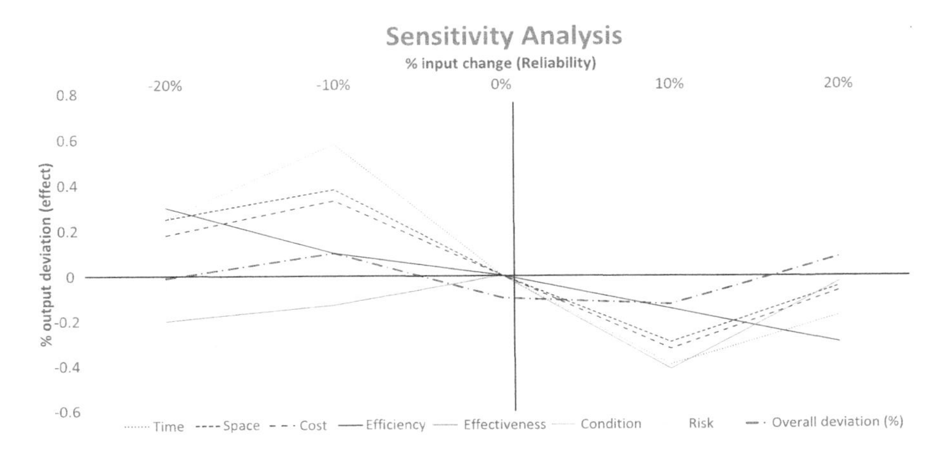

FIGURE 15.16 Spider diagram for sensitivity analysis scenarios.

REFERENCES

Abu-Samra, S. (2015). *Integrated Asset Management System for Highways Infrastructure.* LAPLAMBERT Academic Publishing, Saarbrücken.

Abu-Samra, S. (2017). "How Can Fix Montréal Roads." CityLife TV show - Panel Discussion, Montreal.

Abusamra, S. (2019). "Coordination and Multi-Objective Optimization Framework for Managing Municipal Infrastructure under Performance-Based Contracts." Doctoral of Philosophy Thesis, Concordia University, Montreal.

Amador, L. and Magnuson, S. (2011). "Adjacency modeling for coordination of investments in infrastructure asset management: Case Study of Kindersley, Saskatchewan, Canada." *Transportation Research Board (TRB)*, 2246, 8–15.

Bank of Canada. (2023). "Interest Rate." https://www.bankofcanada.ca/rates/daily-digest/. Mar 15, 2023.

Hachey, N. (2018). "B/C Analysis Project for Montréal Water Network." Interview, Sep 10, 2018, Montreal.

Qin, X. and Cutler, C. (2014). *Review of Road User Costs and Methods.* South Dakota University, Vermillion.

Sabourin, P. and Abusamra, S. (2018). "How Can Montréal Fix Its Roads." Panel Discussion, CityLife TV Show, Nov 14, 2018, Montreal.

TDOT (Texas Department of Transportation). (2015). "Value of Time and User Costs." https://www.researchgate.net/publication/362429765_Calculating_Road_User_Cost_for_Specific_Sections_of_Highway_for_Use_in_Alternative_Contracting_Project. Apr 25, 2023.

Trading Economics. (2023). "Canada Interest Rate." https://tradingeconomics.com/canada/interest-rate. Mar 15, 2023.

Ville de Montreal. (2022a). "Condition of Montréal Roads." https://donnees.ville.montreal.qc.ca/dataset/condition-chaussees-reseau-routier. Apr 22, 2022.

Ville de Montreal. (2022b). "View Performance Indicators." https://ville.montreal.qc.ca/vuesurlesindicateurs/. Apr 22, 2022.

16 Town of Kindersley Case Study

INTRODUCTION

A 53 km stretch from the town of Kindersley roads, water, and combined sewer networks was selected for the analysis. The network comprises 125 corridors of spatially co-located roads, water, and sewer assets. The dataset scale/size, in terms of the number of corridors, was scaled up six times, given the power of MOSEK linear and mixed-integer programming optimization algorithm. The case study was modeled and coded on REMOSOFT software. The condition states of the corridors were adopted from the town of Kindersley case study (Amador and Magnuson 2011). Time value of money has been estimated at 2% interest rate (Trading Economics 2023; Bank of Canada 2023). Furthermore, the study planning horizon was 25 years and could be expanded up to 100 years. The weights of the systems were assumed according to the overall life-cycle costs (LCC) of each system across 100 years, using the longest life method. As highlighted earlier, the results displayed 45%, 25%, and 30% for the roads, water, and sewer systems, respectively. However, those weights are subject to change according to the stakeholders' preferences (e.g., condition, replacement cost, and crews' availability). The physical deterioration curves, financial information (e.g., intervention direct costs and indirect/user costs), temporal information (e.g., intervention time and disruption time), risk probability of failure (POF), and consequences of failure (COF) scoring criteria have been obtained from the town of Kindersley.

ASSET INVENTORY

The asset inventory contains 125 corridors with all the co-located assets including roads, combined sewer/stormwater, and water networks as outlined in Tables 16.1, 16.2, and 16.3. It includes the corridor length, road width, number of lanes, area, average annual daily traffic, water or sewer pipe material and diameter, excavation depth, soil type, demand category, age/year or installation, etc. The asset inventory acts as a central database for the computational models such that all the necessary information about the corridor can be extracted directly from the inventory and used for further analysis.

TEMPORAL DATASET

The temporal dataset includes the unit rates of different intervention actions and unit rate breakdown for each intervention activity as displayed earlier in Table 9.3. A summary of the combined average unit rates is shown in Table 16.4.

DOI: 10.1201/9781003327509-21

TABLE 16.1
Roads' Network

Corridor ID#	Road Type	Structural Design	Traffic	Corridor Length (m)	Age (years)
1	Gravel	Weak	Light	230	18
2	Gravel	Weak	Light	245	18
3	Road	Weak	Light	117	1
4	Road	Weak	Light	111	7
5	Road	Strong	Medium	185	5
6	Road	Weak	Light	108	7
7	Road	Weak	Light	77	9
8	Road	Strong	Light	120	1
9	Gravel	Weak	Light	20	18
10	Gravel	Weak	Light	253	18
11	Gravel	Weak	Light	32	18
12	Gravel	Weak	Light	255	18
13	Gravel	Weak	Light	256	18
14	Road	Weak	Light	103	1
15	Road	Weak	Light	104	7
16	Road	Weak	Light	137	7
17	Road	Weak	Light	140	7
18	Road	Weak	Medium	115	9
19	Road	Weak	Light	143	7
20	Road	Weak	Light	1	8
21	Road	Strong	Medium	186	5
22	Road	Weak	Light	205	7
23	Road	Strong	Medium	101	5
24	Road	Weak	Light	133	3
25	Road	Weak	Light	95	7
26	Road	Weak	Light	89	11
27	Road	Weak	Light	110	1
28	Road	Weak	Light	83	3
29	Road	Weak	Light	106	3
30	Road	Weak	Light	86	11
31	Road	Strong	Light	81	3
32	Road	Weak	Light	142	7
33	Road	Weak	Light	126	16
34	Road	Weak	Light	76	9
35	Road	Strong	Light	37	1
36	Road	Weak	Light	127	7
37	Road	Weak	Light	121	13
38	Road	Weak	Light	123	15
39	Road	Weak	Light	136	13
40	Road	Weak	Light	124	11
41	Road	Weak	Light	125	7
42	Road	Weak	Light	148	7

(Continued)

TABLE 16.1 (*Continued*)
Roads' Network

Corridor ID#	Road Type	Structural Design	Traffic	Corridor Length (m)	Age (years)
43	Road	Weak	Light	68	7
44	Road	Strong	Light	55	9
45	Gravel	Weak	Light	34	18
46	Road	Weak	Light	197	9
47	Road	Weak	Light	153	11
48	Road	Weak	Light	174	10
49	Road	Strong	Light	210	1
50	Road	Weak	Light	158	11
51	Road	Weak	Light	164	13
52	Road	Weak	Light	228	18
53	Gravel	Weak	Light	226	18
54	Road	Weak	Light	64	18
55	Road	Strong	Light	176	7
56	Road	Strong	Light	47	7
57	Road	Strong	Light	177	7
58	Road	Strong	Medium	5	7
59	Road	Strong	Light	52	13
60	Road	Strong	Light	171	7
61	Gravel	Weak	Light	265	18
62	Gravel	Weak	Light	232	18
63	Road	Weak	Light	222	18
64	Road	Weak	Light	221	18
65	Road	Weak	Light	27	18
66	Gravel	Weak	Light	33	18
67	Gravel	Weak	Light	243	18
68	Gravel	Weak	Light	241	18
69	Gravel	Weak	Light	244	18
70	Gravel	Weak	Light	258	18
71	Gravel	Weak	Light	214	18
72	Gravel	Weak	Medium	216	18
73	Gravel	Weak	Light	21	18
74	Road	Strong	Light	162	15
75	Road	Weak	Light	201	18
76	Road	Strong	Light	179	3
77	Gravel	Weak	Light	259	18
78	Road	Strong	Light	61	9
79	Road	Weak	Light	206	3
80	Road	Weak	Light	112	9
81	Road	Weak	Light	116	7
82	Road	Strong	Medium	188	5
83	Road	Weak	Light	118	3

(*Continued*)

TABLE 16.1 (*Continued*)
Roads' Network

Corridor ID#	Road Type	Structural Design	Traffic	Corridor Length (m)	Age (years)
84	Road	Weak	Light	113	7
85	Road	Weak	Light	107	1
86	Road	Weak	Light	109	7
87	Road	Weak	Light	13	1
88	Road	Weak	Light	134	3
89	Road	Strong	Light	128	1
90	Road	Weak	Light	130	13
91	Road	Weak	Medium	215	25
92	Road	Weak	Light	131	3
93	Road	Strong	Light	129	1
94	Road	Strong	Light	93	5
95	Road	Strong	Light	94	5
96	Road	Weak	Light	90	7
97	Road	Weak	Light	84	7
98	Road	Weak	Light	88	3
99	Road	Weak	Light	91	7
100	Road	Weak	Light	96	15
101	Road	Weak	Light	102	7
102	Road	Weak	Light	138	7
103	Road	Weak	Light	135	1
104	Road	Weak	Light	194	16
105	Road	Strong	Light	40	5
106	Road	Weak	Light	78	15
107	Road	Weak	Light	49	16
108	Road	Strong	Light	75	13
109	Road	Weak	Light	139	13
110	Road	Weak	Light	149	11
111	Road	Weak	Light	69	3
112	Road	Weak	Light	150	7
113	Road	Weak	Light	45	7
114	Road	Weak	Light	46	1
115	Road	Strong	Medium	189	5
116	Road	Weak	Light	70	3
117	Road	Weak	Light	147	3
118	Road	Weak	Light	207	3
119	Road	Weak	Light	190	1
120	Road	Weak	Light	168	15
121	Road	Weak	Light	154	1
122	Road	Strong	Medium	169	5
123	Road	Strong	High	159	9
124	Road	Weak	Light	262	18
125	Road	Strong	High	199	9
126	Road	Strong	High	167	9

TABLE 16.2
Sewer Network

Corridor ID#	Pipe Type	Pipe Diameter (inch)	Age (years)	Excavation Depth (feet)	Corridor Length (m)	Demand Category	Demand Age (years)
1	Pvc	S10	26	15	230	Low	26
2	Pvc	S8	26	15	245	Low	26
3	Conc	S15	44	15	117	Medium	44
4	Conc	S21	46	15	111	High	46
5	Conc	S8	46	15	185	Medium	46
6	Conc	S21	46	15	108	Medium	46
7	Conc	S18	44	15	77	High	44
8	Conc	S18	44	15	120	Low	44
9	Pvc	S8	12	15	20	Medium	12
10	Pvc	S6	27	15	253	High	27
11	Pvc	S8	6	15	32	Low	6
12	Conc	S8	35	15	255	Low	35
13	Pvc	S8	35	15	256	Medium	35
14	Conc	S24	46	15	103	High	46
15	Conc	S21	46	15	104	Medium	46
16	Conc	S15	46	15	137	Medium	46
17	Conc	S15	45	15	140	High	45
18	Conc	S18	45	15	115	Low	45
19	Vct	S12	44	15	143	Medium	44
20	Pvc	S8	44	15	1	High	44
21	Ac	S8	46	15	186	High	46
22	Conc	S10	60	15	205	Low	60
23	Conc	S30	46	15	101	Low	46
24	Conc	S15	46	15	133	Medium	46
25	Conc	S15	44	15	95	High	44
26	Conc	S15	44	15	89	Medium	44
27	Conc	S21	46	15	110	Medium	46
28	Conc	S18	44	15	83	High	44
29	Conc	S18	46	15	106	Low	46
30	Conc	S24	44	15	86	Medium	44
31	Conc	S24	44	15	81	High	44
32	Vct	S12	44	15	142	High	44
33	Conc	S15	44	15	126	Low	44
34	Conc	S21	44	15	76	Low	44
35	Pvc	S8	27	15	37	Medium	27
36	Conc	S15	44	15	127	High	44
37	Vct	S12	44	15	121	Medium	44
38	Vct	S12	44	15	123	Medium	44
39	Vct	S12	44	15	136	High	44
40	Vct	S12	44	15	124	Low	44
41	Vct	S12	44	15	125	Medium	44
42	Vct	S12	44	15	148	High	44

(Continued)

TABLE 16.2 (*Continued*)
Sewer Network

Corridor ID#	Pipe Type	Pipe Diameter (inch)	Age (years)	Excavation Depth (feet)	Corridor Length (m)	Demand Category	Demand Age (years)
43	Pvc	S12	20	15	68	High	20
44	Vct	S8	46	15	55	Low	46
45	Pvc	S8	27	15	34	Low	27
46	Conc	S8	44	15	197	Medium	44
47	Vct	S10	44	15	153	High	44
48	Conc	S12	44	15	174	Medium	44
49	Vct	S8	60	15	210	Medium	60
50	Conc	S15	44	15	158	High	44
51	Vct	S12	44	15	164	Low	44
52	Vct	S8	40	15	228	Medium	40
53	Vct	S8	40	15	226	High	40
54	Pvc	S10	20	15	64	High	20
55	Pvc	S18	25	15	176	Low	25
56	Ac	S8	28	15	47	Low	28
57	Pvc	S18	25	15	177	Medium	25
58	Pvc	S10	2	15	5	High	2
59	Pvc	S18	25	15	52	Medium	25
60	Vct	S18	25	15	171	Medium	25
61	Pvc	S10	26	15	265	High	26
62	Pvc	S10	26	15	232	Low	26
63	Vct	S8	50	15	222	Medium	50
64	Conc	S8	50	15	221	High	50
65	Pvc	S8	40	15	27	High	40
66	Pvc	S8	6	15	33	Low	6
67	Pvc	S8	26	15	243	Low	26
68	Pvc	S8	26	15	241	Medium	26
69	Ac	S8	26	15	244	High	26
70	Conc	S8	35	15	258	Medium	35
71	Vct	S8	42	15	214	Medium	42
72	Vct	S8	42	15	216	High	42
73	Pvc	S8	12	15	21	Low	12
74	Conc	S15	44	15	162	Medium	44
75	Vct	S8	52	15	201	High	52
76	Pvc	S18	25	15	179	High	25
77	Vct	S8	35	15	259	Low	35
78	Pvc	S8	35	15	61	Low	35
79	Vct	S8	60	15	206	Medium	60
80	Conc	S21	46	15	112	High	46
81	Conc	S18	45	15	116	Medium	45
82	Ac	S8	46	15	188	Medium	46
83	Conc	S15	44	15	118	High	44
84	Conc	S21	46	15	113	Low	46

(*Continued*)

TABLE 16.2 (*Continued*)
Sewer Network

Corridor ID#	Pipe Type	Pipe Diameter (inch)	Age (years)	Excavation Depth (feet)	Corridor Length (m)	Demand Category	Demand Age (years)
85	Conc	S18	46	15	107	Medium	46
86	Conc	S21	46	15	109	High	46
87	Ac	S8	36	15	13	High	36
88	Conc	S15	46	15	134	Low	46
89	Conc	S15	44	15	128	Low	44
90	Vct	S12	44	15	130	Medium	44
91	Vct	S10	45	15	215	High	45
92	Vct	S12	44	15	131	Medium	44
93	Conc	S15	44	15	129	Medium	44
94	Conc	S15	44	15	93	High	44
95	Conc	S15	44	15	94	Low	44
96	Conc	S15	44	15	90	Medium	44
97	Conc	S18	44	15	84	High	44
98	Conc	S18	44	15	88	High	44
99	Conc	S15	44	15	91	Low	44
100	Conc	S15	44	15	96	Low	44
101	Conc	S30	46	15	102	Medium	46
102	Conc	S12	46	15	138	High	46
103	Conc	S15	44	15	135	Medium	44
104	Conc	S8	44	15	194	Medium	44
105	Pvc	S8	7	15	40	High	7
106	Conc	S18	44	15	78	Low	44
107	Conc	S8	44	15	49	Medium	44
108	Vct	S12	44	15	75	High	44
109	Conc	S12	46	15	139	High	46
110	Vct	S12	44	15	149	Low	44
111	Pvc	S12	20	15	69	Low	20
112	Vct	S12	44	15	150	Medium	44
113	Pvc	S8	50	15	45	High	50
114	Pvc	S8	1	15	46	Medium	1
115	Ac	S8	46	15	189	Medium	46
116	Pvc	S12	20	15	70	High	20
117	Vct	S12	44	15	147	Low	44
118	Conc	S8	50	15	207	Medium	50
119	Ac	S8	46	15	190	High	46
120	Conc	S15	46	15	168	High	46
121	Vct	S10	44	15	154	Low	44
122	Conc	S15	46	15	169	Low	46
123	Conc	S8	44	15	159	Medium	44
124	Ac	S6	36	15	262	High	36
125	Conc	S8	44	15	199	Medium	44
126	Conc	S18	46	15	167	Medium	46

TABLE 16.3
Water Network

Corridor ID#	Pipe Type	Pipe Diameter (inch)	Age (years)	Excavation Depth (feet)	Corridor Length (m)	Demand Category	Demand Age (years)
1	Pvc	S12	21	17	230	Medium	21
2	Pvc	S6	27	17	245	High	27
3	Pvc	S8	20	17	117	Medium	20
4	Pvc	S8	26	17	111	Medium	26
5	Pvc	S8	22	17	185	High	22
6	Pvc	S6	27	17	108	Low	27
7	Ac	S6	44	17	77	Medium	44
8	Uci	S6	50	17	120	High	50
9	Pvc	S6	27	17	20	High	27
10	Pvc	S6	27	17	253	Low	27
11	Ac	S6	37	17	32	Medium	37
12	Ac	S6	37	17	255	High	37
13	Pvc	S6	8	17	256	Medium	8
14	Pvc	S8	26	17	103	Medium	26
15	Pvc	S8	26	17	104	High	26
16	Pvc	S8	26	17	137	Low	26
17	Pvc	S8	9	17	140	Medium	9
18	Pvc	S6	9	17	115	High	9
19	Pvc	S6	9	17	143	High	9
20	Uci	S6	50	17	1	Low	50
21	Pvc	S8	22	17	186	Low	22
22	Pvc	S8	5	17	205	Medium	5
23	Pvc	S8	22	17	101	High	22
24	Ci	S6	58	17	133	Medium	58
25	Uci	S6	47	17	95	Medium	47
26	Ac	S6	44	17	89	High	44
27	Pvc	S8	22	17	110	Low	22
28	Ci	S8	57	17	83	Medium	57
29	Uci	S6	47	17	106	High	47
30	Pvc	S8	16	17	86	High	16
31	Ci	S6	57	17	81	Low	57
32	Pvc	S8	9	17	142	Low	9
33	Ci	S8	55	17	126	Medium	55
34	Ac	S6	36	17	76	High	36
35	Pvc	S6	27	17	37	Medium	27
36	Ci	S6	57	17	127	Medium	57
37	Uci	S6	50	17	121	High	50
38	Ci	S6	57	17	123	Low	57
39	Ci	S8	51	17	136	Medium	51
40	Ci	S8	51	17	124	High	51
41	Ci	S6	52	17	125	High	52
42	Ci	S8	51	17	148	Low	51

(Continued)

TABLE 16.3 (*Continued*)
Water Network

Corridor ID#	Pipe Type	Pipe Diameter (inch)	Age (years)	Excavation Depth (feet)	Corridor Length (m)	Demand Category	Demand Age (years)
43	Pvc	S8	11	17	68	Low	11
44	Ac	S6	46	17	55	Medium	46
45	Ac	S6	42	17	34	High	42
46	Ci	S6	50	17	197	Medium	50
47	Ac	S6	45	17	153	Medium	45
48	Ac	S6	45	17	174	High	45
49	Ac	S6	45	17	210	Low	45
50	Ac	S6	44	17	158	Medium	44
51	Ac	S6	44	17	164	High	44
52	Ac	S6	50	17	228	High	50
53	Ac	S6	40	17	226	Low	40
54	Pvc	S6	29	17	64	Low	29
55	Pvc	S8	29	17	176	Medium	29
56	Pvc	S8	29	17	47	High	29
57	Pvc	S6	29	17	177	Medium	29
58	Ac	S8	34	17	5	Medium	34
59	Ac	S6	31	17	52	High	31
60	Pvc	S6	2	17	171	Low	2
61	Pvc	S16	25	17	265	Medium	25
62	Pvc	S12	21	17	232	High	21
63	Pvc	S6	26	17	222	High	26
64	Pvc	S6	50	17	221	Low	50
65	Pvc	S6	29	17	27	Low	29
66	Pvc	S6	27	17	33	Medium	27
67	Pvc	S6	27	17	243	High	27
68	Ac	S6	32	17	241	Medium	32
69	Pvc	S6	27	17	244	Medium	27
70	Pvc	S6	8	17	258	High	8
71	Ac	S6	32	17	214	Low	32
72	Plastic	S6	26	17	216	Medium	26
73	Pvc	S6	50	17	21	High	50
74	Ac	S6	42	17	162	High	42
75	Ac	S6	45	17	201	Low	45
76	Pvc	S8	25	17	179	Low	25
77	Pvc	S6	27	17	259	Medium	27
78	Ac	S6	35	17	61	High	35
79	Ac	S8	40	17	206	Medium	40
80	Pvc	S8	20	17	112	Medium	20
81	Pvc	S12	20	17	116	High	20
82	Ci	S8	51	17	188	Low	51
83	Pvc	S8	20	17	118	Medium	20
84	Ci	S6	60	17	113	High	60

(*Continued*)

TABLE 16.3 (*Continued*)
Water Network

Corridor ID#	Pipe Type	Pipe Diameter (inch)	Age (years)	Excavation Depth (feet)	Corridor Length (m)	Demand Category	Demand Age (years)
85	Pvc	S8	22	17	107	High	22
86	Pvc	S8	27	17	109	Low	27
87	Pvc	S8	5	17	13	Low	5
88	Ci	S6	58	17	134	Medium	58
89	Ci	S8	55	17	128	High	55
90	Uci	S6	58	17	130	Medium	58
91	Ac	S6	42	17	215	Medium	42
92	Ci	S6	58	17	131	High	58
93	Uci	S6	50	17	129	Low	50
94	Ci	S8	55	17	93	Medium	55
95	Pvc	S6	50	17	94	High	50
96	Ci	S8	57	17	90	High	57
97	Ac	S6	44	17	84	Low	44
98	Pvc	S6	16	17	88	Low	16
99	Ci	S6	50	17	91	Medium	50
100	Uci	S6	47	17	96	High	47
101	Pvc	S8	16	17	102	Medium	16
102	Pvc	S8	11	17	138	Medium	11
103	Steel	S8	51	17	135	High	51
104	Ac	S6	44	17	194	Low	44
105	Ac	S6	36	17	40	Medium	36
106	Uci	S6	50	17	78	High	50
107	Uci	S6	50	17	49	High	50
108	Pvc	S6	50	17	75	Low	50
109	Pvc	S8	11	17	139	Low	11
110	Ci	S8	51	17	149	Medium	51
111	Uci	S6	60	17	69	High	60
112	Pvc	S6	9	17	150	Medium	9
113	Pvc	S6	9	17	45	Medium	9
114	Ci	S6	1	17	46	High	1
115	Ci	S8	51	17	189	Low	51
116	Pvc	S8	11	17	70	Medium	11
117	Ci	S6	50	17	147	High	50
118	Ac	S6	46	17	207	High	46
119	Pvc	S8	11	17	190	Low	11
120	Ac	S6	45	17	168	Low	45
121	Ac	S6	46	17	154	Medium	46
122	Ci	S6	54	17	169	High	54
123	Ac	S6	45	17	159	Medium	45
124	Ac	S6	42	17	262	Medium	42
125	Pvc	S16	26	17	199	High	26
126	Pvc	S8	28	17	167	Low	28

TABLE 16.4
Average Unit Rates and Costs (Direct and Indirect)

Intervention Name	Assets			Coordination	Average Unit Rate (hr/unit)	Average Unit Cost ($/unit)	Unit	Notes
	Road	Water	Sewer					
Crack sealing				Conventional	0.5	$0.75	Linear meter	Varies according to the number of lanes, road structural type, and traffic
Micro surfacing				Conventional	4	$8.00	m^2	Varies according to the number of lanes, road structural type, and traffic
Patching				Conventional	9	$60.00	m^2	Varies according to the number of lanes, road structural type, and traffic
Resurfacing				Conventional	15	$90.00	m^2	Varies according to the number of lanes, road structural type, and traffic
Reconstruction				Conventional	25	$145.00	m^2	Varies according to the number of lanes, road structural type, and traffic
Water pipelining				Conventional	1.5	$1,200.00	Linear meter	Varies according to the pipe material, pipe diameter, and excavation depth
Water pipe replacement				Conventional	2	$1,750.00	Linear meter	Varies according to the pipe material, pipe diameter, and excavation depth
Sewer pipelining				Conventional	2.5	$1,450.00	Linear meter	Varies according to the pipe material, pipe diameter, and excavation depth
Sewer pipe replacement				Conventional	3	$2,200.00	Linear meter	Varies according to the pipe material, pipe diameter, and excavation depth
Roads and water coordination				Partial coordination	3.5	$1,800.00	Linear meter	Varies according to the pipe material, pipe diameter, and excavation depth
Roads and sewer coordination				Partial coordination	4.5	$2,250.00	Linear meter	Varies according to the pipe material, pipe diameter, and excavation depth
Water and sewer coordination				Partial coordination	5	$2,800.00	Linear meter	Varies according to the pipe material, pipe diameter, and excavation depth
Full coordination (roads, water, and sewer)				Full coordination	5	$2,900.00	Linear meter	Varies according to the pipe material, pipe diameter, and excavation depth

Note: The black shade implies that this activity is applicable to this asset class.

Spatial Dataset

The town of Kindersley is a triangular-shaped urban area as displayed in Figure 16.1. As shown in Figure 16.2, there are two hydrologic models for the combined sewer and stormwater network. The area savings for the partial and full coordination scenarios as opposed to the conventional scenario were computed for each corridor, as displayed in Table 15.2, depending on several factors such as excavation depth, pipe diameter, site conditions, etc.

Financial Dataset

The financial dataset includes the costs of different intervention actions, cost breakdown for each intervention activity, indirect/user costs associated with disrupting the public, and intervention cost in partial and full coordination scenarios. Similar intervention direct and indirect costs were utilized for the two case studies. A summary of the combined average direct and indirect costs is displayed in Table 16.4.

Physical Dataset

The physical dataset includes the condition, age, physical characteristics, etc. The deterioration curves were similar for both the cases.

Resilience Preparedness Dataset

The resilience preparedness dataset includes the demand and capacity data necessary to compute the resilience preparedness of the water and sewer assets within each corridor. The estimation of the current demand is based on the estimation of

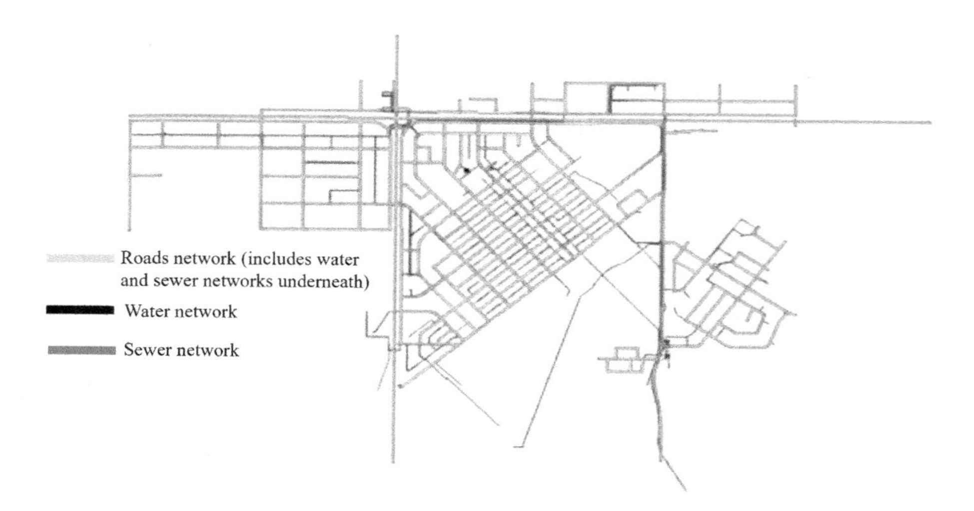

FIGURE 16.1 Extract from the GIS maps for roads, water, and sewer network.

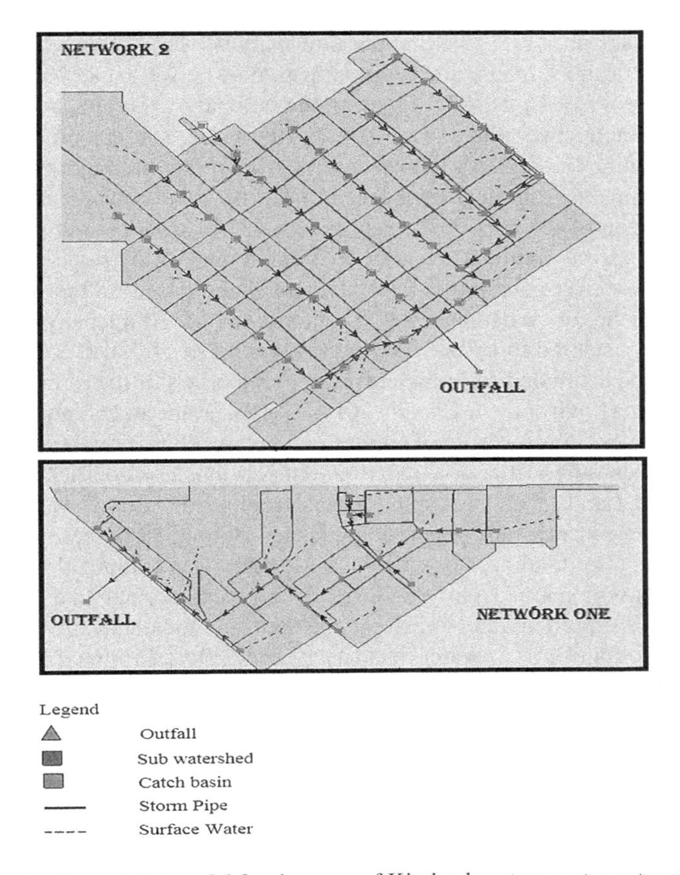

FIGURE 16.2 StormCAD model for the town of Kindersley stormwater network.

the water flow from the rational method. For the combined sewer and stormwater network, two networks with two different outfalls were analyzed, as displayed in Figure 16.2, and accordingly, two hydrological models were built in StormCAD to estimate the current and future flow demand and capacity of each pipe in the network (Bentley 2018). The modeling started by defining the catchment area and breaking it down into sub-catchments. Thenceforth, land use categories and topography information were used to identify and determine the drainage area for each land type and associated impervious area for each sub-basin. Discharge water moves to the related inlet through the sub-catchment. For each sub-catchment, the rational runoff coefficient was assigned based on the composite runoff index. Hence after, the concentration time (e.g., the time needed for the stormwater to flow from the most remote point in the sub-catchment to the inlet) was defined for each sub-catchment and intensity duration frequency data were used to compute the future demand. It is worth noting that the catch-basin/manhole component per sub-catchment was assumed as the main point to get discharged water into the pipes system. From each main inlet, water goes to a pipe and the flow demand was estimated and analyzed. The catchment runoff, inlets, junctions, gutters, pipe networks, and outfalls computations were provided by StormCAD using the rational method to compute the peak flow of combined sewer and stormwater. Furthermore, sub-catchment areas were modeled to define the region's influence within the urban area that tributes to each series of catchments. The slope of the terrain was the main consideration for modeling the direction of the runoff. Current demand flow capacity and future demand capacity ratios were obtained from the hydrologic models by dividing the hydraulic demand flow over each pipe's capacity, calculated based on the pipe's physical attributes. For the future demand, a 12% increase in rainfall intensity due to climate change was considered to update the runoff (Mailhot et al. 2012). The runoff coefficient could be estimated from the corresponding values of the return period from the IDF curves (Environment Canada 2014). For the water and combined sewer and stormwater pipes, the future demand was computed. Thus, population growth and future land uses were utilized to update the imperviousness coefficient (*C*). Spatial data from the Landsat of the United States Geological Survey was used to analyze the imperviousness and vegetation changes from 1988 to 2013 in Kindersley, Saskatchewan. The total area of each land use class was computed with respect to the total study region area from 1988 to 2013, and the historical data trend was used to predict the future demand of each pipe. Runoff coefficients were selected based on each type of land cover (Water Security Agency 2014). For the demand capacity prediction, land use/cover modeling was used to display the trend for each pipe from 1988 to 2013, as outlined in Table 16.5. In summary, water and combined sewer and stormwater pipes were classified into three groups as follows: (1) low demand pipes that are receiving very little changing demand across the future, (2) medium demand pipes that are receiving a moderate changing demand across the future, and (3) high demand pipes that are receiving a large amount of increasing flow demand across the future. The concept of pipe's apparent age was used to estimate the flow demand capacity predication trend, as displayed in Figure 16.3. For each pipe, the pipe's apparent age was estimated through matching its current flow demand capacity ratio with the assigned demand trend (e.g., low, medium, and high). Finally, the future prediction of demand was computed based on the developed prediction curves.

TABLE 16.5
Pipe Land Use/Cover Statistics

Year	Impervious	Green	Bare Soil	Water
1988	17	60.5	20	2.5
1993	19	60.3	18	2.7
1998	27	53.4	17	2.6
2003	29	46.7	22	2.3
2008	38	42.4	17	2.6
2013	40	29.6	28	2.4

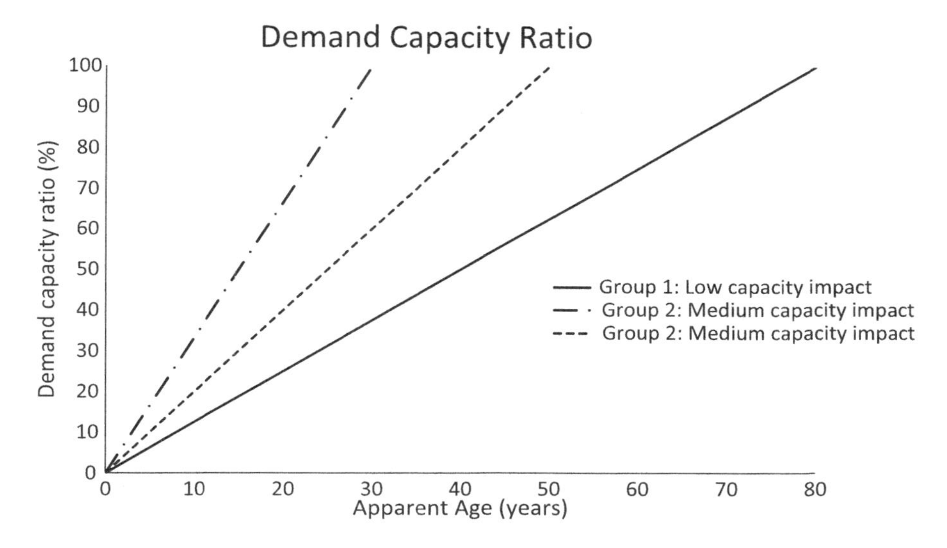

FIGURE 16.3 Demand capacity ratio of different pipe groups.

Risk Dataset

The risk dataset includes the probability and consequences of failure. The same POF and COF scoring criteria were similar for both the cases.

TOWN OF KINDERSLEY CASE STUDY

In this case study, the model was run for the maintenance planning optimization and as such, the penalties and incentives were set to "0". The optimization model was run for 25 years planning horizon using REMSOFT software integrated with MOSEK linear programming optimization engine. Given the power of MOSEK, the optimization reaches an exact solution, using linear goal optimization, and attains the intervention schedule/plan for five scenarios as outlined in Table 16.6. Furthermore, the sensitivity analysis analyzes the impact of changing the reliability key performance indicator (KPI) threshold on the other KPIs, and comparing it with the baseline scenario.

Optimization Results

The results of each scenario will be discussed and analyzed separately in the following sections. Time value of money was applied with a 2% interest rate. Net Present Worth (NPW) and Equivalent Uniform Annual Cost (EUAC) calculations were conducted to represent the life-cycle costs (LCC) and compare the financial outcome of the different scenarios (Abusamra 2019).

TABLE 16.6
Town of Kindersley Optimization Scenarios

Scenario ID #	Coordination	Scenario ID#	Scenario Name	Outcome Description
Scenario 1	Conventional (Silo)	Scenario 1	Combined conventional for roads, water, and sewer networks	Exact solution for the intervention plan of the roads, water, and sewer networks each asset run separately
Scenario 2	Partial coordination	Scenario 2	Roads and sewer networks partial coordination optimization	Exact solution for the intervention plan of the partially coordinated roads and sewer networks along with the conventional water network
		Scenario 3	Roads and water networks partial coordination optimization	Exact solution for the intervention plan of the partially coordinated roads and water networks along with the conventional sewer network
Scenario 4		Scenario 4	Water and sewer networks partial coordination optimization	Exact solution for the intervention plan of the partially coordinated water and sewer networks along with the conventional roads network
Scenario 5	Full coordination	Scenario 5	Roads, water, and sewer networks full coordination optimization	Exact solution for the intervention plan of the full coordination of the co-located assets

Scenario 1 – Combined Conventional for Roads, Water, and Sewer Networks

This scenario aims at reaching an exact optimal intervention plan for each asset separately across the 25 years planning horizon. Three separate optimizations were run for each asset separately and the results were combined in this scenario. It is the basis of computing the improvement deviational variables in the partial and full coordination scenarios as it acts as the baseline scenario that represents the conventional approach most of the municipalities are currently operating.

MOSEK linear optimization engine was used to reach an exact optimal intervention plan that meets the pre-defined contractual KPIs' thresholds throughout the planning horizon. The objective of the combined conventional optimization was maximizing the improvement deviational variables across the planning horizon. The constraints were meeting the unacceptable performance and demand capacity ratio thresholds. The variables were the intervention actions that need to be taken for each system within each corridor at every single point of time across the planning horizon. The optimization results are summarized in Table 16.7 and Figure 16.4. The results displayed a total of 560 intervention actions split into 183 for roads, 144 for sewer (67 replacement with bigger diameter), and 233 for water (103 replacements with

TABLE 16.7
Scenario 1 – Conventional Optimization Results

KPI	Combined Conventional (Total)	Roads	Water	Sewer
Time (hours)	2,673,608	2,052,245	577,970	43,392
Space (m²)	397,069	157,276	222,038	17,757
Cost – Equivalent Uniform Annual Cost (EUAC) ($)	$2,918,743	$591,343	$1,904,109	$423,290
Cost – Net Present Worth (NPW) ($)	$56,983,949	$11,545,060	$37,174,798	$8,264,092
Average condition (%)	66%	57%	63%	79%
Average risk (%)	36%	19%	11%	5%
Average resilience (%)	80%	N/A	65%	95%
# of intervention actions	560	183	233	144
Time per km per year (hours/km/year)	2,018	1,549	436	33
Cost per km per year ($//km/year)	$55,070.62	$11,157.42	$35,926.59	$7,986.61
Average repair length per year (km/year)	5.2	2.2	1.7	1.3
Average number of interventions per year	22.4	7.32	9.32	5.76
Ratio of interventions per year – number of annual interventions/number of corridors (%)	18%	6%	7%	5%

bigger diameter). This distribution is because the water and sewer networks were in a very good condition state and poor resilience preparedness. Thus, undertaking replacement actions for bigger diameter improved the resilience preparedness by 14%, dropping from 75% to 61% demand capacity ratio. The average number of revisits for each corridor was four times, which represents two road activities, and the other two activities are for the water and sewer pipes. The average number of interventions per year was 22 interventions for the 125 corridors, which results in an average disruption ratio of 18%. The fact that there is no coordination among the three spatially located assets increased the public nuisance and results in more repair time as opposed to coordinated interventions. The overall network was in a very good initial reliability of 84%. After running the optimization for 25 years, the reliability improved to 98% because of the replacement actions undertaken throughout the planning horizon. Furthermore, the risk index dropped from 17% to 2% because of the improved reliability as displayed in Figure 16.4B. The intervention program resulted in 2.7 million repair hours over the 25 years with an average of 2,000 repair hours per km per year. The intervention program disrupted an area of 397,000 m² over the 25 years with an average of 5.2 km per year, which equates to around 10% of the network. The intervention program resulted in NPW of $57 million, equivalent to an EUAC of $3 million, for undertaking the intervention actions for the 53 km of the town's road, water, and sewer networks. Those costs were broken down to 22% for roads repair and rehabilitation as well as pipelining, amounting to $12.5 million over the 25 years planning horizon, and 78% for water and sewer pipes replacement and road reconstruction, amounting to $44.5 million over the 25 years planning horizon. The average annual expenditure was $55,000 $/year/km.

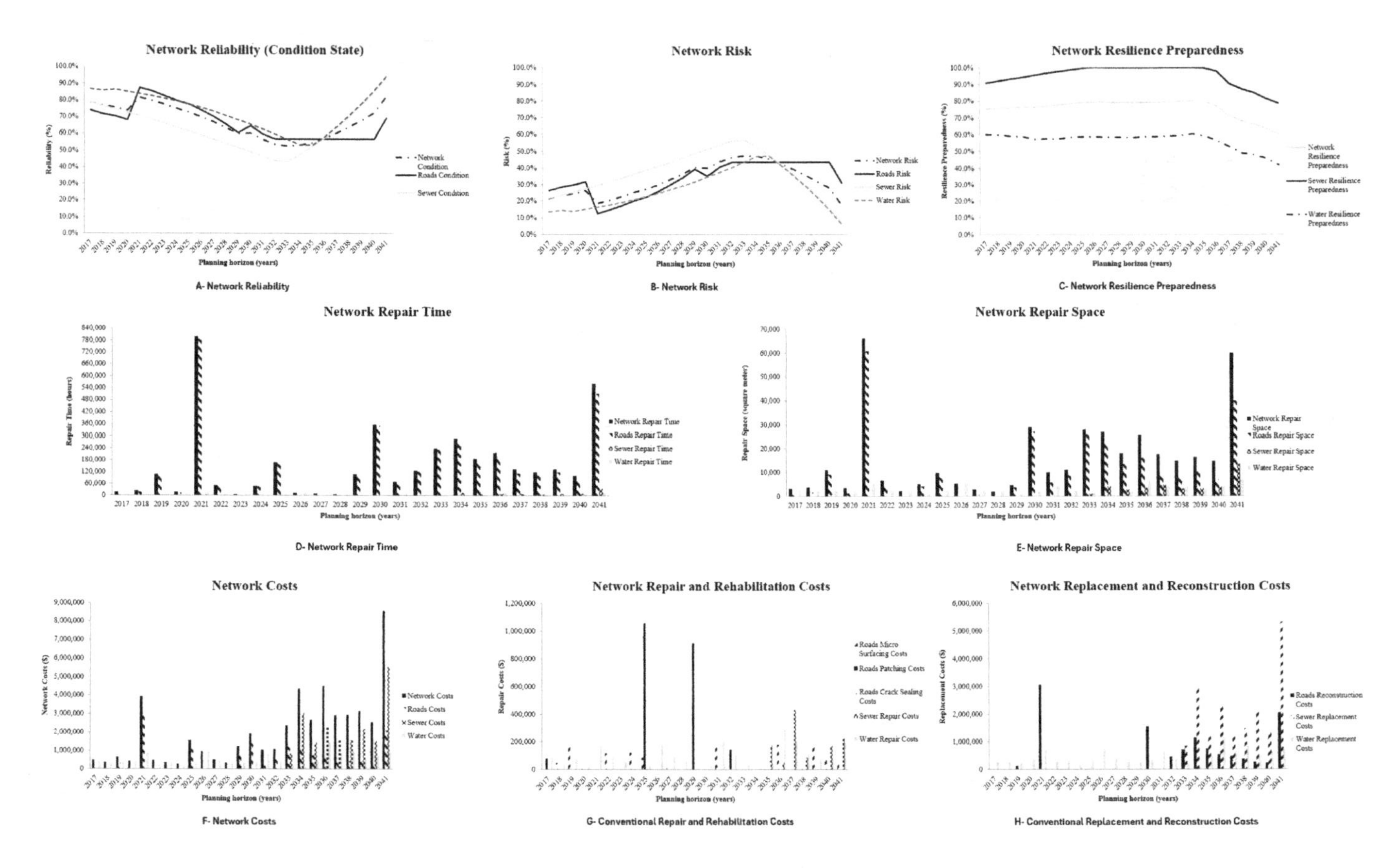

FIGURE 16.4 Scenario 1 – Conventional optimization results.

Partial Coordination Scenarios Optimization

The partial coordination optimization scenarios aim at reaching an exact optimal intervention plan for the network based on partially coordinating the interventions of two out of the three networks together and undertaking conventional interventions for the third system. Like the conventional scenario, the study was carried out across 25 years planning horizon. Scenarios 2 through 4 were conducted for the partially coordinated roads and water and conventional sewer, partially coordinated roads and sewer with conventional water, and partially coordinated water and sewer with conventional roads, respectively. Finally, the partially coordinated optimization results were compared with the baseline combined conventional (Scenario 1) and improvements were computed accordingly to showcase the benefits of partial coordinated planning and execution as opposed to conventional silo planning and execution.

Scenario 2 – Partial Coordination (Roads and Sewer)

This scenario was carried out on partially coordinated roads and sewer networks along with the conventional water network. Two separate optimizations were run: (1) roads and sewer networks and (2) water network. MOSEK linear optimization engine was used to reach an exact optimal intervention plan that meets the pre-defined contractual KPIs' thresholds throughout the planning horizon. The objective of the partial coordination optimization was maximizing the improvement deviational variables across the planning horizon. The constraints were meeting the unacceptable performance and demand capacity ratio thresholds. The variables were the intervention actions that need to be taken for each system within each corridor at every single point of time across the planning horizon. The optimization results are summarized in Tables 16.8, 16.9, and Figure 16.5. The results displayed a total

TABLE 16.8
Scenario 2 – Partially Coordinated (Roads and Sewer) Optimization Results

KPI	Combined Network (Total)	Roads and Sewer	Water	Roads Conventional
Time (hours)	1,093,561	91,449	494,194	507,914
Space (m^2)	282,758	19,283	190,943	72,530
Cost – Equivalent Uniform Annual Cost (EUAC) ($)	$3,569,007	$1,653,598	$1,717,999	$197,409
Cost – Net Present Worth (NPW) ($)	$69,679,348	$32,283,956	$33,541,282	$3,854,110
# of intervention actions	271	122	112	37
Time per km per year (hours/km/year)	825	69	373	383
Cost per km per year ($//km/year)	$67,339.75	$31,199.97	$32,415.08	$3,724.70
Average repair length per year (km/year)	3.4	1.5	1.4	0.5
Average number of interventions per year	10.84	4.88	4.48	1.48
Ratio of interventions per year – number of annual interventions/number of corridors (%)	9%	4%	4%	1%

TABLE 16.9
Scenario 2 – Partially Coordinated (Roads and Sewer) KPIs

KPI	Combined Network (Total) (%)	Roads (%)	Water (%)	Sewer (%)
Average condition (%)	67	61	74	66
Average risk (%)	34	18	8	18
Average resilience (%)	75	N/A	64	86

of 271 intervention actions split into 37 for conventional road actions, 112 for water (13 replacement with bigger diameter), and 122 for partially coordinated road and sewer (112 replacements with bigger diameter). This distribution is because the water and sewer networks were in a very good condition state and poor resilience preparedness. Thus, undertaking replacement actions for bigger diameter improved the resilience preparedness by 33%, dropping from 75% to 37% demand capacity ratio as displayed in Figure 16.5C. The average number of revisits for each corridor was two times, which shows the potential savings compared to the combined conventional scenario. The average number of interventions per year was ten interventions for the 125 corridors, which results in an average disruption ratio of 9%. The overall network was in a very good initial reliability of 84%. After running the optimization for 25 years, the reliability improved to 98% because of the undertaken replacement actions. Furthermore, the risk index dropped from 20% to 2% because of the improved reliability. The intervention program resulted in 1 million repair hours over the 25 years with an average of 825 repair hours per km per year, which reveals a network coordination ratio (NCR) of 59% as opposed to the baseline combined conventional scenario. Furthermore, it disrupted an area of 282,000 m^2 repair space over the 25 years with an average of 3.4 km per year, which equates to around 6% of the network and is 4% less conventional scenario. The intervention program resulted in NPW of $69 million, equivalent to an EUAC of $3.5 million, for undertaking the conventional and partially coordinated roads and sewer intervention actions for the 53 km of the town's road, water, and sewer networks. Those costs were broken down to 46% for partially coordinated intervention actions, amounting to $32 million over the 25 years planning horizon, and 54% for conventional roads and water, amounting to $37 million over the 25 years planning horizon. The average annual expenditure was $67,000 $/year/km.

The coordination scenarios are compared with the baseline combined conventional to compute the potential savings in terms of the pre-defined multi-dimensional performance assessment indicators. As such, the partially coordinated roads and sewer scenario was compared with the combined conventional scenario and the results are outlined in Tables 16.10 and 16.11. The results displayed huge temporal and spatio-temporal savings represented through a 59% NCR and 29% Spatio-Temporal Improvement Factor (STIF). However, this coordination scenario was not cost-effective as it revealed a life-cycle costs improvement factor (LIF) of −22%, which represents extra costs as opposed to the conventional scenario. Furthermore, the results displayed a slightly improved condition, resilience preparedness, and risk.

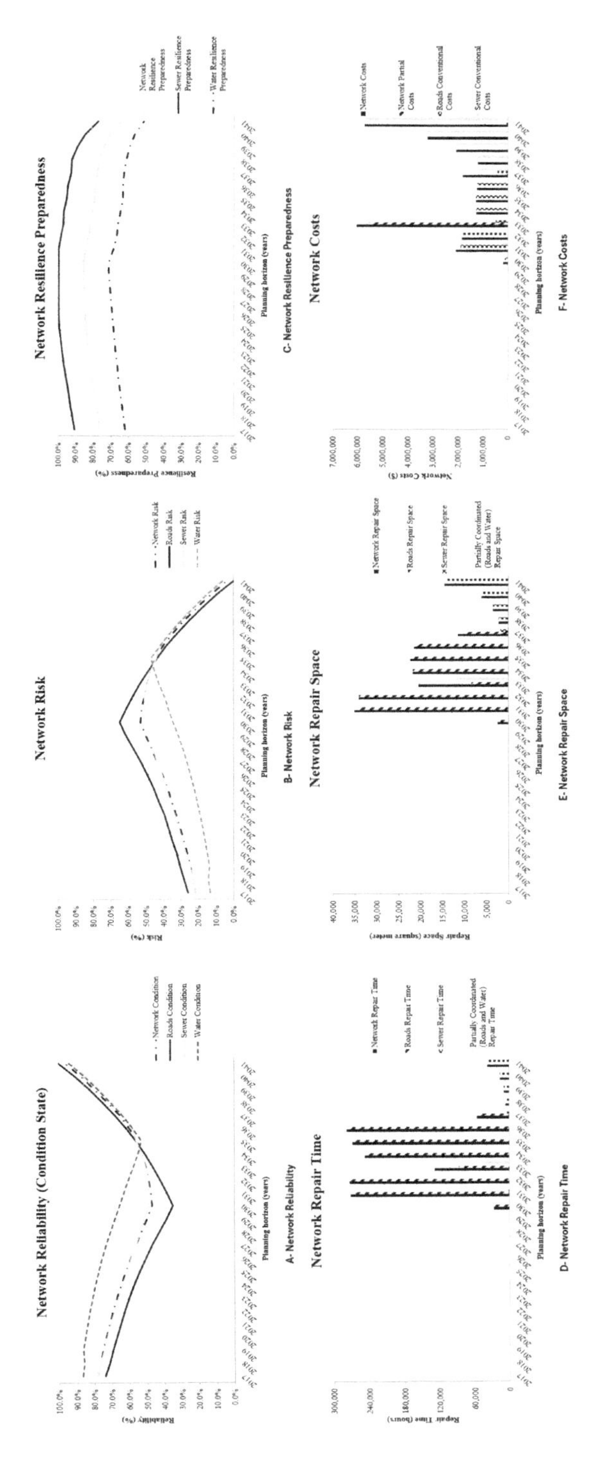

FIGURE 16.5 Scenario 2 – Partially coordinated (roads and sewer) optimization results.

TABLE 16.10
Results Comparison – Scenario 2 vs Baseline

KPI	Scenario 1 – Combined Conventional (Baseline)	Scenario 2 – Roads and Sewer Partial Coordination	Difference (%)
Time (hours)	2,673,608	1,093,561	59
Space (m^2)	397,069	282,758	29
Cost – Equivalent Uniform Annual Cost (EUAC) ($)	$2,918,743	$3,569,007	−22
Cost – Net Present Worth (NPW) ($)	$56,983,949	$69,679,348	−22
Average condition (%)	66%	67%	1
Average risk (%)	36%	34%	5
Average resilience (%)	80%	75%	6
# of intervention actions	560	271	52
Time per km per year (hours/km/year)	2,018	825	59
Cost per km per year ($//km/year)	$55,070.62	$67,339.75	−22
Average repair length per year (km/year)	5.2	3.4	34
Average number of interventions per year	22.4	10.84	52
Ratio of interventions per year – number of annual interventions/number of corridors (%)	18%	9%	52

TABLE 16.11
Scenario 2 – Improvement Deviational Variables

Performance Indicator	KPI	Weights of Importance (%)	Scenario 2 – Roads and Sewer (%)
Time	NCR	10	59
Space	STIF	10	29
Cost	LIF	20	−22
Efficiency	IEF	5	52
Effectiveness	IFF	5	2
Condition	CIF	20	1
Resilience preparedness	RPIF	10	6
Risk	RIF	20	5
Overall improvement (%)			9

Those savings were represented through a 1% condition improvement factor (CIF), 6% resilience preparedness improvement factor (RPIF), and 5% risk improvement factor (RIF) reflecting minor improvements in terms of condition, resilience preparedness, and risk. For the efficiency and effectiveness, the results displayed an intervention efficiency factor (IEF) of 52% and intervention effectiveness factor

(IFF) of 2%, which reflects fewer public disruptions (e.g., less disruption time and a fewer number of interventions) with longer corridor/asset operating times. Through combining the above-mentioned coordination savings, the partially coordinated roads and sewer scenario revealed an overall improvement of 9% as opposed to the baseline combined conventional one.

Scenario 3 – Partial Coordination (Roads and Water)

This scenario was carried out on the partially coordinated roads and water networks along with the conventional sewer network. Two separate optimizations were run: (1) roads and water networks and (2) sewer network. MOSEK linear optimization engine was used to reach an exact optimal intervention plan that meets the pre-defined contractual KPIs' thresholds throughout the planning horizon. The objective of the partial coordination optimization was maximizing the improvement deviational variables across the planning horizon. The constraints were meeting the unacceptable performance and demand capacity ratio thresholds. The variables were the intervention actions that need to be taken for each system within each corridor at every single point of time across the planning horizon. The optimization results could be summarized in Tables 16.12, 16.13, and Figure 16.6. The results displayed a total of 316 intervention actions split into 82 for conventional road actions, 168 for sewer (42 replacement with bigger diameter), and 66 for partially coordinated road and water (18 replacements with bigger diameter). This distribution is because the water and sewer networks were in a very good condition state and poor resilience preparedness. Thus, undertaking replacement actions for bigger diameter improved the resilience preparedness by 11% dropping from 75% to 64% demand capacity ratio as displayed in Figure 16.6C. The average number of revisits for each corridor was

TABLE 16.12
Scenario 3 – Partially Coordinated (Roads and Sewer) Optimization Results

KPI	Combined Network (Total)	Roads and Water	Sewer	Roads Conventional
Time (hours)	1,614,876	42,122	74,356	1,498,398
Space (m²)	195,432	12,610	28,965	153,858
Cost – Equivalent Uniform Annual Cost (EUAC) ($)	$1,436,639	$295,928	$699,736	$440,975
Cost – Net Present Worth (NPW) ($)	$28,048,165	$5,777,528	$13,661,271	$8,609,364
# of intervention actions	316	5	229	82
Time per km per year (hours/km/year)	1,219	32	56	1,131
Cost per km per year (S//km/year)	$27,106.40	$5,583.54	$13,202.57	$8,320.29
Average repair length per year (km/year)	3.4	0.1	2.2	1.2
Average number of interventions per year	12.64	0.2	9.16	3.28
Ratio of interventions per year – number of annual interventions/number of corridors (%)	10%	0%	7%	3%

TABLE 16.13
Scenario 3 – Partially Coordinated (Roads and Sewer) KPIs

KPI	Combined Network (Total) (%)	Roads (%)	Water (%)	Sewer (%)
Average condition (%)	67	61	74	65
Average risk (%)	34	18	8	9
Average resilience (%)	80	N/A	65	95

two times, which shows the potential savings compared to the combined conventional scenario. The average number of interventions per year was 12 interventions for the 125 corridors, which results in an average disruption ratio of 10%. The overall network was in a very good initial reliability of 84%. After running the optimization for 25 years, the reliability improved to 95% because of the undertaken replacement actions. Furthermore, the risk index dropped from 20% to 4% because of the improved reliability. Furthermore, the intervention program resulted in 1.6 million repair hours over the 25 years with an average of 1,200 repair hours per km per year, which reveals an NCR of 40% as opposed to the baseline combined conventional scenario. Furthermore, it disrupted an area of 195,000 m^2 repair space over the 25 years with an average of 2.6 km per year, which equates to around 5% of the network and is 5% less conventional scenario. The intervention program resulted in NPW of $28 million, equivalent to an EUAC of $1.4 million, for undertaking the conventional and partially coordinated roads and water intervention actions for the 53 km of the town's road, water, and sewer networks. Those costs were broken down to 20% for partially coordinated intervention actions, amounting to $5.7 million over the 25 years planning horizon, and 80% for conventional roads and sewer, amounting to $22.3 million over the 25 years planning horizon. The average annual expenditure was $27,000 $/year/km.

The coordination scenarios are compared with the combined conventional one to compute the potential savings in terms of the pre-defined multi-dimensional performance assessment indicators. As such, the partially coordinated roads and water scenario was compared with the combined conventional scenario and the results are outlined in Tables 16.14 and 16.15. The results displayed huge temporal, spatio-temporal, and cost savings represented through a 40% NCR, 51% STIF, and 51% LIF. However, this coordination scenario displayed no improvement in terms of condition (CIF = 0) and resilience preparedness (RPIF = 0) as it revealed similar results compared to the combined conventional one. Furthermore, the results displayed a slight improvement in the risk represented through a 4% RIF. For the efficiency and effectiveness, the results displayed an IEF of 44% and IFF of 5%, which reflects fewer public disruptions (e.g., less disruption time and a fewer number of interventions) with longer corridor/asset operating times. Through combining the above-mentioned coordination savings, the partially coordinated roads and water scenario revealed an overall improvement of 22% as opposed to the baseline combined conventional one.

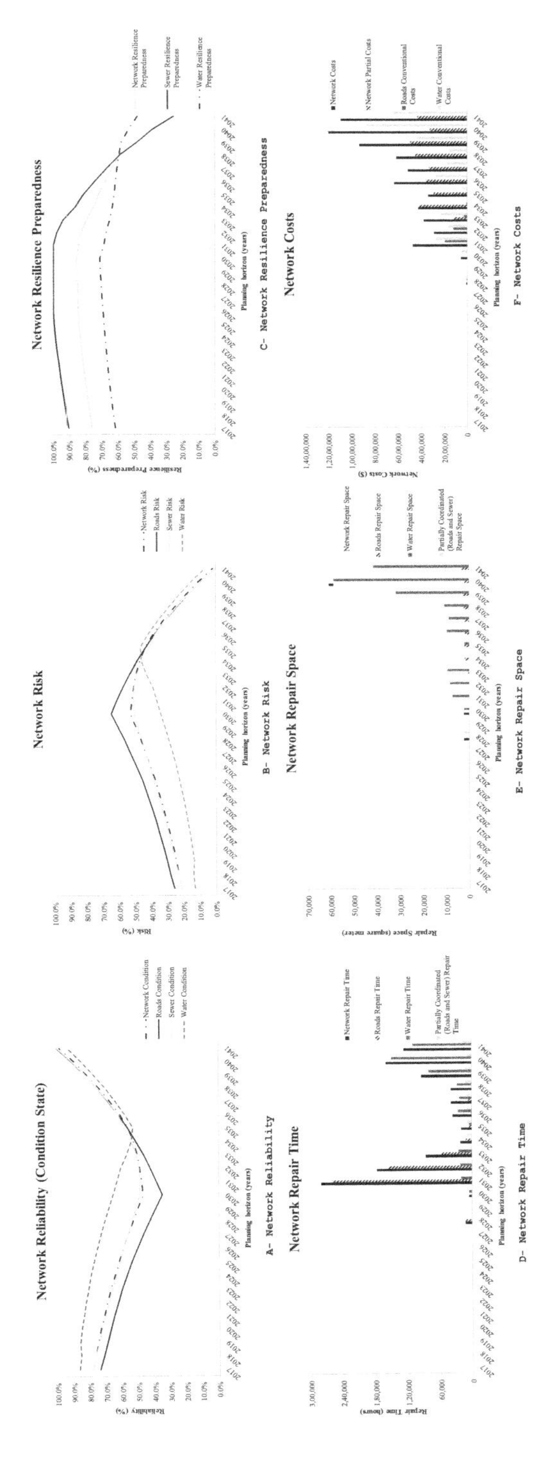

FIGURE 16.6 Scenario 3 – Partially coordinated (roads and water) optimization results.

TABLE 16.14
Results Comparison – Scenario 3 vs Baseline

KPI	Scenario 1 – Combined Conventional (Baseline)	Scenario 3 – Roads and Water Partial Coordination	Difference (%)
Time (hours)	2,673,608	1,614,876	40
Space (m²)	397,069	195,432	51
Cost – Equivalent Uniform Annual Cost (EUAC) ($)	$2,918,743	$1,436,639	51
Cost – Net Present Worth (NPW) ($)	$56,983,949	$28,048,165	51
Average condition (%)	66%	67%	0
Average risk (%)	36%	34%	4
Average resilience (%)	80%	80%	0
# of intervention actions	560	316	44
Time per km per year (hours/km/year)	2,018	1,219	40
Cost per km per year ($//km/year)	$55,070.62	$27,106.40	51
Average repair length per year (km/year)	5.2	3.4	34
Average number of interventions per year	22.4	12.64	44
Ratio of interventions per year – number of annual interventions/number of corridors (%)	18%	10%	44

TABLE 16.15
Scenario 3 – Improvement Deviational Variables

Performance Indicator	KPI	Weights of Importance (%)	Scenario 3 – Roads and Water (%)
Time	NCR	10	40
Space	STIF	10	51
Cost	LIF	20	51
Efficiency	IEF	5	44
Effectiveness	IFF	5	5
Condition	CIF	20	0
Resilience preparedness	RPIF	10	0
Risk	RIF	20	4
Overall improvement (%)			*22*

Scenario 4 – Partial Coordination (Water and Sewer)

This scenario was carried out on the partially coordinated water and sewer networks along with the conventional roads network. Two separate optimizations were run: (1) water and sewer networks and (2) roads network. MOSEK linear optimization engine was used to reach an exact optimal intervention plan that meets the pre-defined contractual KPIs' thresholds throughout the planning horizon.

The objective of the partial coordination optimization was maximizing the improvement deviational variables across the planning horizon. The constraints were meeting the unacceptable performance and demand capacity ratio thresholds. The variables were the intervention actions that need to be taken for each system within each corridor at every single point of time across the planning horizon. The optimization results could be summarized in Tables 16.16, 16.17, and Figure 16.7. The results displayed a total of 338 intervention actions split into 89 for conventional road actions, 249 for partially coordinated water and sewer (163 replacements with bigger diameter). This distribution is because the water and sewer networks were in a very good condition state and poor resilience preparedness. Thus, undertaking replacement actions for bigger diameter improved the resilience preparedness by 67%, dropping from 75% to 8% demand capacity ratio. The average number of revisits for each corridor was two times, which shows the potential savings compared to the combined conventional scenario. The average number of interventions per year was 13 interventions for the 125 corridors, which results in an average disruption ratio of 11%. The overall network was in a very good initial reliability of 84%. After running the optimization for 25 years, the reliability improved to 98% because of the undertaken replacement actions. Furthermore, the risk index dropped from 20% to 2% because

TABLE 16.16
Scenario 4 – Partially Coordinated (Water and Sewer) Optimization Results

KPI	Combined Network (Total)	Water and Sewer	Roads Conventional
Time (hours)	1,636,767	85,953	1,550,814
Space (m^2)	180,755	21,845	158,910
Cost – Equivalent Uniform Annual Cost (EUAC) ($)	$1,621,048	$1,168,893	$452,155
Cost – Net Present Worth (NPW) ($)	$31,648,453	$22,820,822	$8,827,630
# of intervention actions	338	249	89
Time per km per year (hours/km/year)	1,235	65	1,170
Cost per km per year ($//km/year)	$30,585.80	$22,054.58	$8,531.23
Average repair length per year (km/year)	3.5	2.3	1.2
Average number of interventions per year	13.52	9.96	3.56
Ratio of interventions per year – number of annual interventions/number of corridors (%)	11%	8%	3%

TABLE 16.17
Scenario 4 – Partially Coordinated (Water and Sewer) KPIs

KPI	Combined Network (Total) (%)	Roads (%)	Water (%)	Sewer (%)
Average condition (%)	74	11	21	66
Average risk (%)	30	32	28	25
Average resilience (%)	73	N/A	60	86

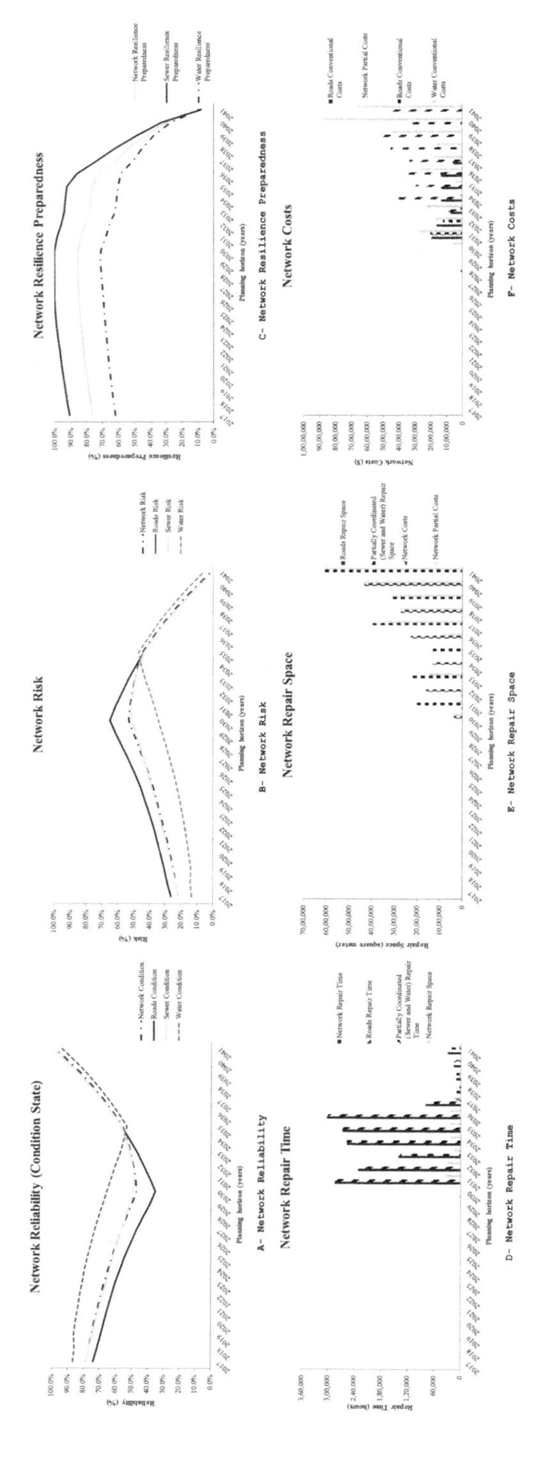

FIGURE 16.7 Scenario 4 – Partially coordinated (water and sewer) optimization results.

of the improved reliability. The intervention program resulted in 1.6 million repair hours over the 25 years with an average of 1,200 repair hours per km per year, which reveals an NCR of 39% as opposed to the baseline combined conventional scenario. Furthermore, it disrupted an area of 180,000 m^2 repair space over the 25 years with an average of 2.4 km per year, which equates to around 4.5% of the network and is 5.5% less conventional scenario. The intervention program resulted in NPW of $31.6 million, equivalent to an EUAC of $1.6 million, for undertaking the conventional and partially coordinated water and sewer intervention actions for the 53 km of the town's road, water, and sewer networks. Those costs were broken down to 72% for partially coordinated intervention actions, amounting to $22.8 million over the 25 years planning horizon, and 28% for conventional roads, amounting to $8.8 million over the 25 years planning horizon. The average annual expenditure was $30,000 $/year/km.

The coordination scenarios are compared with the combined conventional one to compute the potential savings in terms of the pre-defined multi-dimensional performance assessment indicators. Accordingly, the partially coordinated water and sewer scenario was compared with the combined conventional scenario and the results are outlined in Tables 16.18 and 16.19. The results displayed huge temporal, spatio-temporal, and cost savings represented through a 39% NCR, 54% STIF, and 44% LIF. Furthermore, this coordination scenario displayed decent improvement in terms of condition (CIF = 10%), resilience preparedness (RPIF = 9%), and risk (RIF = 10%). For the efficiency and effectiveness, the results displayed an IEF of 40% and IFF of 7%, which reflects fewer public disruptions (e.g., less disruption time and a fewer

TABLE 16.18
Results Comparison – Scenario 4 vs Baseline

KPI	Scenario 1 – Combined Conventional (Baseline)	Scenario 4 – Roads and Water Partial Coordination	Difference (%)
Time (hours)	2,673,608	1,636,767	39
Space (m^2)	397,069	180,755	54
Cost – Equivalent Uniform Annual Cost (EUAC) ($)	$2,918,743	$1,621,048	44
Cost – Net Present Worth (NPW) ($)	$56,983,949	$31,648,453	44
Average condition (%)	66%	74%	10
Average risk (%)	36%	30%	10
Average resilience (%)	80%	73%	9
# of intervention actions	560	338	40
Time per km per year (hours/km/year)	2,018	1,235	39
Cost per km per year ($//km/year)	$55,070.62	$30,585.80	44
Average repair length per year (km/year)	5.2	3.5	32
Average number of interventions per year	22.4	13.52	40
Ratio of interventions per year – number of annual interventions/number of corridors (%)	18%	11%	40

TABLE 16.19
Scenario 4 – Improvement Deviational Variables

Performance Indicator	KPI	Weights of Importance (%)	Scenario 4 – Water and Sewer (%)
Time	NCR	10	39
Space	STIF	10	54
Cost	LIF	20	44
Efficiency	IEF	5	40
Effectiveness	IFF	5	7
Condition	CIF	20	10
Resilience preparedness	RPIF	10	9
Risk	RIF	20	10
Overall improvement (%)			*27*

number of interventions) with longer corridor/asset operating times. Through combining the above-mentioned coordination savings, the partially coordinated water and sewer scenario revealed an overall improvement of 27% as opposed to the baseline combined conventional one.

Full Coordination Scenario Optimization

The full coordination optimization scenario aims at reaching an exact optimal intervention plan for the network based on full coordination of the interventions for the three co-located networks altogether and allowing conventional interventions for the roads only, given its shorter service life and thus, requires more frequent intervention actions. Like the conventional scenario, the study was carried out across 25 years planning horizon. Scenario 5 was conducted for the full coordination of roads, water, and sewer networks. The full coordination optimization results were compared with the baseline combined conventional (Scenario 1) and improvements were computed to showcase the benefits of fully coordinated planning and execution as opposed to conventional silo planning and execution.

Scenario 5 (Full Coordination – Roads, Water, and Sewer)

This scenario was carried out on the full coordination roads, water, and sewer networks. One optimization was run for the whole network including all the co-located assets (roads, water, and sewer networks). MOSEK linear optimization engine was used to reach an exact optimal intervention plan that meets the pre-defined contractual KPIs' thresholds throughout the planning horizon. The objective of the partial coordination optimization was maximizing the improvement deviational variables across the planning horizon. The constraints were meeting the unacceptable performance and demand capacity ratio thresholds. The variables were the intervention actions that need to be taken for each system within each corridor at every single point of time across the planning horizon. The optimization results are summarized in Tables 16.20, 16.21, and Figure 16.8. The results displayed a total of 186 intervention actions split into 18 for conventional road actions and 168 full coordination actions (22 replacement with bigger diameter). This distribution is because the water and sewer networks were in a very good condition state and poor resilience preparedness. Thus, undertaking replacement actions for bigger diameter improved the resilience preparedness by 29%, dropping from 75% to 16% demand capacity ratio. The average number of revisits for each corridor was one time, which shows huge savings compared to the combined conventional scenario. The average number of interventions per year was seven interventions for the 125 corridors, which results in the lowest average disruption ratio of 6%. The overall network was in a very good initial reliability of 84%. After running the optimization for 25 years, the reliability improved to 98% because of the undertaken replacement actions. Furthermore, the risk index dropped from 20% to 2% because of the improved reliability. The intervention program resulted in 748,000 repair hours over the 25 years with an average of 565 repair hours per km per year, which reveals an NCR of 72% as opposed to the baseline combined conventional scenario. Furthermore, it disrupted an area of 145,000 m^2 repair space over the 25 years with an average of 1.9 km per year, which equates to around 3.6% of the network and is 6.4% less conventional scenario.

TABLE 16.20
Scenario 5 – Full Coordination Optimization Results

KPI	Combined Network (Total)	Roads, Water and Sewer	Roads Conventional
Time (hours)	748,074	244,467	503,604
Space (m^2)	145,467	60,518	84,948
Cost – Equivalent Uniform Annual Cost (EUAC) ($)	$1,529,741	$1,456,329	$73,412
Cost – Net Present Worth (NPW) ($)	$29,865,841	$28,432,579	$1,433,261
# of intervention actions	186	168	18
Time per km per year (hours/km/year)	565	185	380
Cost per km per year ($//km/year)	$28,863.05	$27,477.91	$1,385.14
Average repair length per year (km/year)	2.4	2.2	0.2
Average number of interventions per year	7.44	6.72	0.72
Ratio of interventions per year – number of annual interventions/number of corridors (%)	6%	5%	1%

TABLE 16.21
Scenario 5 – Full Coordination KPIs

KPI	Combined Network (Total) (%)	Roads (%)	Water (%)	Sewer (%)
Average condition (%)	67	61	74	66
Average risk (%)	34	18	8	8
Average resilience (%)	69	N/A	57	81

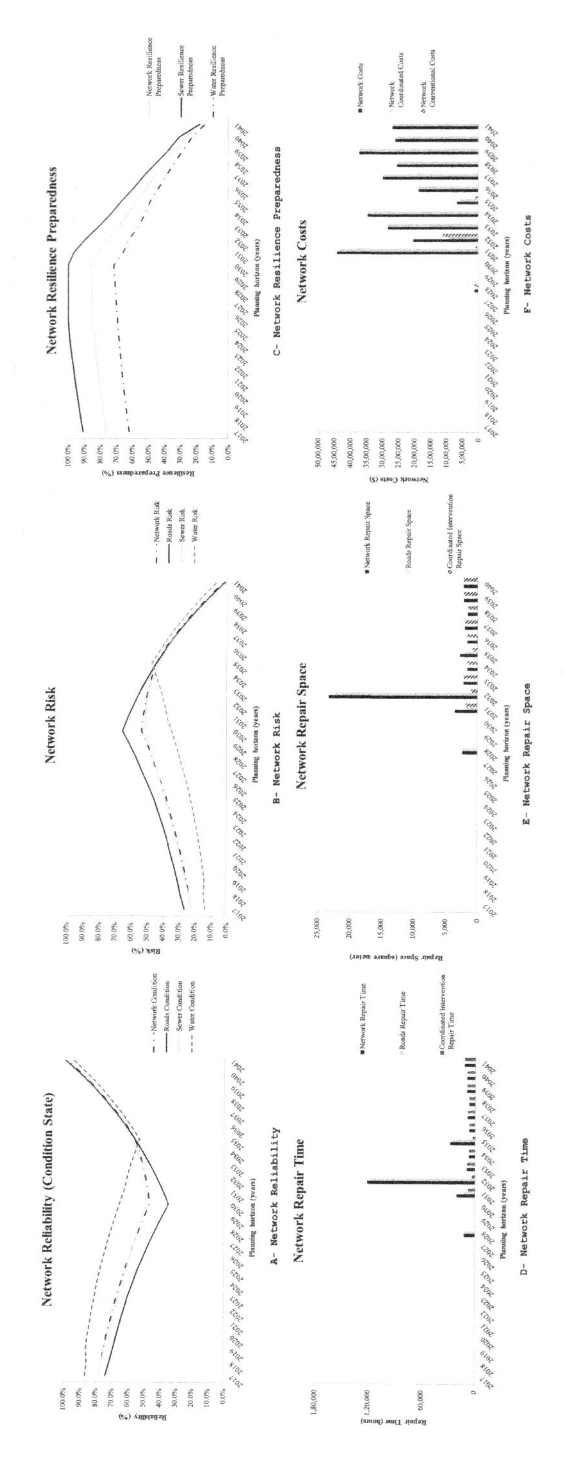

FIGURE 16.8 Scenario 5 – Full coordination optimization results.

The intervention program resulted in NPW of $29 million, equivalent to an EUAC of $1.5 million, for undertaking the conventional roads and full coordination road, water, and sewer intervention actions for the 53 km of the town's road, water, and sewer networks. Those costs were broken down to 95% for full coordination intervention actions, amounting to $27.6 million over the 25 years planning horizon, and 5% for conventional roads, amounting to $1.4 million over the 25 years planning horizon. The average annual expenditure was $27,000 $/year/km.

The coordination scenarios are compared with the combined conventional one to compute the potential savings in terms of the pre-defined multi-dimensional performance assessment indicators. Accordingly, the full coordination road, water, and sewer scenario was compared with the baseline combined conventional scenario and the results are outlined in Tables 16.22 and 16.23. The results displayed huge temporal, spatio-temporal, and cost savings represented through a 72% NCR, 63% STIF, and 48% LIF. Furthermore, this coordination scenario displayed fair improvement in terms of condition (CIF = 1%), resilience preparedness (RPIF = 14%), and risk (RIF = 5%). For the efficiency and effectiveness, the results displayed an IEF of 67% and IFF of 9%, which reflects fewer public disruptions (e.g., less disruption time and a fewer number of interventions) with longer corridor/asset operating times. Through combining the above-mentioned coordination savings, the full coordination roads, water, and sewer scenario revealed an overall improvement of 29% as opposed to the baseline combined conventional one.

TABLE 16.22
Results Comparison – Scenario 5 vs Baseline

KPI	Scenario 1 – Combined Conventional (Baseline)	Scenario 5 – Full Coordination	Difference (%)
Time (hours)	2,673,608	748,074	72
Space (m^2)	397,069	145,467	63
Cost – Equivalent Uniform Annual Cost (EUAC) ($)	$2,918,743	$1,529,741	48
Cost – Net Present Worth (NPW) ($)	$56,983,949	$29,865,841	48
Average condition (%)	66%	67%	1
Average risk (%)	36%	34%	5
Average resilience (%)	80%	69%	14
# of intervention actions	560	186	67
Time per km per year (hours/km/year)	2,018	565	72
Cost per km per year ($//km/year)	$55,070.62	$28,863.05	48
Average repair length per year (km/year)	5.2	2.4	54
Average number of interventions per year	22.4	7.44	67
Ratio of interventions per year– number of annual interventions/number of corridors (%)	18%	6%	67

TABLE 16.23
Scenario 5 – Improvement Deviational Variables

Performance Indicator	KPI	Weights of Importance (%)	Scenario 5 – Full Coordination (%)
Time	NCR	10	72
Space	STIF	10	63
Cost	LIF	20	48
Efficiency	IEF	5	67
Effectiveness	IFF	5	9
Condition	CIF	20	1
Resilience preparedness	RPIF	10	14
Risk	RIF	20	5
Overall improvement (%)			*29*

Summary of Optimization Results

The objective of this section is to summarize the key results of all the optimization scenarios and combine them to visualize the outcome of convention versus partial and full coordination. The summary results are outlined in Tables 16.24, 16.25, and 16.26. The conventional optimization scenarios displayed the outcome of undertaking single asset-based intervention and the results have displayed a larger number of interventions compared to the coordinated scenarios. The condition, risk, and resilience were slightly better for the coordinated scenario as opposed to the conventional one. The time, space consumption, and cost experienced a major change as the common or duplicated activities were carried out once instead of n_s times. Furthermore, the performance assessment improvement results showed the full coordination scenario as the best scenario with 29% savings, followed by the partially coordinated water and sewer with 27%, and ending with the partially coordinated roads and sewer with only 9%.

Sensitivity Analysis

Sensitivity analysis is performed to study and verify the sensitivity of the increasing or decreasing the minimally acceptable reliability thresholds on the other KPIs'. It can answer many what-if questions such as: "Should we pay more for enhancing the network reliability?" And if the answer is yes, "what is the cost premium between the proposed intervention program and the optimal intervention program?" The sensitivity analysis was undertaken only for the full coordination scenario with four new optimization cases ranging between −20% and +20% with 10% increments. The results are outlined in Tables 16.27 and 16.28. The improvement deviation variables of the sensitivity analysis scenarios were compared with the baseline case (Scenario 5) and the difference was plotted in Figure 16.9. The system showed to be very sensitive to changes in the reliability threshold. For instance, if the municipality decreased their acceptable reliability threshold by 20%, it would potentially result in 86% less

TABLE 16.24
Combined Conventional and Coordinated Optimization Summary Results

KPI	Scenario 1 – Combined Conventional (Baseline)	Scenario 2 – Roads and Sewer	Scenario 3 – Roads and Water	Scenario 4 – Water and Sewer	Scenario 5 – Full Coordination
Time (hours)	2,673,608	1,093,561	1,614,876	1,636,767	748,074
Space (m^2)	397,069	282,758	195,432	180,755	145,467
Cost – Equivalent Uniform Annual Cost (EUAC) ($)	$2,918,743	$3,569,007	$1,436,639	$1,621,048	$1,529,741
Cost – Net Present Worth (NPW) ($)	$56,983,949	$69,679,348	$28,048,165	$31,648,453	$29,865,841
Average condition (%)	66%	67%	67%	74%	67%
Average risk (%)	36%	34%	34%	12%	34%
Average resilience (%)	80%	75%	80%	73%	69%
# of intervention actions	560	271	316	338	186
Time per km per year (hours/km/year)	2,018	825	1,219	1,235	565
Cost per km per year ($//km/year)	$55,070.62	$67,339.75	$27,106.40	$30,585.80	$28,863.05
Average repair length per year (km/year)	5.2	3.4	3.4	3.5	2.4
Average number of interventions per year	22.4	10.84	12.64	13.52	7.44
Ratio of interventions per year – number of annual interventions/number of corridors (%)	18%	9%	10%	11%	6%

TABLE 16.25
Combined Conventional and Coordinated Improvement (%)

KPI	Scenario 1 – Combined Conventional (Baseline) (%)	Scenario 2 – Roads and Sewer (%)	Scenario 3 – Roads and Water (%)	Scenario 4 – Water and Sewer (%)	Scenario 5 – Full Coordination (%)
Time (hours)	0	59	40	39	72
Space (m^2)	0	29	51	54	63
Cost – Equivalent Uniform Annual Cost (EUAC) ($)	0	−22	51	44	48
Cost – Net Present Worth (NPW) ($)	0	−22	51	44	48
Average condition (%)	0	1	0	10	1
Average risk (%)	0	5	4	65	5
Average resilience (%)	0	6	0	9	14
# of intervention actions	0	52	44	40	67
Time per km per year (hours/km/year)	0	59	40	39	72
Cost per km per year ($//km/year)	0	−22	51	44	48
Average repair length per year (km/year)	0	34	34	32	53
Average number of interventions per year	0	52	44	40	67
Ratio of interventions per year – number of annual interventions/number of corridors (%)	0	52	44	40	67

TABLE 16.26
KPIs' Summary Results for Partial and Full Coordination Scenarios

Performance Indicator	KPI	Weights of Importance (%)	Scenario 2 – Roads and Sewer (%)	Scenario 3 – Roads and Water (%)	Scenario 4 – Water and Sewer (%)	Scenario 5 – Full Coordination (%)
Time	NCR	10	59	40	39	72
Space	STIF	10	29	51	54	63
Cost	LIF	20	−22	51	44	48
Efficiency	IEF	5	52	44	40	67
Effectiveness	IFF	5	2	5	7	9
Condition	CIF	20	1	0	10	1
Resilience preparedness	RPIF	10	6	0	9	14
Risk	RIF	20	5	4	10	5
Overall improvement (%)			*9*	*22*	*27*	*29*

disruption time, 85% less space, 7% less cost, 33% improved efficiency, 3% effectiveness, 10% decline in the average network reliability and resilience preparedness with a 24% higher risk exposure as a result of the declined reliability, resulting in a 3% decline in the overall improvement compared to the baseline case (Scenario 5). Similarly, the other sensitivity analysis scenarios were carried out and the results are outlined in Tables 16.27, 16.28, and Figure 16.9. In summary, slight changes in the reliability drastically affect the other KPIs. The repair time and efficiency showed to be the most sensitive items to the changes in the reliability thresholds. However, the intervention effectiveness showed to be the least sensitive item to the changes in the reliability thresholds as it is not directly related to the reliability but rather the selection of the effective intervention action as well as the skills of the crews' undertaking the intervention.

TABLE 16.27
Sensitivity Analysis Scenarios' Optimization Results

KPI	Case 1 (−20%)	Case 2 (−10%)	Baseline Case (Scenario 5)	Case 3 (10%)	Case 4 (20%)
Time (hours)	643,874	594,514	0	−1,540,131	17,571
Space (m²)	124,335	115,053	0	−193,302	21,714
Cost – Equivalent Uniform Annual Cost (EUAC) ($)	$108,433.00	$74,466.00	$0.00	−$35,921.00	−$177,940.00
Cost – Net Present Worth (NPW) ($)	$2,116,982.00	$1,453,837.00	$0.00	−$701,305.00	−$3,474,013.00
Average condition (%)	−7%	−5%	0%	18%	8%
Average risk (%)	−8%	−5%	0%	19%	8%
Average resilience (%)	−7%	−5%	0%	8%	6%
# of intervention actions	52	43	0	−396	−159
Time per km per year (hours/km/year)	486	449	0	−1,162	13
Cost per km per year ($//km/year)	$2,045.90	$1,405.02	$0.00	−$677.76	−$3,357.37
Average repair length per year (km/year)	0.32	0.28	–	−1.40	−0.79
Average number of interventions per year	2.08	1.72	0	−15.84	−6.36
Ratio of interventions per year – number of annual interventions/number of corridors (%)	2%	1%	0%	−13%	−5%

TABLE 16.28
KPIs' Summary Results for Sensitivity Analysis Scenarios

Performance Indicator	KPI	Weights of Importance (%)	Case 1 (−20%)	Case 2 (−10%)	Baseline Case (Scenario 5)	Case 3 (10%)	Case 4 (20%)
Time	NCR	10	86%	79%	0%	−206%	2%
Space	STIF	10	85%	79%	0%	−133%	15%
Cost	LIF	20	7%	5%	0%	−2%	−12%
Efficiency	IEF	5	33%	17%	0%	−217%	−83%
Effectiveness	IFF	5	3%	5%	0%	8%	10%
Condition	CIF	20	−10%	−7%	0%	27%	12%
Resilience preparedness	RPIF	10	−10%	−7%	0%	12%	9%
Risk	RIF	20	−24%	−15%	0%	56%	24%
Overall deviation (%)			*−3%*	*3%*	*0%*	*7%*	*0%*

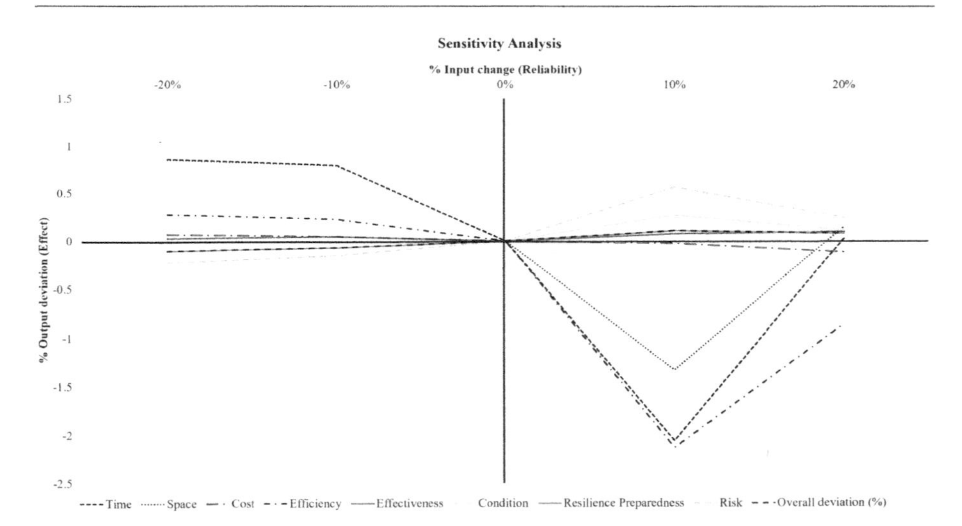

FIGURE 16.9 Spider diagram for the sensitivity analysis scenarios.

REFERENCES

Abusamra, S. (2019). "Coordination and Multi-Objective Optimization Framework for Managing Municipal Infrastructure under Performance-Based Contracts." Doctoral of Philosophy Thesis, Concordia University, Montreal.

Amador, L. and Magnuson, S. (2011). "Adjacency modeling for coordination of investments in infrastructure asset management: Case study of Kindersley, Saskatchewan, Canada." *Transportation Research Board (TRB)*, 2246, 8–15.

Bank of Canada. (2023). "Interest Rate." https://www.bankofcanada.ca/rates/daily-digest/. Mar 15, 2023.

Bentley (2018). "StormCAD, Storm Sewer Analysis and Design Software." https://www.bentley.com/en/products/product-line/hydraulics-and-hydrology-software/stormcad. Sep 7, 2018.

Environment Canada. (2014). "Engineering Climate Datasets." https://climate.weather.gc.ca/prods_servs/engineering_e.html. Mar 14, 2022.

Mailhot, A., Beauregard, I., Talbot, G., Caya, D., and Binerr, S. (2012). "Future changes in intense precipitation over Canada assessed from multi-model NARCCAP ensemble simulations." *International Journal of Climatology*, 32 (8), 1151–1163.

Trading Economics. (2023). "Canada Interest Rate." https://tradingeconomics.com/canada/interest-rate. Mar 15, 2023.

Water Security Agency. (2014). "Stormwater Guidelines: EPB 322." https://pdf4pro.com/view/stormwater-guidelines-saskh20-ca-5704c6.html. Apr 12, 2022.

Part 6

Future of Cities and Infrastructure in 2050

17 Vision for 2050

Sustainable Cities and Coordinated Infrastructure

INTRODUCTION

The world of infrastructure stands as a cornerstone, supporting the very foundations of our society. Roads that connect us, buildings that shelter us, and energy systems that power us are the arteries and bones of our modern civilization. This chapter summarizes the current state of the infrastructure, challenges and inefficiencies faced by municipalities while managing their assets, and the coordination framework and its benefits. Thenceforth, it moves to the future of our cities and infrastructure as we peer into the horizon of 2050. A future promises not only profound transformation but also an imperative for innovation like never before. The world we know today is on the verge of unprecedented change. The convergence of cutting-edge technologies, environmental imperatives, and shifting demographics is reshaping our expectations of what infrastructure means and how it should function. As we enter this new era, it becomes critical to envision the future of infrastructure, one that is not only sustainable and resilient but also capable of adapting to a rapidly evolving world.

SUMMARY AND FINDINGS

Municipalities are facing significant inefficiencies and financial burdens due to underperforming infrastructure. This increases the risk of sudden failures and service disruptions, necessitating immediate corrective actions to maintain deteriorating assets. Additionally, aging infrastructure systems are putting immense pressure on governments, leading to growing deficits for repairs and replacements. Infrastructure projects encounter numerous challenges and risks throughout their lifecycles, such as demand fluctuations, uncertainties, natural disasters, and criticality. Crucial intervention decisions are made not only at the beginning of the lifecycle but are also regularly revised to ensure acceptable service levels while adhering to tight budgets and maintaining minimum acceptable conditions. Therefore, various alternatives must be considered to optimize the use of available expenditures and resources within budget constraints.

The need for asset management has been amplified by several challenges, compelling governments to act proactively and optimize expenditures. Urbanization, for example, requires new infrastructure or increased demand on existing infrastructure, such as more traffic on roads and higher demand for processed water. This forces asset managers to consider resilience in their rehabilitation and replacement

DOI: 10.1201/9781003327509-23

decisions, such as expanding roads, enlarging water and sewer pipes, and building additional facilities like water pumping stations and treatment plants.

In light of these issues, this book introduces a coordination and optimization asset management framework for municipal infrastructure. The framework supports both the structuring of performance-based contracts (PBC) and the maintenance planning phases. For PBC structuring, it helps asset managers establish optimal thresholds and penalties-incentives (P/I) systems for key performance indicators (KPIs). During the maintenance planning phase, it aids municipalities or private partners in selecting intervention plans that meet predefined KPI thresholds and make optimal use of expenditures. The models are adaptable to various stakeholder preferences, such as enhanced performance, minimized costs, and reduced disruption.

The framework can be applied to intervention scheduling and fund allocation for municipal co-located infrastructure. The process begins by defining KPIs along with their thresholds and associated P/I. Subsequently, multi-dimensional performance assessment models are developed to quantify temporal, financial, spatial, risk, reliability, resilience preparedness, efficiency, effectiveness, and health savings indicators for various coordination scenarios. These are compared with conventional scenarios to compute potential coordination savings, which arise from the spatial interdependency among municipal co-located infrastructure systems.

The temporal dimension calculates the corridor coordination ratio to compare conventional intervention scenarios, based on asset-based maintenance, with full and partial coordination scenarios, which coordinate interventions of right-of-way assets. The spatial dimension assesses the spatial and temporal savings of full and partial coordination scenarios compared to conventional scenarios. The financial dimension evaluates the monetary savings of full and partial coordination scenarios, incorporating both direct costs (e.g., manpower, equipment, materials) and indirect costs (e.g., disruption – user costs), along with the time value of money during trade-off analysis.

The intervention efficiency and effectiveness dimensions measure the efficiency and effectiveness of coordinating intervention actions versus independent actions for each asset. Efficiency represents utility cut costs (e.g., longer disruption time), while effectiveness represents the duration when the corridor is disruption-free. Computations focus on disruption and operating durations. The reliability dimension assesses the corridor condition/reliability to reflect the impact of intervention actions on the corridor condition/reliability for different coordination scenarios. The risk dimension calculates the corridor's probability and consequences of failure for various coordination scenarios, comparing coordinated and conventional intervention scenarios.

The resilience preparedness dimension evaluates the corridor's resilience concerning climate change and urbanization, focusing on the replacement of water and sewer pipes due to their long service lives and significant public disruptions. It considers the impact of urbanization (e.g., land use change, population growth) and climate change (e.g., increased rainfall intensity and frequency) on water and combined sewer and stormwater systems. The corridor health dimension integrates indicator scores into a priority index to help asset managers prioritize corridors and make critical intervention decisions.

The PBC structuring optimization provides municipalities with near-optimal KPIs and P/I systems that minimize maintenance contractors' contingencies without compromising service levels. The maintenance planning optimization ensures that intervention plans meet predefined KPI thresholds and make the best use of available resources. It features both a near-optimal hierarchical optimization technique using genetic algorithms (GAs) and an exact optimization technique using the MOSEK linear programming optimization engine. The innovative aspect of the multi-objective optimization technique lies in its integration of metaheuristics, binary coding, integer programming, and non-preemptive goal optimization procedures. This combination effectively balances the scheduling of various intervention alternatives, significantly reducing the search space. Consequently, the framework can be scaled up to include more than three infrastructure systems or to extend the planning horizon. Optimal utilization of expenditures is crucial in the decision-making process for infrastructure systems. This system can be used for planning and scheduling interventions across all coordination scenarios (e.g., conventional, partial, and full coordination). Additionally, it considers the effects of urbanization and climate change, which may necessitate expanding some assets to meet increased demand.

In summary, this research develops a novel coordination and optimization framework for municipal co-located infrastructure. The methodology was applied to roads, water, and sewer networks, but it can be extended to other infrastructure systems that are spatially co-located and may disrupt traffic, such as oil and gas pipelines, electricity networks, bridges, and telecommunication networks. The key outcomes of this system are (1) quantifying and demonstrating potential coordination savings due to existing spatial interdependencies and (2) integrating contractual KPIs and P/I systems into the decision-making process.

The system was applied to a 9 km area in the city of Montreal and expanded to a 53 km area in the town of Kindersley. In Montreal, a GA-based optimization engine was used, modeled with sophisticated spreadsheets integrated with Evolver. In Kindersley, the system was modeled using the REMSOFT software package, featuring a linear programming MOSEK optimization engine to find an exact solution to the problem.

The results from both case studies showed significant savings in favor of full coordination over conventional asset management (i.e., single asset-based management) in terms of cost, time, space, risk, resilience preparedness, reliability, efficiency, and effectiveness. For the city of Montreal, the PBC structuring achieved a near-optimal set of KPI thresholds and their associated penalties and incentives. The maintenance planning optimization revealed an overall improvement of 15% in favor of the full coordination program compared to the conventional one. These savings reflect a 10% improved reliability and less risk exposure, 18% reduced costs, and 12% reduced disruption time, along with other savings detailed earlier.

Additionally, the coordinated intervention program proved to be more efficient in over 70% of the corridors, with fewer interventions per corridor, more temporal and financial savings, and less public disruption. A sensitivity analysis was conducted to analyze the impacts of changing the reliability threshold on the other indicators. The results showed that the system is sensitive to changes in the reliability threshold. For instance, if the municipality increased its acceptable reliability threshold by 10%, it would potentially result in 42% more disruption time, 31% additional space, 33% extra cost, etc.

Similarly, the town of Kindersley maintenance planning optimization showed 29% overall improvement in the performance assessment indicators as opposed to the conventional scenario. Likewise, improvements were computed for the partial coordination scenarios, and the overall improvement to the performance assessment indicators ranged between 9% and 27%, with the partial coordination of water and sewer being the optimal partial coordination scenario compared with the other two scenarios. A sensitivity analysis was conducted to analyze the impacts of changing the reliability threshold on the other indicators. The results showed that the system is sensitive to changes in the reliability threshold. For instance, if the municipality decreased its acceptable reliability threshold by 20%, it would potentially result in 86% less repair manhours, 85% less space, 7% less cost, etc.

In summary, the framework provides decision-makers with a computational tool that (1) defines KPI thresholds and their associated P/I in the PBC structuring phase and (2) optimally allocates the limited budget in the maintenance planning phase to enhance the performance of right-of-way corridor assets while minimizing public disruption and maximizing intervention program efficiency.

CITIES AND INFRASTRUCTURE IN 2050: CHALLENGES AND OPPORTUNITIES

To better manage our infrastructure, we need to envision the future design and layout of our cities. In this book, we have chosen the year 2050 as a key milestone, anticipating significant transformations in infrastructure from various perspectives. In the following sections, we will explore the factors, challenges, and opportunities that will shape the future of our infrastructure. Each aspect of this transformation aims to create a more sustainable, connected, and resilient infrastructure.

We examine how cities are evolving to accommodate growing populations while minimizing their environmental impact. This includes transitioning from fossil fuels to renewable energy and developing new transportation systems that are faster, safer, and more sustainable. Our focus is on designing infrastructure that is environmentally responsible and resilient to climate change. Additionally, we address the challenges of funding and policy, balancing innovation, financing, and governance to support infrastructure development.

URBANIZATION: OPPORTUNITY FOR SMARTER CITIES

Rapid urbanization is reshaping the world's landscape. By 2050, nearly 70% of the global population is projected to live in cities (UN 2018). This poses an unprecedented challenge for infrastructure. However, it also offers an opportunity to create smarter, more efficient urban environments through the integration of advanced technologies such as the Internet of Things (IoT), Artificial intelligence (AI), and sustainable urban planning. The cities of the future can evolve to meet the needs of a growing populace while reducing their environmental footprint through the following initiatives:

- **Smart City Initiatives**: Cities are investing in smart technologies to improve services, reduce waste, and enhance the quality of life for residents. IoT sensors will monitor traffic, energy usage, and waste management in real time, allowing for data-driven decision-making. As of 2021, over 1,000 cities worldwide have initiated smart city projects. By 2050, it is projected that these initiatives will collectively save over 2.3 billion MWh of electricity and reduce CO_2 emissions by 1.9 billion metric tons annually (IEA 2021).
- **Sustainable Urban Planning**: Urban planners are adopting eco-friendly designs, emphasizing green spaces, efficient public transportation, and mixed-use zoning to reduce congestion and emissions. For instance, cities like Copenhagen are investing heavily in cycling infrastructure, resulting in over 45% of the population commuting by bicycle (C40 Cities 2016). Such initiatives aim to reduce congestion and carbon emissions significantly.
- **Infrastructure Expansion**: Governments are investing in the expansion and maintenance of critical urban infrastructure, such as public transit systems, bridges, and water supply networks, to accommodate growing populations. To cope with urbanization, China has embarked on an extensive high-speed rail network expansion, with over 38,000 miles of high-speed rail by 2025, fostering connectivity and economic growth (SCMP 2022).

Energy Evolution: From Fossil Fuels to Renewable Power

The energy sector is at the center of the fight against climate change. As we approach 2050, the transition from fossil fuels to renewable energy sources becomes imperative. We will discuss the challenges of phasing out traditional power generation and the opportunities that lie in harnessing solar, wind, and other sustainable energy sources and how we can build resilient energy grids and storage systems to ensure a stable energy supply for future generations.

- **Renewable Energy Investment**: Governments and private companies are investing heavily in renewable energy sources such as solar and wind. Advanced technologies in energy storage, such as improved batteries, are being developed to ensure a continuous power supply. In 2020, global investment in renewable energy exceeded $300 billion. By 2050, it is projected that renewable energy sources will provide more than 70% of the world's electricity (IRENA 2020).
- **Grid Modernization**: Electric grids are being upgraded to accommodate distributed energy sources, allowing for efficient integration of renewables and reducing energy wastage. The European Union's Clean Energy for all Europeans package aims to modernize electricity grids, facilitating the integration of renewable sources and reducing transmission losses by up to 40% (European Commission 2019).
- **Carbon Capture and Storage**: Innovations in carbon capture and storage (CCS) technologies aim to mitigate emissions from industries that are hard to decarbonize. The Quest CCS facility in Canada, which began operations

in 2015, has already captured and stored over 5 million metric tons of CO_2, showcasing the potential of this technology (Shell Canada 2015).

Transportation Transformation: The Road to Autonomous Mobility

The way we move people and goods is on the edge of a revolution. Autonomous vehicles, hyperloop systems, and high-speed rail are changing the face of transportation. We will discuss the challenges of integrating these innovations into existing infrastructure, addressing safety concerns, and ensuring equitable access. We will also discuss the potential benefits of reducing traffic congestion, lowering emissions, and enhancing the overall efficiency of transportation networks.

- **Infrastructure Adaptation**: Roads and highways are being equipped with smart infrastructure, such as sensors and communication networks, to support autonomous vehicles. High-speed rail and hyperloop projects are under development to offer sustainable long-distance transportation options (Liu et al. 2023). As of 2021, over 60 cities worldwide had initiated autonomous vehicle testing programs (Rana and Hossain 2021). The hyperloop project between Dubai and Abu Dhabi aims to reduce travel time from 90 minutes to just 12 minutes (Intellias 2019).
- **Regulatory Frameworks**: Governments are working on regulations to ensure the safe deployment of autonomous vehicles while addressing ethical and legal concerns. In 2020, the U.S. Department of Transportation issued a set of guidelines for autonomous vehicles, fostering innovation while ensuring safety (U.S. DOT 2020).
- **Accessibility and Equity**: Planners are focusing on equitable transportation access, especially in underserved areas, to ensure that the benefits of transportation innovations are widely distributed. Programs like the "Vision Zero" initiative in New York City, launched in 2014, aim to eliminate traffic-related deaths by 2050, prioritizing safety and equitable access to transportation (New York City DOT 2024).

Sustainability in Design: Building for the Long Term

The infrastructure of the future must be designed with sustainability at its core. Whether it is resilient buildings that can withstand extreme weather events or eco-friendly materials that reduce environmental impact, sustainable design principles are paramount. We will discuss the challenges faced by architects, engineers, and planners in creating infrastructure that can endure for decades while minimizing its ecological footprint.

- **Green Building Standards**: Architects and builders are adhering to strict green building standards, using materials that reduce carbon emissions and designing structures that optimize energy use. The Leadership in Energy and Environmental Design (LEED) certification, established in 1998, has

now been applied to over 100,000 projects in more than 165 countries (U.S. Green Building Council 2023).
- **Resilient Design**: Infrastructure is being designed to withstand natural disasters and climate change impacts. This includes elevated structures, flood barriers, and energy-efficient retrofits of existing buildings. The Netherlands, with its Delta Works project initiated in the 1950s, showcases how resilient infrastructure can protect against rising sea levels and storm surges (TouristSecrets 2023).
- **Circular Economy Principles**: A shift toward a circular economy involves designing infrastructure with materials that can be reused and recycled, reducing waste and environmental impact. The European Union's Circular Economy Action Plan aims to make sustainable products the norm and reduce waste generation by 30% by 2030 (European Commission 2020).

Resilience in the Face of Uncertainty: Adapting to Climate Change

Climate change is a global reality that will continue to impact our infrastructure. Rising sea levels, extreme weather events, and shifting climate patterns pose significant challenges. Yet, these challenges also underscore the importance of building resilient infrastructure that can withstand the unpredictable forces of nature. We will discuss the strategies and technologies that can enhance resilience and ensure our infrastructure can weather the storms of the future.

- **Climate Resilient Infrastructure**: Engineering practices are evolving to incorporate climate resilience, including sea level rise protections, stormwater management, and drought-resistant designs. After Hurricane Katrina in 2005, New Orleans invested over $14.5 billion in rebuilding and improving its flood defenses and water management systems (Columbia Climate School 2024).
- **Nature-Based Solutions**: Implementing nature-based solutions like wetland restoration and green infrastructure can provide natural defenses against climate-related hazards. Singapore's "ABC Waters" program, initiated in 2006, transforms drains and canals into beautiful and functional streams, incorporating nature-based solutions for flood management (FEMA 2021).
- **Early Warning Systems**: Advanced monitoring and early warning systems help anticipate and mitigate the impact of extreme weather events on infrastructure. In 2019, India implemented an advanced Cyclone Warning System, resulting in a significant reduction in casualties during cyclones like Amphan in 2020 (World Meteorological Organization 2020).

Funding and Policy: Navigating the Path Forward

Turning these visions of future infrastructure into reality will require significant investments and thoughtful policy decisions. We will discuss the financial and political challenges associated with funding large-scale infrastructure projects and creating regulatory frameworks that encourage innovation while safeguarding the public interest.

- **Public–Private Partnerships**: Governments are partnering with private companies to finance and manage large-scale infrastructure projects, sharing both the risks and rewards. The California High-Speed Rail project, with an estimated cost of $80 billion, shows how public and private entities can collaborate to fund and develop transformative infrastructure (California High-Speed Rail Authority 2023).
- **Incentives for Innovation**: Governments are providing incentives and grants for research and development in infrastructure technologies, promoting innovation in the sector. Germany's Research and Innovation Program for Sustainable Mobility (Funding Directive II), established in 2020, supports innovative transportation projects with grants totaling €2 billion (Federal Ministry of Transport and Digital Infrastructure 2020).
- **Sustainable Finance**: The financial sector is increasingly integrating sustainability criteria into investment decisions, directing capital toward projects that align with environmental and social goals. As of 2021, the total value of green bonds issued worldwide exceeded $1 trillion, providing a growing pool of capital for sustainable infrastructure projects (Climate Bonds Initiative 2021).

We are already moving forward. These facts and figures underscore the tangible progress and ambitious targets set for addressing the challenges and seizing the opportunities that lie ahead in shaping the future of infrastructure in 2050. As we navigate and seize the opportunities that lie ahead, the future of infrastructure in 2050 emerges as a dynamic landscape where technology, sustainability, and resilience converge to shape a world that is not only well connected but also better equipped to address the pressing challenges of our times.

REFERENCES

C40 Cities. (2016). "C40 Good Practice Guides: Copenhagen - City of Cyclists." https://www.c40.org/case-studies/c40-good-practice-guides-copenhagen-city-of-cyclists/. Jul 13, 2023.

California High-Speed Rail Authority. (2023). "California High-Speed Rail Project." https://www.hsr.ca.gov/. Feb 21, 2024.

Climate Bonds Initiative. (2021). "Green Bonds Market Summary." https://www.climatebonds.net/resources/reports/green-bonds-market-summary-2021. Jul 22, 2024.

Columbia Climate School. (2024). "The Case for Climate-Resilient Infrastructure." https://news.climate.columbia.edu/2024/07/22/the-case-for-climate-resilient-infrastructure/. Aug 21, 2024.

European Commission. (2019). "Clean Energy for All Europeans Package." Publications Office of the European Union. https://energy.ec.europa.eu/topics/energy-strategy/clean-energy-all-europeans-package_en. July 15, 2023.

European Commission. (2020). "Circular Economy Action Plan." https://environment.ec.europa.eu/strategy/circular-economy-action-plan_en. Jul 20, 2023.

Federal Ministry of Transport and Digital Infrastructure. (2020). "Research and Innovation Program for Sustainable Mobility (Funding Directive II)." https://www.bmvi.de/EN/Home/home.html. Jul 21, 2024.

FEMA. (2021). "Building Community Resilience with Nature-Based Solutions: A Guide for Local Communities." https://www.fema.gov/sites/default/files/documents/fema_risk-map-nature-based-solutions-guide_2021.pdf. Jul 21, 2023.

IEA. (2021). "Empowering Cities for a Net Zero Future." https://www.iea.org/reports/empowering-cities-for-a-net-zero-future. Jul 12, 2023.

Intellias. (2019). "How Will Urban Infrastructure Change with Autonomous Driving? https://intellias.com/how-will-urban-infrastructure-change-with-autonomous-driving/. Jul 15, 2023.

International Renewable Energy Agency (IRENA). (2020). "Global Landscape of Renewable Energy Finance 2020." International Renewable Energy Agency. https://www.irena.org/-/media/Files/IRENA/Agency/Publication/2020/Apr/IRENA_GRO_Summary_2020.pdf?la=en&hash=1F18E445B56228AF8C4893CAEF147ED0163A0E47. Jul 14, 2023.

Liu, H., Yang, M., Guan, C., Chen, Y. S., Keith, M., You, M., and Menendez, M. (2023). "Urban infrastructure design principles for connected and autonomous vehicles: A case study of Oxford, UK". *Computational Urban Science*, Springer, 3(1), 34.

New York City Department of Transportation. (2024). "New York City Streets Are Safer and More Equitable after 10 Years of Vision Zero. https://www.nyc.gov/html/dot/html/pr2024/10-years-vision-zero.shtml. Jul 16, 2024.

Rana, M. M. and Hossain, K. (2021). "Connected and autonomous vehicles and infrastructures: A literature review". *International Journal of Pavement Research and Technology*, Springer, 16, 264–284.

SCMP. (2022). "China Plans Expansion of High-Speed Railway Equal to Combined Length of Next 5 Largest Countries by Network Size by 2025." https://www.scmp.com/economy/china-economy/article/3163957/china-plans-expansion-high-speed-rail-network-equal-combined?module=perpetual_scroll_0&pgtype=article. Jul 13, 2023.

Shell Canada. (2015). "Quest Carbon Capture and Storage Project." https://www.shell.ca/en_ca/about-us/projects-and-sites/quest-carbon-capture-and-storage-project.html. Jul 16, 2023.

TouristSecrets. (2023). "Netherlands Delta Works: Wonder of the Modern World." https://www.touristsecrets.com/destinations/europe/netherlands/netherlands-delta-works-wonder-of-the-modern-world-guide/. Jan 18, 2024.

U.S. Department of Transportation. (2020). "Automated Vehicles Comprehensive Plan." https://www.transportation.gov/AV. Jul 16, 2023.

U.S. Green Building Council. (2023). "LEED Rating System." https://www.usgbc.org/leed. Jul 18, 2023.

United Nations (UN). (2018). "68% of the World Population Projected to Live in Urban Areas by 2050, Says UN." https://www.un.org/development/desa/en/news/population/2018-revision-of-world-urbanization-prospects.html. Jul 12, 2023.

World Meteorological Organization. (2020). "Cyclone Amphan Highlights the Value of Multi-Hazard Early Warnings." https://wmo.int/media/news/cyclone-amphan-highlights-value-of-multi-hazard-early-warnings-0. Jul 20, 2023.

18 Transportation, Sustainability, and Resilience in the Infrastructure of 2050

INTRODUCTION

As we progress, infrastructure remains vital, supporting our roads, buildings, and energy systems. Looking ahead to 2050, we face a future that demands significant changes and innovative ideas. Today, we are on the edge of major shifts. Advanced technologies, environmental needs, and changing populations are redefining infrastructure. It is essential to envision a future where infrastructure is sustainable, resilient, and adaptable.

In the next four chapters, we will share our perspectives on the future of infrastructure, examining the challenges and opportunities ahead. We will explore key aspects shaping the infrastructure landscape in 2050, including smart cities, sustainability, renewable energy, transportation changes, sustainable design, governance, digital technology, and AI. Each chapter will illustrate how these elements contribute to a more sustainable, connected, and resilient world, guiding us toward a brighter future in 2050. Table 18.1 provides an overview of the upcoming chapters and their topics.

This chapter explores the transformative potential of future transportation systems and their profound impact on infrastructure by 2050. As advancements in transportation promise faster, safer, cleaner, and more accessible mobility solutions, the necessary infrastructure will evolve to support these innovations. Key areas of focus include the integration of autonomous vehicles, which will drive the development of smart roads and optimized traffic management, and the emergence of high-speed transportation options such as the Hyperloop and supersonic travel, which will require dedicated networks and specialized facilities. Additionally, urban mobility will see improvements through enhanced public transit, active transportation initiatives, and innovative last-mile solutions.

The evolution of infrastructure will be driven by the dual imperatives of sustainability and resilience. Addressing climate change and environmental challenges will require infrastructure that not only minimizes ecological impact but also endures and adapts to future uncertainties. This will involve integrating green infrastructure such as green roofs and urban forests to enhance stormwater management, support biodiversity, and improve air quality. Circular economy principles will guide the lifecycle of infrastructure, emphasizing waste reduction, material reuse, and resource

DOI: 10.1201/9781003327509-24

TABLE 18.1
An Overview of Chapters 18–21

Chapter	Theme	Topic
Chapter 18	Transportation, Sustainability, and Resilience in the Infrastructure of 2050	Future of Transportation Sustainability Resilience
Chapter 19	Role of Digital Technology and Asset Management in the Infrastructure of 2050	Digital Technology Asset Management Innovation and Disruption
Chapter 20	Adapting Infrastructure for Inclusivity, Security, and Well-Being by 2050	Social and Demographic Changes Security and Privacy Health and Well-Being
Chapter 21	Governance, Stakeholder Engagement, and Circular Economy Principles in 2050 Infrastructure	Governance and Regulation Financing, Funding and Collaboration Stakeholder Engagement and Participation Circular Economy

efficiency. Concurrently, resilience will be built into designs to withstand and recover from natural disasters and climate extremes, ensuring continuity of essential services.

FUTURE OF TRANSPORTATION

The future of transportation is set to be a driver in shaping the infrastructure of 2050. As we approach this horizon, transportation systems will undergo a profound transformation, offering faster, safer, cleaner, and more accessible mobility solutions for people and goods. This evolution will have far-reaching implications for the infrastructure that supports and enables these advancements.

Autonomous Vehicles

Autonomous vehicles, often referred to as self-driving cars, will play a central role in the future of transportation. These vehicles are equipped with sensors, cameras, and advanced AI systems, enabling them to navigate without human intervention. The impact on infrastructure will be substantial.

- **Smart Roads:** Infrastructure will include smart road systems with sensors embedded in roadways to communicate with autonomous vehicles. These systems will help vehicles navigate, optimize traffic flow, and enhance safety.
- **Traffic Management:** Traffic management will become more data-driven, leveraging real-time data from autonomous vehicles to optimize traffic flow, reduce congestion, and improve road safety.

- **Parking:** Autonomous vehicles will require fewer parking spaces as they can drop off passengers and find parking spaces efficiently. This will free up valuable urban land previously used for parking lots.

High-Speed Transportation

In the quest for faster, more efficient transportation, high-speed options will reshape the infrastructure:

- **Hyperloop:** The development of Hyperloop systems, using vacuum tubes to transport pods at near-supersonic speeds, will require the creation of dedicated infrastructure networks. These systems will significantly reduce travel times between cities, potentially transforming regional economies. For instance, the Hyperloop project currently under development between Los Angeles and San Francisco aims to reduce travel time between these cities to just 30 minutes, marking a significant leap in transportation efficiency (HyperloopTT 2023).
- **Supersonic Travel:** Supersonic aircraft provide rapid long-distance travel, necessitating the development of specialized airports and flight corridors to accommodate these high-speed aircraft (Meier 2020).

Urban Mobility

In urban areas, transportation infrastructure will evolve to be more integrated, reliable, and user-friendly.

- **Public Transit:** Public transit systems will undergo substantial upgrades, incorporating digital technology to offer real-time information and seamless payment systems. Electric buses and trains will become the norm, reducing emissions and noise pollution.
- **Active Transportation:** Cities will prioritize walking and cycling infrastructure to encourage active transportation. Dedicated lanes, pedestrian-friendly streets, and bike-sharing programs will become commonplace.
- **Last-Mile Solutions:** To address the "last-mile" challenge of getting commuters from transit hubs to their final destinations, cities will continue evolving innovative solutions such as electric scooters, shared bikes, and micromobility options.

Delivery and Logistics

The future of transportation will also impact the movement of goods.

- **Delivery Drones:** Drones and autonomous delivery robots will revolutionize the delivery of goods. Infrastructure will include drone landing pads and storage facilities for automated delivery hubs.

- **Logistics Centers:** Urban logistics centers will be strategically located to optimize the delivery of goods within cities, reducing congestion and emissions.

Sustainable Transportation

Sustainability will be at the forefront of transportation infrastructure.

- **Electric Charging Infrastructure:** Widespread adoption of electric vehicles (EVs) will require a dense network of EV charging stations, both in urban areas and along highways.
- **Sustainable Fuels:** Infrastructure will support the use of sustainable fuels, such as hydrogen for fuel cell vehicles and biofuels for conventional vehicles.
- **Infrastructure for Micromobility:** Dedicated lanes and parking spaces for electric scooters, bicycles, and other micromobility solutions will be integrated into urban infrastructure.

Accessibility and Equity

Future transportation infrastructure will prioritize accessibility and equity.

- **Public Transportation Accessibility:** Infrastructure will focus on ensuring that public transportation is accessible to all, including people with disabilities and underserved communities.
- **Affordable Transit:** Policies and infrastructure will work together to make public transportation more affordable and inclusive for all income levels.

Safety and Regulation

As transportation technology advances, infrastructure and regulations must keep pace.

- **Safety Infrastructure:** Infrastructure will incorporate advanced safety features, such as intelligent traffic lights and crash avoidance systems, to reduce accidents and save lives.
- **Regulatory Frameworks:** Governments will establish regulatory frameworks to ensure the safe operation of autonomous vehicles and new transportation technologies.

Real-World Example: The Hyperloop Project

The Hyperloop project serves as a compelling real-world example of how high-speed transportation innovations will reshape infrastructure in 2050 (HyperloopTT 2023). This transformative concept envisions a revolutionary mode of transportation that combines cutting-edge technology with dedicated infrastructure to redefine the way we travel between cities. It is designed to transport passenger pods through vacuum tubes at near-supersonic speeds. The key advantages include:

- **Reduced Travel Times:** The Hyperloop's design aims to dramatically reduce travel times between major urban centers. For example, the initial proposed Hyperloop route between Los Angeles and San Francisco seeks to reduce the journey time from several hours to just 30 minutes.
- **Sustainability:** The system aims to be energy-efficient, with the potential for renewable energy sources to power the system. This focus on sustainability aligns with the broader goals of reducing emissions and environmental impact.
- **Minimal Land Footprint:** The elevated or underground tube infrastructure minimizes the land footprint required for the transportation system, making it an efficient use of urban and rural space.

The development and implementation of the Hyperloop will have significant infrastructure implications.

- **Dedicated Infrastructure:** Hyperloop transportation requires the construction of dedicated vacuum tubes and infrastructure networks. These tubes will travel long distances, potentially spanning several regions or even countries.
- **Specialized Stations:** To accommodate high-speed travel, specialized Hyperloop stations will be required at various points along the route. These stations will serve as departure and arrival points for passengers and cargo.
- **Tunneling and Construction:** Building the vacuum tubes and stations will necessitate extensive construction and engineering efforts, including tunneling through varied terrains and coordinating with local authorities.
- **Economic Impact:** Successful Hyperloop projects can potentially transform regional economies by enabling rapid commuting between major cities and encouraging business development along the transportation corridors.

The Hyperloop project is an example of how future transportation systems will drive infrastructure development. As cities and regions plan and invest in these high-speed networks, they position themselves for a more interconnected, efficient, and sustainable future in 2050 and beyond. The success of initiatives like the Hyperloop highlights the critical role that innovation plays in shaping the infrastructure landscape of the future.

The future of transportation will reshape infrastructure in profound ways. The integration of autonomous vehicles, high-speed transportation, sustainable solutions, and equitable access will drive the evolution of infrastructure to meet the needs of a rapidly changing world. The result will be a transportation network that is faster, more efficient, and more sustainable, ultimately enhancing the quality of life for people around the globe while reducing the environmental footprint of our cities.

SUSTAINABILITY

As we move toward 2050, sustainability stands as a paramount force that will fundamentally reshape the landscape of infrastructure. The challenges posed by climate change and environmental degradation are now impossible to ignore, demanding a fundamental shift in how we design and construct the infrastructure that underpins modern society. Sustainability will be the key that guides us toward infrastructure that is both resilient and environmentally responsible.

Green Infrastructure

One of the cornerstones of future infrastructure will be green infrastructure. This approach harnesses the power of natural systems, such as wetlands, forests, and green roofs, to provide a multitude of ecosystem services that enhance the quality of urban living.

- **Stormwater Management:** Green infrastructure absorbs and filters rainwater, reducing the risk of urban flooding and water pollution. Sustainable urban design will include permeable pavements, rain gardens, and green corridors that act as natural sponges.
- **Biodiversity Conservation:** Incorporating green spaces within urban areas will promote biodiversity. Urban planners will set aside areas for parks, green belts, and wildlife corridors to support local ecosystems.
- **Air Quality Improvement:** Green infrastructure, such as urban forests and green walls, will play a pivotal role in improving air quality. These natural elements absorb pollutants and release oxygen, contributing to cleaner, healthier urban environments.

Circular Economy Principles

In the infrastructure of 2050, circular economy principles will guide every phase of a project's lifecycle, from design and construction to operation and decommissioning.

- **Reducing Waste:** Designing infrastructure with an eye toward reducing waste will be paramount. Innovative construction techniques, such as modular and prefabricated elements, will minimize waste generation during the building process.
- **Reuse and Recycling:** Materials used in infrastructure will be carefully selected for their ability to be reused or recycled. Concrete, steel, and other materials will be sourced sustainably and designed for disassembly and repurpose.
- **Resource Efficiency:** Infrastructure will be constructed to maximize resource efficiency, utilizing sustainable materials and technologies that minimize resource consumption over time. Energy-efficient buildings and renewable energy systems will be the norm.

Resilient Infrastructure

Resilience in infrastructure will be a cornerstone for addressing the growing threats posed by natural disasters, including floods, earthquakes, and wildfires, and the uncertainties of a changing climate.

- **Adaptability:** Resilient infrastructure will be designed to adapt to changing conditions. Buildings will be constructed to withstand extreme weather events, and transportation networks will incorporate features that minimize disruption during disasters.
- **Redundancy:** The infrastructure of 2050 will prioritize redundancy and backup systems, ensuring the continuity of essential services in the face of adversity.
- **Flexibility:** Flexible infrastructure will be designed to evolve as conditions change. For example, flood defenses will be adjustable to account for rising sea levels, and transportation systems will be modular to accommodate changing mobility needs.

Real-World Example: Copenhagen's Green Roof Policy

The city of Copenhagen serves as a real-world example of how sustainability principles are already shaping future infrastructure (City of Copenhagen 2022). Copenhagen has adopted an ambitious green roof policy that promotes the installation of green roofs on new buildings. This initiative provides numerous benefits, including:

- **Climate Resilience:** Green roofs provide insulation and reduce the urban heat island effect, helping the city adapt to changing climate conditions.
- **Improved Air Quality:** The vegetation on green roofs captures air pollutants, enhancing overall air quality.
- **Biodiversity Enhancement:** Green roofs serve as habitats for birds, insects, and plants, fostering biodiversity within the city.
- **Aesthetic and Social Value:** Green roofs contribute to the aesthetic appeal of the city, providing green spaces for relaxation and recreation.

As we defy the pressing challenges of climate change and environmental degradation, the integration of green infrastructure, circular economy principles, and resilience will be non-negotiable. These principles will not only protect the environment but also enhance the quality of life for urban populations, ensuring a future where infrastructure is both robust and ecologically responsible.

RESILIENCE

In 2050, the future of infrastructure will be driven by the need for resilience. As the world faces climate change, natural disasters, and pandemics, resilient infrastructure will stand as a bedrock of stability and continuity. Resilient infrastructure is designed to not only withstand but also recover swiftly from shocks and stresses, ensuring the uninterrupted provision of essential services and minimizing disruptions to communities. Resilience will be central to shaping infrastructure in 2050 in various ways.

Climate Resilience

- **Sea Level Rise Protections:** Infrastructure in coastal regions will incorporate measures to combat rising sea levels and storm surges. Protective barriers, elevated structures, and improved drainage systems will become standard features.
- **Extreme Weather Preparedness:** Infrastructure will be designed and retrofitted to withstand extreme weather events, including hurricanes, typhoons, and droughts. Robust materials and construction techniques will be employed to reduce vulnerability.

Disaster–Resilient Infrastructure

- **Disaster–Resilient Design:** Infrastructure projects will prioritize resilience against natural disasters such as earthquakes, floods, and wildfires. Flexible and adaptable designs will allow for swift recovery after such events.
- **Early Warning Systems:** Advanced monitoring and early warning systems will provide real-time alerts, allowing for proactive responses to disasters and minimizing their impact on infrastructure.

Infrastructure Redundancy

- **Built-in Redundancy:** Resilient infrastructure will feature built-in redundancy, ensuring that critical services can continue even in the event of failures. Redundant power sources, data centers, and transportation routes will become common.
- **Decentralization:** Decentralized infrastructure systems will reduce the risk of single points of failure. Distributed energy generation, for example, will ensure a stable power supply.

Real-World Example: Netherlands' Delta Works

The Netherlands' Delta Works project serves as an iconic example of resilient infrastructure (Delta Programme 2024).

- **Storm Surge Barriers:** The Delta Works features a network of storm surge barriers and dams that protect low-lying regions from flooding during severe storms.
- **Adaptive Planning:** The project continues to evolve and adapt to changing conditions, demonstrating a long-term commitment to infrastructure resilience.

Resilience will be at the forefront of shaping the infrastructure of 2050. In a world marked by increasing uncertainty and unpredictability, infrastructure that can withstand and recover from shocks and stresses will be crucial for maintaining societal stability and well-being. By prioritizing resilience in infrastructure design, construction, and operation, communities can face the challenges of the future with confidence, knowing that their essential services and critical assets are prepared to endure and thrive in the face of adversity. Resilience in infrastructure is not just a technical consideration; it is a promise of a safer and more secure tomorrow.

REFERENCES

City of Copenhagen. (2022). "Green Roofs in Copenhagen." https://interlace-hub.com/green-roof-policy-copenhagen. Jul 20, 2023.

Delta Programme. (2024). "2024 Delta Programme." https://english.deltaprogramma.nl/delta-programme/2024-delta-programme. Aug 15, 2024.

Hyperloop Transport Technologies (HyperloopTT). (2023). "Progress towards First US Route." https://www.hyperlooptt.com/projects/feasibility-study/. Jul 10, 2024.

Meier, R. (2020). "United States Establishes "Supersonic Corridor" to Test New Civilian Jets That Fly Faster than the Speed of Sound." *ADN*. https://www.airdatanews.com/united-states-establishes-supersonic-corridor-to-test-new-civilian-jets-that-fly-faster-than-the-speed-of-sound/. Jul 10, 2023.

19 Role of Digital Technology and Asset Management in the Infrastructure of 2050

INTRODUCTION

In 2050, digital technology will transform infrastructure, making it more connected, intelligent, and responsive. The integration of data, sensors, AI, and advanced analytics will revolutionize how we design, build, and manage our built environment from individual buildings to entire cities. This digital evolution promises to enhance efficiency, sustainability, and quality of life. As we move forward, the power of digital transformation will be evident in several areas. Data-driven decision-making and asset management will be at the forefront, allowing for real-time insights that improve resource utilization and prevent sudden failures. AI and advanced analytics will continuously refine infrastructure operations, predicting maintenance needs and optimizing performance. The path ahead promises a future where digital technology and asset management not only meet the demands of modern infrastructure but also adapt to the evolving needs of societies.

DIGITAL TECHNOLOGY

In 2050, digital technology will transform infrastructure, making it more connected, intelligent, and responsive. This transformation will impact every facet of our built environment, from individual buildings to entire cities, as we harness the power of data, sensors, AI, and advanced analytics to optimize efficiency, sustainability, and quality of life.

The Power of Digital Transformation

- **Data-Driven Decision-Making:** In 2050, infrastructure decisions will be guided by vast amounts of real-time data collected from sensors embedded in buildings, roads, bridges, and utilities. This data will empower stakeholders to make informed decisions that optimize the utilization of resources, reduce waste and sudden failures, and enhance asset performance.
- **AI and Advanced Analytics:** AI and advanced analytics will play pivotal roles in infrastructure optimization. AI-driven algorithms will continuously

DOI: 10.1201/9781003327509-25

analyze data to identify patterns, predict maintenance needs, and enhance resource allocation.

- **Enhanced User Experience:** Infrastructure will be more responsive to the needs and preferences of users. Smart buildings, for example, will adjust lighting, temperature, and security based on occupants' behavior and preferences, resulting in a more comfortable and energy-efficient environment.

Smart Buildings

Smart buildings will be on the frontline of infrastructure innovation. Equipped with sensors, these structures will continuously collect and analyze data on various aspects of their operation. This data will be used to enhance energy efficiency, comfort, security, and maintenance.

- **Energy Efficiency:** Sensors will monitor factors such as occupancy, temperature, and lighting conditions. AI algorithms will then adjust heating, cooling, and lighting systems in real time to minimize energy consumption, reducing both costs and environmental impact.
- **Comfort:** Smart buildings will personalize comfort for occupants. They will adapt lighting, heating, and cooling to individual preferences, creating an environment that promotes well-being and productivity.
- **Security:** Integrated security systems will employ AI to detect and respond to threats, such as unauthorized access or anomalies in building operations, ensuring the safety of occupants.
- **Maintenance:** Predictive maintenance powered by AI will continuously assess the condition of building components and systems. By anticipating maintenance needs and preventing breakdowns, this approach will reduce downtime and extend the lifespan of infrastructure.

Smart Grids

The energy sector will transform with smart grids. These grids will not only deliver electricity but will also intelligently balance supply and demand while integrating diverse energy sources, including renewables, into the mix.

- **Balancing Supply and Demand:** Smart grids will use real-time data on electricity usage and production to optimize distribution. During periods of high renewable energy production, excess power can be stored or routed to where it's needed most, ensuring a stable and efficient energy supply.
- **Renewable Energy Integration:** By seamlessly incorporating solar, wind, and other renewable sources, smart grids will accelerate the transition away from fossil fuels. This will significantly reduce greenhouse gas emissions, making energy production more sustainable.
- **Resilience:** Smart grids will enhance the resilience of energy infrastructure. In the event of disruptions, such as extreme weather or system failures, the grid can rapidly reroute power and restore service to affected areas.

Smart Cities

At the scale of cities, the concept of smart cities will redefine urban living. These cities will serve as integrated digital platforms that coordinate an array of services, from transportation to healthcare, education, and public safety, all aimed at improving the quality of life for residents.

- **Transportation:** Smart transportation systems will provide real-time data on traffic conditions, enabling efficient routing and reducing congestion. Autonomous vehicles will communicate with traffic infrastructure to enhance safety and efficiency.
- **Healthcare:** Digital health systems will facilitate remote monitoring and telemedicine, improving healthcare access and outcomes for urban populations.
- **Education:** Smart education platforms will provide personalized learning experiences, leveraging AI to tailor curriculum and resources to individual student needs.
- **Public Safety:** Advanced data analytics will enhance public safety by predicting and preventing accidents, crimes, and disasters.
- **Sustainability:** Smart cities will prioritize sustainability by optimizing resource usage, reducing waste, and lowering emissions through data-driven decision-making.

Real-World Example: Singapore's Smart Nation Initiative

Singapore's Smart Nation initiative is a tangible embodiment of the potential of digital technology in infrastructure (Smart Nation Programme Office 2016):

- **Data-Driven Urban Management:** Singapore employs data analytics and sensors to manage various aspects of urban life. Traffic flow is optimized through real-time data analysis, reducing congestion and travel times. Energy consumption is minimized through smart grid technologies, contributing to sustainability efforts.
- **Public Service Enhancement:** The initiative enhances the quality of life for residents. Public services such as healthcare and public transport are integrated into a seamless digital ecosystem, offering convenience and efficiency.
- **Sustainable Urban Living:** Digital technology enables Singapore to embrace sustainability. The city-state uses data to monitor environmental parameters and make informed decisions regarding conservation and resource management.

The integration of digital technology into infrastructure in 2050 promises to revolutionize our cities and the way we live, work, and move within them. It will not only enhance efficiency and convenience but also contribute to sustainability, resilience, and well-being of urban populations. As we embrace this digital revolution, our infrastructure will be equipped to meet the needs of a rapidly evolving world, ultimately resulting in a more connected, intelligent, and responsive future.

ASSET MANAGEMENT

In 2050, asset management will become crucial for maintaining efficiency, ensuring the longevity of assets, and promoting sustainability. As the demands for our infrastructure grow, the need to optimize its performance throughout its entire lifecycle becomes paramount. Asset management practices in 2050 will be characterized by cutting-edge technologies, AI, data-driven decision-making, and comprehensive insights into the true costs and benefits of infrastructure projects.

Digital Twins

These digital replicas of physical infrastructure assets will revolutionize how we monitor, maintain, and optimize their performance:

- **Real-Time Monitoring:** Digital twins will enable real-time monitoring of infrastructure assets, from bridges and water treatment plants to transportation networks and energy grids. These virtual representations will continuously collect and process data on asset conditions, usage patterns, and performance.
- **Proactive Maintenance:** With the aid of data analytics and machine learning, asset managers will anticipate maintenance needs before they become critical issues. Predictive maintenance algorithms will identify trends and anomalies in asset performance, enabling timely interventions to prevent failures and extend the lifespan of critical infrastructure.
- **Optimization:** Digital twins will empower asset managers to make data-driven decisions about the operation and maintenance of infrastructure. By simulating different scenarios and testing various interventions within the digital twin environment, asset managers can identify the most cost-effective strategies for optimizing asset performance.

Predictive Maintenance and Data-Driven Decision-Making

AI will transform asset management into a data-driven discipline, enabling more precise and informed decisions. By analyzing vast data sets generated by sensors and monitoring systems, AI algorithms will be able to:

- **Anticipate Failures:** AI algorithms will process real-time data from sensors embedded in infrastructure assets, providing up-to-the-minute insights into asset performance, condition, and usage. This will result in the detection of early warning signs of equipment or infrastructure asset degradation, allowing for proactive maintenance and minimizing costly downtime.
- **Optimize Maintenance Scheduling:** AI can assess risk factors associated with asset operation and propose mitigation strategies. This includes

identifying vulnerabilities to natural disasters, cyber threats, or equipment failures. In alignment with risk-based approach, they can optimize maintenance schedules by considering risks, asset conditions, usage patterns, and other factors, ensuring that maintenance interventions are conducted when they are most cost-effective.

Asset Lifecycle Planning

In 2050, infrastructure projects will be evaluated not only for their upfront costs but also for their long-term economic viability. Lifecycle costing will provide comprehensive insights into the total costs and benefits of infrastructure investments. AI will play a pivotal role in comprehensive asset lifecycle planning.

- **Asset Selection:** AI can assist in selecting the most suitable infrastructure assets for specific projects, factoring in their expected lifespan, maintenance requirements, and long-term costs.
- **Lifecycle Costing:** Lifecycle costing will also assess the broader benefits and returns generated by infrastructure projects, including economic, social, and environmental impacts. This holistic approach will guide decision-makers in selecting projects that offer the best long-term value to society.
- **Adaptive Planning:** AI-driven simulations and scenario analyses will help asset managers adapt to changing conditions, such as shifting demographics, climate change, or technological advancements.

Real-World Example: Barcelona's Water Supply Management

The city of Barcelona serves as a compelling real-world example of how advanced asset management practices are shaping the future of infrastructure. In Barcelona, digital twin technology has been harnessed to monitor the city's water supply and distribution systems in real time, resulting in numerous benefits including (Brears 2023):

- **Efficient Operations:** Real-time monitoring enables Barcelona to optimize its water supply operations, ensuring that water is delivered where and when it is needed most efficiently.
- **Proactive Maintenance:** Data analytics and predictive algorithms help identify potential issues before they escalate, allowing for proactive maintenance and minimizing costly disruptions.
- **Resource Conservation:** The ability to monitor water usage patterns and detect leaks promptly contributes to the conservation of water resources and reduces waste.

Asset management in 2050 will shift toward being data-driven, proactive, and sustainable. The infusion of AI, digital twins, and predictive maintenance into asset management practices will redefine how infrastructure assets are monitored, maintained, and optimized and ensure that infrastructure assets are not only built efficiently but also proactively managed throughout their lifecycles.

INNOVATION AND DISRUPTION

The future of infrastructure in 2050 will be tied to innovation and disruption, driven by a convergence of emerging technologies, novel business models, and societal changes. Innovations such as blockchain, biotechnology, nanotechnology, and quantum computing will reshape infrastructure, offering both opportunities and threats that enhance efficiency, security, quality, and resilience.

Blockchain Technology

Blockchain's impact on infrastructure will be profound for infrastructure projects and operations.

- **Project Management:** Blockchain will enhance transparency and traceability in project management by creating records of project milestones, expenditures, and contracts.
- **Smart Contracts:** Smart contracts on blockchain will automate payment processes, releasing funds when predefined conditions are met, and reducing delays and disputes in construction and procurement.
- **Supply Chain Management:** Blockchain will revolutionize supply chain management by providing real-time visibility into the movement of materials and equipment, reducing inefficiencies, and enhancing accountability.

Biotechnology and Nanotechnology

Biotechnology and nanotechnology will play crucial roles in sustainable infrastructure.

- **Biological Infrastructure:** Bioengineered materials and processes will enable the development of self-healing infrastructure materials and eco-friendly construction practices, reducing maintenance and environmental impact.
- **Nanomaterials:** Nanotechnology will produce advanced construction materials with superior strength, durability, and energy efficiency, revolutionizing building design and construction.

Quantum Computing

Quantum computing will tackle complex infrastructure challenges:

- **Simulation and Modeling:** Quantum computers will perform rapid simulations and modeling for complex infrastructure projects, optimizing designs, materials, and energy usage within a digital twin environment.
- **Cybersecurity:** Quantum-resistant cryptography will protect critical infrastructure systems from future cyber threats.

Disruptive Business Models

Innovative business models will redefine infrastructure delivery.

- **As-a-Service Models:** Infrastructure-as-a-Service (IaaS) and Platform-as-a-Service (PaaS) models will shift ownership and management responsibilities to service providers, reducing upfront costs and risks.
- **Sharing Economy:** The sharing economy will expand to include shared infrastructure assets, enabling efficient utilization of resources and reducing the need for new construction.

Real-World Examples: Blockchain in Infrastructure

Blockchain's potential in infrastructure is already being realized in areas like supply chain and infrastructure financing.

- **Transparent Supply Chain:** Walmart uses blockchain to track the source and journey of food products, ensuring food safety and traceability (Harvard Business Review 2022).
- **Infrastructure Financing:** Projects like the Brooklyn Microgrid in New York use blockchain to enable peer-to-peer energy trading among residents, reducing reliance on centralized power grids (Mengelkamp et al. 2018).

The interplay between innovation and disruption will shape our infrastructure in 2050. Emerging technologies, coupled with forward-thinking business models, will create new paradigms for infrastructure development and operation. Our ability to harness innovation while managing disruptions will be a key determinant of success in shaping infrastructure that is efficient, sustainable, resilient, and adaptive to the evolving needs of societies in the mid-21st century.

REFERENCES

Brears, R. C. (2023). "Optimising Water Resource Management: Smart Water Solutions and Success in Barcelona." *Mark and Focus*. https://medium.com/mark-and-focus/optimising-water-resource-management-smart-water-solutions-and-success-in-barcelona-637611941b0d. Aug 2, 2023.

Harvard Business Review. (2022). "How Walmart Canada Uses Blockchain to Solve Supply-Chain Challenges." https://hbr.org/2022/01/how-walmart-canada-uses-blockchain-to-solve-supply-chain-challenges. Aug 5, 2023.

Mengelkamp, E., Gärttner, J., Rock, K., Kessler, S., Orsini, L., and Weinhardt, C. (2018). "Designing microgrid energy markets: A case study: The Brooklyn Microgrid." *Applied Energy*, ScienceDirect, 210, 870–880.

Smart Nation Programme Office. (2016). "Smart Nation: The Way Forward." Singapore Government. https://www.smartnation.gov.sg/files/publications/smart-nation-strategy-nov2018.pdf. Aug 2, 2023.

20 Adapting Infrastructure for Inclusivity, Security, and Well-Being by 2050

INTRODUCTION

As we approach 2050, social and demographic shifts will influence how we design and develop infrastructure systems. Changes such as an aging population, rapid urbanization, the rise of the middle class, and increasing cultural diversification will demand infrastructure that is inclusive, accessible, and adaptable. For instance, the needs of an aging population will drive the creation of age-friendly urban spaces and mobility solutions, while urbanization will push for integrated transportation and affordable housing. The expanding middle class will increase demands for enhanced consumer services and digital connectivity, and cultural diversification will necessitate infrastructure that reflects and respects diverse backgrounds. Accessibility will remain a critical focus, ensuring that all individuals, regardless of ability or socioeconomic status, can benefit from essential services and opportunities. Furthermore, as infrastructure becomes more digital and interconnected, security and privacy will become crucial, requiring robust measures to protect against cyber threats and ensure data protection. Additionally, infrastructure will play a pivotal role in enhancing health and well-being by improving air quality, reducing noise pollution, and promoting access to green spaces and active transportation. These evolving social, demographic, and technological factors will guide the design and implementation of infrastructure to create equitable, efficient, and vibrant urban environments that promote safety, health, and inclusivity for all.

SOCIAL AND DEMOGRAPHIC CHANGES

In 2050, social and demographic changes will greatly influence the design, development, and operation of infrastructure systems. These shifts, including an aging population, urbanization, the rise of the middle class, and cultural diversification, will shape infrastructure in ways that prioritize inclusivity, affordability, accessibility, and diversity.

AGING POPULATION

The aging of the global population will bring about unique challenges and opportunities for infrastructure development.

DOI: 10.1201/9781003327509-26

- **Age-Friendly Infrastructure:** Cities worldwide will need to invest in age-friendly infrastructure that caters to the needs of older citizens. This includes universally accessible public transport, pedestrian-friendly urban planning, and healthcare facilities that address age-related health concerns.
- **Senior Housing:** There will be a growing demand for senior housing and care facilities that provide safe, accessible, and socially engaging environments for the elderly.
- **Mobility Solutions:** Transportation systems will need to accommodate reduced mobility, with features such as low-floor buses, ramps, and priority seating, ensuring that older adults can move freely and comfortably within cities.

Urbanization

Urbanization will keep shaping infrastructure development in 2050 in different ways.

- **Integrated Transportation:** Urban infrastructure will focus on integrated transportation solutions such as efficient public transit systems, pedestrian-friendly streets, and cycling infrastructure. These features will help reduce congestion, improve air quality, and enhance the quality of life in densely populated urban centers.
- **Affordable Housing:** As cities expand, the demand for affordable and inclusive housing will grow. Infrastructure projects will need to include provisions for affordable housing developments, promoting socioeconomic diversity within urban areas.
- **Smart Cities:** Urban infrastructure will integrate digital technologies to create smart cities that coordinate various services, including transportation, healthcare, education, and public safety. These digital platforms will enhance the urban living experience by improving efficiency and reducing the environmental footprint.

Rise of the Middle Class

The expanding middle class will reshape infrastructure requirements and priorities:

- **Consumer Demand:** As the middle-class population keeps growing, there will be greater demand for consumer goods and services. Infrastructure will need to support the transportation and logistics networks required for efficient supply chains and delivery.
- **Quality of Life:** The middle class will seek improved quality of life, driving infrastructure investments in areas like recreational facilities, cultural amenities, and green spaces within urban environments.
- **Digital Connectivity:** With an increasing middle-class population, there will be a surge in demand for high-speed internet access and digital services, prompting investments in digital infrastructure.

Cultural Diversification

Cultural diversification will encourage inclusive infrastructure designs, respect and celebrate cultural differences.

- **Inclusive Public Spaces:** Urban infrastructure will create inclusive public spaces that reflect diverse cultural backgrounds and foster community cohesion. This includes cultural centers, parks, sports facilities, public art facilities, and recreational spaces.
- **Multilingual Services:** Infrastructure services will offer multilingual support, making them more accessible and accommodating for culturally diverse populations.
- **Affordability and Safety:** Infrastructure services should remain affordable to ensure that lower income individuals and communities can access essential services. Infrastructure projects will prioritize safety, particularly in urban areas for all residents. Initiatives like "Complete Streets" will ensure that streets are safe for pedestrians, cyclists, and all road users.

Accessibility

Accessibility is a crucial aspect of modern infrastructure, ensuring that all individuals can benefit from essential services and opportunities. Infrastructure must prioritize universal accessibility, ensuring that people with disabilities can use public spaces, transportation, and facilities.

- **Mobility Options:** Infrastructure should offer diverse mobility options, including public transport, cycling lanes, and pedestrian-friendly streets, to accommodate varying transportation preferences.
- **Healthcare Access:** Accessibility to healthcare facilities, including remote and underserved areas, is vital for equitable healthcare provision.
- **Digital Inclusion:** Ensuring that digital infrastructure reaches underserved populations to bridge the digital divide is crucial for inclusive participation in the modern economy.

Real-World Examples: Tokyo's Age-Friendly Infrastructure and "Complete Streets" in the United States

Tokyo provides a real-world example of how a city is adapting its infrastructure to meet the needs of an aging population (World Bank 2022).

- **Universally Accessible Transport:** Tokyo's transportation system is designed to be universally accessible, with low-floor buses, accessible train stations, and priority seating for seniors and people with disabilities.
- **Age-Friendly Urban Planning:** The city's urban planning prioritizes walkability, green spaces, and access to healthcare facilities, creating an environment that supports the well-being of older residents.

- **Community Engagement:** Tokyo fosters community engagement among older citizens through cultural and recreational programs, creating social connections and enhancing the overall quality of life.

The "Complete Streets" program in the United States exemplifies inclusive infrastructure (Federal Highway Administration 2024).

- **Multi-Modal Design:** "Complete Streets" designs roads to accommodate various modes of transportation, including walking, cycling, and public transit. This promotes safety and accessibility for everyone.
- **Community Engagement:** The program involves community engagement to ensure that the infrastructure aligns with the specific needs and preferences of the local population.

The social and demographic shifts will be pivotal in shaping infrastructure in 2050. Infrastructure will adapt to be more inclusive and accessible to all regardless of age, ability, income, or background. This commitment to social equity will guide the design and implementation of comprehensive and integrated projects, fostering a more equitable society. By 2050, infrastructure will symbolize human progress and our dedication to leaving no one behind.

SECURITY AND PRIVACY

As technology progresses and our dependence on digital systems increases, prioritizing security and privacy becomes essential. Infrastructure systems will need to address vulnerabilities stemming from cyberattacks, physical threats, and natural hazards. Robust cybersecurity measures will safeguard digital infrastructure, while strategies to counter physical threats will protect critical assets. Resilience will be key, ensuring that infrastructure can maintain continuity during disasters.

Cybersecurity Resilience

Infrastructure will adopt advanced cybersecurity measures to protect against evolving cyber threats.

- **Threat Detection:** AI-driven threat detection systems and early warning systems will continuously monitor networks and systems for unusual activity, enabling proactive response to potential cyberattacks.
- **Zero Trust Security:** The Zero Trust model will become the norm, with infrastructure systems requiring verification of all users and devices attempting to access resources, reducing the attack surface.
- **Cyber Hygiene:** Cyber hygiene practices, including regular patching, software updates, and employee training, will be standard to prevent common cyber vulnerabilities.

Physical Threat Mitigation

Infrastructure will incorporate physical security measures to protect against threats such as terrorism and sabotage.

- **Critical Asset Protection:** Infrastructure operators will implement security measures to protect critical assets from physical threats, including surveillance, access control, and perimeter security.
- **Resilient Design:** Infrastructure will be designed with resilience in mind, incorporating redundancy, flexibility, adaptability, and smart material to withstand and recover from natural and man-made disasters such as floods, earthquakes, and wildfires.

Privacy and Data Protection

Privacy and data protection will be paramount, particularly in the context of smart infrastructure.

- **Data Encryption:** Infrastructure systems will implement robust data encryption techniques to protect user data and ensure privacy in an interconnected environment.
- **User Consent:** Strict regulations will govern data collection and usage, requiring explicit user consent for the gathering and utilization of personal information.

Real-World Example: U.S. Department of Homeland Security

The U.S. Department of Homeland Security (DHS) serves as a real-world example of a comprehensive strategy to safeguard critical infrastructure (Department of Homeland Security 2024).

- **Cybersecurity Initiatives:** The DHS collaborates with public and private sector partners to enhance cybersecurity for critical infrastructure, conducting assessments, providing guidance, and promoting best practices.
- **Physical Security:** The DHS offers guidance and support to protect physical infrastructure against terrorist threats, conducting risk assessments and offering training programs.

Security and privacy will be paramount for the infrastructure of 2050. As technology advances and infrastructure becomes more digital, strong cybersecurity will be essential to protect against online threats. At the same time, infrastructure must be designed to withstand physical risks and ensure it remains functional during emergencies. Privacy will also be key to maintaining public trust, especially as data becomes more central to how infrastructure operates.

HEALTH AND WELL-BEING

In 2050, the health and well-being of communities will be a paramount consideration in the infrastructure design and planning to redefine how communities interact with the built environments. Infrastructure projects will extend beyond their conventional roles, actively contributing to improving air and water quality, reducing noise pollution, and enhancing the overall quality of life for residents.

Air Quality and Health

As cities and infrastructure evolve, improving air quality will be a key focus, with projects aimed at reducing emissions and creating green spaces to enhance public health.

- **Emission Reduction:** Infrastructure projects will prioritize reducing emissions from transportation and industry, leading to improved air quality, and reduced respiratory illnesses.
- **Green Spaces:** Urban areas will be transformed with the creation of green spaces, parks, and urban forests that not only enhance aesthetics but also act as natural air filters, improving overall air quality.

Noise Reduction

Noise reduction will become a priority in urban planning, with infrastructure projects integrating acoustic designs to minimize noise pollution.

- **Acoustic Design:** Urban infrastructure will incorporate noise-reduction measures, from quieter road surfaces to the strategic placement of noise barriers, reducing noise pollution that negatively impacts mental health.

Mental and Physical Well-Being

Enhancing mental and physical well-being will be central to infrastructure development, with a focus on increasing access to natural spaces and supporting active transportation to foster healthier lifestyles and improved quality of life.

- **Access to Nature:** Infrastructure will promote access to natural spaces, positively impacting mental health. Green corridors, urban gardens, and park networks will provide residents with opportunities for relaxation and recreation.
- **Active Transportation:** Infrastructure will support active transportation modes such as walking and cycling, promoting physical fitness and well-being.

Real-World Example: Greening of Urban Spaces: Singapore

Singapore is a great example of a city that has successfully implemented greening of urban spaces. The city has integrated extensive green spaces, such as parks, gardens, and green roofs, into its urban planning (World Economic Forum 2022).

- **Mental Health Benefits:** Access to green spaces in Singapore helps reduce stress, anxiety, and depression and improves well-being among residents.
- **Physical Health Benefits:** Green spaces in Singapore encouraged residents to do physical activities and resulted in a reduction of lifestyle-related diseases such as obesity and diabetes.

Health and well-being will be key in shaping infrastructure in 2050. Infrastructure projects will go beyond serving functional roles; they will become integral to enhancing the overall quality of life for communities. As infrastructure design incorporates health and well-being principles, the built environment will play a pivotal role in promoting healthier, happier, and more vibrant societies.

REFERENCES

Department of Homeland Security. (2024). "Strategic Guidance and National Priorities for U.S. Critical Infrastructure Security and Resilience." https://www.dhs.gov/publication/strategic-guidance-and-national-priorities-us-critical-infrastructure-security-and. Jul 10, 2024.

Federal Highway Administration. (2024). "Complete Streets." https://highways.dot.gov/complete-streets. Jul 10, 2024.

World Bank. (2022). "Rethinking Silver: Lessons from Japan's Age-Ready Cities." https://blogs.worldbank.org/en/voices/rethinking-silver-lessons-japans-age-ready-cities. Aug 8, 2023.

World Economic Forum. (2022). "How to Reap the Benefits of Urban Greening." https://www.weforum.org/agenda/2022/02/urban-greening-extreme-weather-climate-change-cities/. Aug 20, 2023.

21 Governance, Stakeholder Engagement, and Circular Economy Principles in 2050 Infrastructure

INTRODUCTION

By 2050, governance and regulation will be pivotal in sculpting the landscape of infrastructure development, ensuring it is both effective and equitable. The intricate interplay between public and private sectors will become more pronounced, with governance structures evolving to accommodate varying levels of public-private involvement. In some regions, the public sector's role will remain central, particularly for essential services such as water supply and national defense, ensuring widespread access and equity despite the necessity for substantial public funding. Conversely, PPPs will increasingly drive innovation and efficiency by harnessing private sector expertise and capital. The challenge for governance will be to balance these roles effectively, defining clear responsibilities, sharing risks, and safeguarding public interests to foster a collaborative environment conducive to sustainable infrastructure development.

The future of governance will require balancing decentralization with centralized coordination. Local decision-making will enable communities to address specific needs, while centralized planning will ensure alignment with national objectives. Effective coordination among agencies and transparent procurement processes will be essential for fairness and efficiency. Rigorous safety and environmental regulations will safeguard public welfare and encourage innovation. Additionally, integrating circular economy principles into governance will be crucial. This will involve minimizing waste and maximizing resource efficiency through practices like using recycled materials, adopting zero-waste methods, and designing for longevity. Embedding these principles into regulatory standards and procurement processes will not only tackle environmental challenges but also enhance economic viability, driving a shift toward sustainable infrastructure that meets the needs of society.

GOVERNANCE AND REGULATION

In 2050, governance and regulations will serve as the cornerstone upon which efficient, effective, and equitable infrastructure development and operation will rest.

DOI: 10.1201/9781003327509-27

Policies and institutions governing infrastructure will play a pivotal role in shaping the future of our infrastructure.

Level of Public-Private Sector Involvement

Collaboration between the public and private sectors will be central to infrastructure development.

- **Public Sector Dominance:** In some regions or for certain types of infrastructure (e.g., water supply, defense), the public sector may remain dominant. This can ensure equity and access for all citizens, but it may require significant public funding.
- **Public-Private Partnerships (PPPs):** PPPs can promote innovation and efficiency by involving private sector expertise and capital. However, the challenge lies in finding the right balance to ensure that the public interest is safeguarded, and equity is maintained.
- **Risk Allocation:** Effective governance will involve clearly defined roles and responsibilities for public and private partners, as well as mechanisms for risk sharing and mitigation.

Decentralization

Balancing decentralization with centralized coordination will be critical.

- **Local Decision-Making:** Decentralization of infrastructure decision-making can lead to more tailored solutions that meet local needs. Local governments will play an active role in infrastructure development, aligning projects with their communities' unique needs and priorities, and coordination to align local efforts with broader national objectives.
- **Centralized Planning:** Effective governance will ensure centralized coordination to harmonize infrastructure efforts, promoting regional connectivity and optimizing resource allocation.
- **National Coordination:** A centralized approach can ensure consistency and alignment with national priorities but may overlook local nuances.

Coordination

Effective infrastructure management requires both inter-agency coordination to prevent duplication of efforts and resource wastage, and regional collaboration to tackle transboundary projects and shared challenges.

- **Inter-Agency Coordination:** Effective coordination among various government agencies responsible for different infrastructure systems (e.g., transportation, wastewater and sewer, energy, and water) is crucial to avoid duplication of efforts and resource wastage.
- **Regional Collaboration:** Regional collaboration can be important for transboundary projects and to address shared infrastructure challenges.

Quality of Planning and Procurement

Governance structures will prioritize high-quality planning and procurement processes.

- **Transparent Procurement:** Transparent and competitive procurement processes will become the norm. These processes will ensure fairness, efficiency, and cost-effectiveness in infrastructure development.
- **Strategic Planning:** Long-term strategic planning will guide infrastructure investments, considering factors such as long-term needs, environmental sustainability, social inclusion, and economic growth. It ensures that projects are well-conceived and meet future demands.
- **Efficient Procurement:** Transparent and competitive procurement processes can lead to cost-effective infrastructure development by attracting qualified bidders and reducing corruption risks.

Enforcement of Standards and Regulations

Governance and regulations will ensure adherence to standards and regulations.

- **Safety and Environmental Compliance:** Strict enforcement of safety and environmental standards ensures that infrastructure is designed, constructed, and operated with the highest standards, reducing risks to the public and minimizing negative impact of infrastructure projects on the environment.
- **Innovation-Friendly Regulations:** Forward-thinking regulations will encourage innovation in infrastructure technologies, materials, and construction methods while maintaining safety and performance standards.

Transparency and Accountability

Transparency in governance and accountability of public officials and project developers fosters public trust and guarantees the efficient and effective execution of infrastructure projects.

- **Transparency:** Transparent governance processes and accessible information about infrastructure projects build public trust and confidence in the decision-making process.
- **Accountability:** Holding public officials and project developers accountable for their actions ensures that infrastructure projects are carried out efficiently and effectively.

Real-World Example: Germany's Procurement Excellence

Germany's approach to transparent and competitive procurement processes serves as a great real-world example of how governance can impact infrastructure development (Essig et al. 2009).

- **Efficiency:** Transparent and competitive procurement processes have earned Germany a reputation for efficiency and cost-effectiveness in public infrastructure development.
- **Competition:** Competitive bidding ensures that the most qualified and cost-effective contractors are selected for projects, promoting value for money.
- **Innovation:** Germany's governance framework encourages innovation in infrastructure construction and design while ensuring that projects adhere to the highest quality and safety standards.

By 2050, governance and regulation will be key in shaping infrastructure. They will foster public-private collaboration, guide decentralization, ensure transparent procurement, and uphold high standards. Efficient governance and regulation will create a sustainable and safe infrastructure that meets society's needs.

FINANCING, FUNDING, AND COLLABORATION

The future of infrastructure in 2050 will be highly influenced by the mechanisms and sources of financing and funding. Large-scale infrastructure projects will demand substantial capital from a diverse range of sources, including governments, the private sector, multilateral institutions, and civil society. The availability, affordability, and sustainability of financing and funding will play a pivotal role in determining the feasibility and viability of these projects.

Public-Private Collaboration

Public-private partnerships (PPPs) and risk mitigation strategies will be crucial in infrastructure financing, leveraging private sector expertise while ensuring investor confidence through government guarantees and insurance.

- **Public-Private Partnerships (PPPs):** PPPs will remain central to infrastructure financing, leveraging private sector expertise and capital to develop and operate public assets. Governments will increasingly seek private sector participation to share risks and responsibilities.
- **Risk Mitigation:** Governments and multilateral institutions will play a key role in mitigating risks for private investors, offering guarantees and insurance to make infrastructure projects more attractive.

Multilateral Institutions

International funding from multilateral institutions and a strong sustainability focus will drive infrastructure projects, especially in developing countries, aligning investments with climate and sustainability goals.

- **International Funding:** Multilateral institutions like the World Bank, Asian Development Bank, and others will continue to provide financing and technical assistance for infrastructure projects, particularly in developing countries.

- **Sustainability Focus:** Multilateral institutions will increasingly prioritize environmentally sustainable infrastructure projects, aligning their funding with climate and sustainability goals.

Sustainable Finance

Green bonds and social impact bonds will play pivotal roles in financing sustainable and socially beneficial infrastructure projects, reflecting strong investor interest and addressing critical social challenges.

- **Green Bonds:** The use of green bonds will surge, mobilizing capital for environmentally sustainable infrastructure projects. The market value of green bonds has already surpassed $1 trillion, demonstrating strong investor interest in sustainable infrastructure (Climate Bonds Initiative 2020).
- **Social Impact Bonds:** Social impact bonds will gain prominence, channeling funding into projects that address social challenges, such as affordable housing and healthcare infrastructure.

Technology-Driven Financing

Blockchain-based financing and crowdfunding will revolutionize infrastructure investment, enabling broader investor participation and democratizing capital contributions.

- **Blockchain-Based Financing:** Blockchain technology will enable innovative financing mechanisms, such as tokenized securities and digital asset exchanges, making it easier for a broader range of investors to participate in infrastructure projects.
- **Crowdfunding:** Crowdfunding platforms will democratize infrastructure investment, allowing individuals to contribute small amounts of capital to projects they support.

Innovation Ecosystems

- **Research and Development Collaborations:** Public and private entities will collaborate on research and development efforts to drive innovation in infrastructure technologies and practices. These collaborations will accelerate the adoption of cutting-edge solutions.
- **Technology Transfer:** International partnerships will facilitate the transfer of technology and knowledge, enabling developing nations to leapfrog traditional infrastructure development stages.

Real-World Examples: Green Bonds Market and The World Bank's Global Infrastructure Facility (GIF)

The green bonds market and World Bank's GIF serve as great examples for sustainable finance in infrastructure and collaboration (Climate Bonds Initiative 2020; World Bank 2024).

- **Market Growth with Public and Private Sectors:** The green bonds market has witnessed exponential growth, with the market value exceeding $1 trillion in recent years. These bonds raise capital specifically for projects with positive environmental impacts. GIF brings together public and private sector stakeholders to develop sustainable infrastructure projects, aligning their interests and resources.
- **Investor Demand:** Institutional investors, as well as individual investors, are increasingly prioritizing green bonds due to their alignment with environmental sustainability goals. The World Bank assists in mitigating risks associated with infrastructure investments, making projects more attractive to private sector investors.

By 2050, financing, funding, and collaboration will be at the heart of shaping the future of our infrastructure. As infrastructure projects grow in complexity and scale, the collective efforts of governments, businesses, communities, and international organizations will be essential for their success. Public-private collaboration, support from multilateral institutions, and sustainable financing like green bonds will be key. Innovative approaches will ensure infrastructure is economically viable, environmentally friendly, and socially responsible, improving quality of life. Top of Form

STAKEHOLDER ENGAGEMENT AND PARTICIPATION

The future of infrastructure in 2050 will be significantly shaped by the degree and quality of stakeholder engagement and participation. The involvement of diverse stakeholders, including communities, individuals, and organizations, will have a profound impact on infrastructure project design, implementation, and evaluation. Robust stakeholder engagement processes will be instrumental in enhancing project acceptance, legitimacy, accountability, and overall success.

Inclusive Decision-Making

Infrastructure planning and development will involve extensive community consultation and robust citizen feedback mechanisms to ensure local voices are heard and incorporated into project design, with public meetings, surveys, and forums playing a crucial role.

- **Community Consultation:** Infrastructure planning and development will involve extensive community consultation, ensuring that local voices are heard and incorporated into project design. Public meetings, surveys, and forums will be essential in gathering input and feedback.
- **Citizen Feedback Mechanisms:** Robust mechanisms for citizen feedback will become standard practice, allowing individuals to provide ongoing input and hold infrastructure providers accountable for performance.

Co-Design and Co-Creation

Infrastructure projects will increasingly involve stakeholders and communities in the design and co-creation process to ensure they meet local needs and foster a sense of ownership.

- **Collaborative Design:** Infrastructure projects will increasingly embrace co-design principles, involving stakeholders in the design process from inception. This approach will ensure that infrastructure meets the unique needs and preferences of the communities it serves.
- **Co-Creation of Solutions:** Communities and end-users will actively participate in the co-creation of infrastructure solutions, fostering a sense of ownership and responsibility.

Transparent Decision-Making

Infrastructure providers will ensure transparency and accountability by making project data accessible to the public and regularly reporting on progress, budgets, and outcomes.

- **Open Data and Information Sharing:** Infrastructure providers will make project information and data accessible to the public, promoting transparency and informed decision-making.
- **Public Reporting:** Regular reporting on project progress, budgets, and outcomes will be a standard practice, ensuring accountability and public trust.

Real-World Example: "Bristol Approach" in the UK

The "Bristol Approach" in the UK serves as a real-world example of comprehensive stakeholder engagement (KWMC 2024).

- **Community Participation:** Bristol's approach involves extensive engagement with local communities, allowing residents to shape the development of infrastructure projects.
- **Coordinated Planning:** The city actively collaborates with communities, local authorities, and organizations to co-design solutions that align with community needs and aspirations.

Stakeholder engagement and active participation will be pivotal in shaping the infrastructure landscape of 2050. In an era of increasing complexity and interconnectedness, involving those who are directly impacted by infrastructure projects is not only essential for project success but also for fostering trust and ensuring that infrastructure serves the best interests of society. By focusing on inclusivity, transparency, and collaboration, future infrastructure can reflect diverse community aspirations and promote a fair and sustainable future.

CIRCULAR ECONOMY

By 2050, infrastructure will undergo a major transformation through the adoption of circular economy principles. At its core, the circular economy model emphasizes the minimization of waste and the maximization of resource efficiency to substantially reduce the environmental impact of the built environment.

Sustainable Materials and Waste Reduction Practices

Infrastructure projects will increasingly use recycled and reused materials, reduce construction waste through innovative methods, and aim for zero waste by adopting recycling and upcycling practices.

- **Recycled and Reused Materials:** Infrastructure projects will increasingly incorporate recycled and reused materials, reducing the demand for new resources. Materials like reclaimed timber, recycled plastics, and repurposed metals will become standard.
- **Construction Waste Reduction:** Circular economy principles will drive significant reductions in construction waste. Innovative construction methods, such as modular construction, will minimize on-site waste, and surplus materials will be repurposed or recycled.
- **Zero-Waste Goals:** Infrastructure projects will aim for zero waste generation, diverting waste from landfills and adopting recycling and upcycling practices.

Longevity and Durability

Infrastructure will be designed for longevity using low-maintenance materials, reducing the need for frequent replacements and minimizing resource consumption and upkeep.

- **Design for Longevity:** Infrastructure will be designed with a focus on longevity, ensuring that structures have extended lifespans. This reduces the need for frequent replacements and minimizes resource consumption.
- **Low-Maintenance Materials:** Low-maintenance materials will be favored, reducing the energy and resources required for ongoing upkeep and repairs.

Real-World Example: The Use of Recycled Plastics in Roads: The Plastic Man of India

The use of recycled plastics in road construction is an exemplary application of circular economy principles in infrastructure. An Indian scientist developed an innovative method to use plastic waste in road construction through mixing shredded plastic with bitumen resulting in more durable and cost-effective roads. This technique not only helps in managing plastic waste but also enhances the quality of roads, making them more resistant to damage from heavy rains (Government of India 2019). India has around 40,000 km of rural roads constructed using plastic waste across India (TOI News Desk 2024).

- **Resource Conservation:** Recycled plastics, such as plastic bottles and bags, are transformed into a durable and flexible material that can be used in road surfaces. This conserves valuable resources and diverts plastic waste from landfills and oceans.
- **Reduced Environmental Impact:** The use of recycled plastics in road construction reduces the carbon footprint of the infrastructure sector and contributes to sustainability goals.

The circular economy will be a driving force in shaping infrastructure in 2050. By minimizing waste, maximizing resource efficiency, and prioritizing sustainable materials and practices, the infrastructure of the future will not only be environmentally responsible but also economically viable. Circular economy principles will enable infrastructure to meet the needs of growing populations while mitigating environmental challenges, ensuring that our built environment aligns with the broader goals of sustainability and resource conservation.

CONCLUDING REMARKS

In conclusion, infrastructure stands as the bedrock upon of our modern society, forming the essential framework that connects communities, supports industries, and drives progress. Whether in the form of the roads that link us, the buildings that shelter us, or the energy systems that power us, infrastructure is the backbone of our civilization.

As we look ahead to 2050, we stand on the brink of a transformative era, one that promises profound changes and demands unprecedented innovation. The convergence of cutting-edge technologies, ecological imperatives, and evolving demographics is reshaping our understanding of infrastructure and its role. This book has presented a coordination and optimization framework to effectively manage the co-located infrastructure systems. It has also explored the future landscape of infrastructure, revealing a complex interplay of smart cities, renewable energy, transportation revolutions, and sustainable design, all vital to creating a more efficient and harmonious infrastructure for 2050.

The challenges ahead are significant, from reimagining cities to accommodate growing populations and minimizing environmental impact, to transitioning from fossil fuels to renewable energy, and developing transformative transportation systems. We face a profound need for resilient infrastructure that can withstand the uncertainties of climate change while addressing the intricate balance of funding, policy, and innovation.

The key facets shaping infrastructure in 2050 are like a complex symphony, where technology, sustainability, governance, societal change, innovation, security, financing, stakeholder engagement, resilience, circular economy principles, climate adaptation, collaboration, inclusivity, and health and well-being all play interconnected roles. Together, they shape the future of infrastructure, leading us toward a sustainable, adaptable, and inclusive future.

Ultimately, this journey into the future of infrastructure is not just about what might be; it shows our potential to create a world where infrastructure symbolizes human progress and hope for future generations. As we move into the unknown of 2050, let's carry the lessons learned and aim for a future of infrastructure that is brighter, more sustainable, and more harmonious than we can imagine today.

REFERENCES

Knowle West Media Centre (KWMC). (2024). "The Bristol Approach." The Bristol Approach - KWMC. Dec 23, 2024.

Climate Bonds Initiative. (2020). "$1 Trillion Mark Reached in Global Cumulative Green Issuance." https://kwmc.org.uk/projects/bristolapproach/. Aug 21, 2023.

Essig, M., Dorobek, S., Glas, A., and Leuger, S. (2009). "Public Procurement in Germany." In: Khi V. Thai (ed.). *International Handbook of Public Procurement* (pp. 253–270). Boca Raton, FL: Taylor & Francis.

Government of India. (2019). "Guidelines for the Use of Plastic Waste in Road Construction." *Indian Roads Congress*. https://www.scribd.com/document/556289926/WKS-G-16. Aug 20, 2023.

TOI News Desk. (2024). "ODF Plus Model': 40,000 km of rural roads constructed using plastic waste." *The Times of India*. https://timesofindia.indiatimes.com/india/odf-plus-model-40000-km-of-rural-roads-constructed-using-plastic-waste/articleshow/113869779.cms. Dec 23, 2024.

World Bank. (2024). "Global Infrastructure Facility (GIF)." https://fiftrustee.worldbank.org/en/about/unit/dfi/fiftrustee/fund-detail/gif. Jul 20, 2024.

Glossary

Accuracy: The precision level needed while measuring an indicator and the difficulty level, represented either by the frequency or easiness of measurement.

Action Level: The level at which specific actions or interventions are planned and executed.

Active Transportation: Infrastructure prioritizing walking and cycling, including dedicated lanes, pedestrian-friendly streets, and bike-sharing programs.

Age-Friendly Infrastructure: Infrastructure designed to cater to the needs of older citizens, including universally accessible public transport, pedestrian-friendly urban planning, and healthcare facilities that address age-related health concerns.

Aging Infrastructure: Many infrastructure systems are old and need repair or replacement, making them less reliable, more expensive to operate, and more vulnerable to natural disasters.

Aging Population: Refers to the increasing number of older individuals in the global population, which brings unique challenges and opportunities for infrastructure development.

Artificial Neural Networks (ANN): A model-free method that uses historical maintenance records to identify internal relationships among decision factors and their impacts. It is used in data-driven approaches for deterioration and failure models.

Asset Inventory: A comprehensive database containing details about the corridors, including road width, traffic data, and pipe specifications.

Asset Management Specialists: Professionals appointed to manage each asset type due to differences in how co-located assets evolve throughout their lifecycles.

Autonomous Vehicles: Often referred to as self-driving cars, these vehicles are equipped with sensors, cameras, and advanced artificial intelligence (AI) systems, enabling them to navigate without human intervention.

Bidding Process: The process in which the proponent with the highest score is awarded the contract.

Binary Coding Rules: Rules used in optimization models to represent constraints as binary values (0 or 1), indicating whether a constraint is met or not.

Biotechnology and Nanotechnology: Fields that contribute to sustainable infrastructure through bioengineered materials, self-healing infrastructure materials, and advanced construction materials with superior strength and energy efficiency.

Block Funding: The approval of budgets on a program level for various infrastructure services, allowing for flexibility in changing priorities among individual projects.

Blockchain-Based Financing: Innovative financing mechanisms enabled by blockchain technology, allowing broader investor participation in infrastructure projects.

Blockchain Technology: Technology that enhances transparency and traceability in project management, automates payment processes through smart contracts, and revolutionizes supply chain management.

Budget: This is the financial plan that outlines the expected income and expenditure for a specific period. It is crucial for managing resources and ensuring that funds are allocated efficiently to meet the organization's goals.

Carbon Capture and Storage (CCS): Technologies aimed at capturing and storing carbon emissions from industries that are hard to decarbonize.

Centralized Planning: Coordination at a national level to harmonize infrastructure efforts, promoting regional connectivity and optimizing resource allocation.

CIF (Condition Improvement Factor): A metric used to compare the condition improvement of different intervention scenarios with the conventional one.

Circular Economy: An economic model that emphasizes the minimization of waste and the maximization of resource efficiency, particularly in infrastructure development.

Circular Economy Principles: Designing infrastructure with materials that can be reused and recycled, reducing waste and environmental impact.

Circular Economy Principles: Principles guiding the lifecycle of infrastructure projects, focusing on reducing waste, reusing and recycling materials, and maximizing resource efficiency.

Climate Resilience: Infrastructure incorporating measures to combat rising sea levels and storm surges and designed to adapt to changing conditions.

Climate-Resilient Infrastructure: Engineering practices that incorporate climate resilience, including sea level rise protections, stormwater management, and drought-resistant designs.

Co-Design: Collaborative design processes involving stakeholders from the inception of infrastructure projects to ensure they meet community needs.

Combined Sewer: A sewer system that carries both sewage and stormwater.

Condition: This term describes the current state or quality of an asset. It is often assessed through inspections and evaluations to determine the need for maintenance or replacement.

Condition KPI: A metric used to track the condition state or reliability of each asset or system within a corridor. It is computed using an integrated deterioration model that accounts for various factors such as aging, traffic, weather, maintenance, and extreme events.

Condition State: The current state of an asset, such as roads or pipes, based on various factors like age and physical characteristics.

Consequences of Failure (COF): The impact or outcome resulting from the failure of a system or component, which can include economic, operational, social, and environmental effects.

Consortium: A group of municipal departments that work together to define common PBC asset-based performance metrics, identify asset-based thresholds, set the annual intervention budget, specify inspection frequency and techniques, classify asset interdependencies, outline performance metrics and their associated penalties and incentives, and determine the optimal contractual period.

Contract: A legally binding agreement between two parties, usually a government agency and a contractor, that outlines the terms and conditions for maintaining infrastructure such as roads, water, and sewer pipes.

Coordination Benefits: The advantages gained from coordinating infrastructure renewal programs, including financial savings, reduced disruptions, and improved decision-making.

Coordination Framework: A structured approach to managing and planning the interdependencies among various municipal infrastructure systems to enhance efficiency and reduce disruptions.

Coordination Level: The level at which different systems or activities are coordinated to achieve optimal performance.

Coordination Scenarios: Different scenarios in which maintenance and intervention actions are coordinated among various networks (roads, water, sewer) to optimize performance and reduce costs.

Corridor Condition State (CCS): A measure of the overall condition of all systems within a corridor, compiled based on the weights of the importance of each system.

Corridor Priority Index (CPI): An index that represents the urgency of undertaking an intervention based on the key performance indicators (KPIs) such as time, space, cost, efficiency, effectiveness, condition, resilience preparedness, and risk.

Cost Reimbursable Contracts: Contracts that involve payments to contractors for all legitimate actual costs incurred for completed work, plus a fee representing the contractor's profit. Variations include cost plus fixed fee (CPFF), cost plus incentive fee (CPIF), and cost plus award fee (CPAF).

Cultural Diversification: The inclusion and celebration of diverse cultural backgrounds in infrastructure design, including inclusive public spaces and multilingual services.

Cyber Interdependency: A type of interdependency where the state of one infrastructure system depends on information transmitted through the information infrastructure.

Cybersecurity: With the increasing dependence on digitalization and automation in various critical infrastructure, the risk of cyber-attacks is getting higher. For instance, cyber-attacks on traffic systems or autonomous vehicles can cause fatal accidents.

Cybersecurity Resilience: Advanced cybersecurity measures adopted to protect against evolving cyber threats, including AI-driven threat detection systems and the Zero Trust security model.

Decentralization: The distribution of decision-making authority to local governments to create tailored solutions that meet local needs.

Decision-Making: The process of making choices by identifying a decision, gathering information, and assessing alternative resolutions. It is essential for effective and proactive management of public municipal infrastructure.

Decision-Making Framework: A systematic process for making informed decisions regarding the management and coordination of interdependent infrastructure systems.

Delivery Drones: Drones and autonomous delivery robots that will revolutionize the delivery of goods, requiring infrastructure like drone landing pads and storage facilities.

Demand Capacity Ratio: A measure of the current and future flow demand over a pipe's capacity.

Demand Management Gap: The need to better manage the demand for infrastructure services through behavioral change and new technologies. For example, using motion-sensor detectors to switch lights on/off based on people's movement.

Deviational Variables: Variables that represent the deviation from the desired performance levels or thresholds in an optimization model.

Digital Twins: Digital replicas of physical infrastructure assets that enable real-time monitoring, maintenance, and optimization of their performance.

Direct Costs: These represent the costs of the intervention activities needed to be undertaken throughout the planning horizon to deliver the services in an "acceptable" manner without interruption.

Disaster-Resilient Design: Infrastructure projects prioritizing resilience against natural disasters like earthquakes, floods, and wildfires.

Disruption Time (DT): The time to undertake the intervention.

Dynamic Programming: A method used to solve complex problems by breaking them down into simpler subproblems. It is used in multi-objective optimization to address extended planning horizons by dividing them into smaller segments.

Early Warning Systems: Advanced monitoring systems providing real-time alerts for proactive responses to disasters.

Econometric Approaches: Methods that use economic data and statistical techniques to study infrastructure interdependencies, such as input-output models.

Efficiency Gap: The need for managing infrastructure more cost-efficiently. For instance, preventive maintenance for roads saves ten times more compared to rehabilitation.

Electric Charging Infrastructure: A dense network of EV charging stations required for the widespread adoption of electric vehicles.

Environmental Sustainability: Many countries' infrastructure still lacks the capacity to incorporate greener and more sustainable technologies, making it harder to reduce emissions and combat climate change. Examples include electric/hybrid vehicles and solar panel roads.

Excavation of Entrance and Exit Pits: The process of digging entrance and exit pits for trenchless rehabilitation of water and sewer systems.

Financial: The costs needed for frequently measuring and controlling the asset.

Financial KPI: This KPI calculates the intervention's direct and indirect costs and quantifies the lifecycle costs improvement arising from coordinating the interventions versus undertaking them in silos.

Financial Penalties and Incentives (P/I): Monetary values or contractual period extensions used to encourage or discourage certain behaviors or performance levels in a contract.

Fixed Price Contracts: Contracts that involve setting a fixed total price for a defined scope of work. They are used when the scope of work is well-defined, and no significant changes are expected during execution. Variations include firm fixed price (FFP), fixed price with economic price adjustment (FPEPA), and fixed price incentive fee (FPIF).

Frequency: The rate at which an indicator tracks the assets' performance throughout the planning horizon.

Functional Interdependency: The operation of one infrastructure system is necessary for the operation of another infrastructure system.

Genetic Algorithms (GAs): A search heuristic that mimics the process of natural selection to generate high-quality solutions for optimization and search problems. It is used in multi-objective optimization models for prioritizing corridors for renewal, rehabilitation, and preventive maintenance.

Geographical Interdependency: A type of interdependency where a local environmental event can cause state changes in two or more infrastructure systems.

Goal Optimization: A technique used to balance multiple competing objectives by combining all weighted deviations from thresholds along with their relative weights to form an overall deviational goal.

Governance and Regulations: The framework of policies and institutions that guide infrastructure development and operation, ensuring efficiency, effectiveness, and equity.

Green Bonds: Financial instruments that raise capital specifically for environmentally sustainable infrastructure projects.

Green Building Standards: Standards that architects and builders adhere to, using materials that reduce carbon emissions and designing structures that optimize energy use.

Green Infrastructure: Infrastructure that harnesses natural systems like wetlands, forests, and green roofs to provide ecosystem services, enhancing urban living.

Green Spaces: Urban areas transformed with parks, urban forests, and green corridors that enhance aesthetics, act as natural air filters, and improve overall air quality.

Grid Modernization: Upgrading electric grids to accommodate distributed energy sources, allowing for efficient integration of renewables and reducing energy wastage.

Hyperloop: A high-speed transportation system using vacuum tubes to transport pods at near-supersonic speeds, significantly reducing travel times between cities.

IEF (Intervention Efficiency Factor): Proxy for the nuisance caused by repeated interventions executed for the same corridor within a short time span. A metric to assess the overall efficiency of the intervention program for all systems in terms of inter-disruption time extent.

IFF (Intervention Effectiveness Factor): Proxy of time without any disruption once the intervention program is completed. A metric to compare the interventions' effectiveness of the partially coordinated and fully coordinated intervention scenarios with the conventional one.

Inadequate Capacity: Some infrastructure systems may not have enough capacity to meet the needs of a growing population or economy. For example, water systems may not have enough capacity to meet the needs of a growing community.

Incentives and Penalties: Financial rewards or penalties tied to the achievement or failure to meet performance metrics under a PBC.

Inclusive Infrastructure: Infrastructure that ensures the benefits of projects reach all members of society, with a focus on marginalized and vulnerable populations, emphasizing accessibility, affordability, and safety.

Indirect Costs: Sometimes referred to as "social" or "user" costs, these reflect all the costs that are not directly related to the intervention.

Inequalities: Infrastructure gaps can disproportionately affect low-income communities and communities of color, who may have less access to transportation, clean water, and other essential resources.

Infrastructure Interdependency: The mutual reliance between infrastructure systems, where the state or operation of one system affects or is affected by another system.

Infrastructure Systems: The fundamental facilities and systems serving a country, city, or area, including transportation, communication, sewage, water, and electric systems.

Infrastructure-as-a-Service (IaaS): Business model that shifts ownership and management responsibilities of infrastructure to service providers, reducing upfront costs and risks.

Innovation Ecosystems: Collaborative environments where public and private entities work together on research and development to drive innovation in infrastructure.

Installation of Sewer Manholes: The process of installing manholes for the sewer system.

Integrated Transportation: Urban infrastructure that emphasizes efficient public transit systems, pedestrian-friendly streets, and cycling infrastructure to reduce congestion and improve air quality.

Interdependent Assets: Infrastructure components that rely on each other for optimal functionality, such as roads, water, and sewer networks.

Inter-Disruption Time (IDT): The time between one intervention and another.

Intervention Activities: Specific tasks or actions undertaken during the intervention process, categorized by their applicability to be undertaken in parallel with other activities and their units of measurement.

Intervention Effectiveness: The performance or quality of the intervention, reflecting the amount of operating time expected until the next major intervention without undertaking any disruptive intervention activities.

Intervention Efficiency: The percentage of time consumed to undertake all the systems' intervention activities over the total disruption period to restore all the co-located systems.

Intervention Program: A planned schedule of maintenance and repair actions for the assets over a specified period.

Intervention Scenarios: Different strategies for improving the reliability of systems within a corridor, including conventional, partially coordinated, and full coordination scenarios.

Intervention Schedule: A planned sequence of maintenance activities designed to optimize the performance and lifespan of an asset.

Investment Gap: The need for additional funding sources due to ever-increasing social expenditures and aging infrastructure. This includes increasing infrastructure deficits because of delayed repairs and maintenance.

Joint Backfilling and Compaction: The combined process of filling excavated areas and compacting the soil for multiple systems.

Joint Duration (JD): The duration of intervention actions required for two or more systems that can take place concurrently.

Joint Excavation and Shuttering: The combined process of excavating and installing temporary supports for multiple systems.

Joint Venture: A business entity created by two or more parties, generally characterized by shared ownership, shared returns and risks, and shared governance. In the context of PBC, it involves maintenance contractors working together to share risks and returns, enhancing cooperation to maximize intervention efficiency.

KPIs (Key Performance Indicators): Measurable benchmarks used to evaluate performance and track progress against desired goals.

Lack of Connectivity: Gaps in transportation infrastructure, such as the lack of roads or public transportation, which can make it difficult for people to access jobs, education, and other resources.

LIF (Lifecycle Costs Improvement Factor): Potential cost savings that can be attained due to integrating common activities while undertaking the intervention.

Lifecycle Cost (LCC) Analysis: A technique that considers all direct and indirect cost categories such as planning, design, acquisition, maintenance, ownership, and operation of an asset. It helps in evaluating the cost-effectiveness and savings over the asset's lifecycle.

Lifecycle Costing: Evaluation of infrastructure projects based on their upfront costs, operating costs, and long-term benefits and returns.

Lifecycle Costs: The total cost of owning, operating, maintaining, and disposing of an infrastructure asset over its entire lifespan.

Lifecycle Costs (LCC): The total cost of ownership of an asset over its entire life, including acquisition, operation, maintenance, and disposal costs.

Linear Optimization: A mathematical method used to find the best outcome in a model whose requirements are represented by linear relationships.

Location Parameter (γ): A parameter in the Weibull distribution that represents the distribution along the planning horizon.

Logical Interdependency: A type of interdependency where the state of one infrastructure system depends on the state of another via a mechanism that is not physical, cyber, or geographic.

Maintenance Planning Optimization: The process of planning and scheduling maintenance activities to achieve the best performance and cost-effectiveness.

Manage Your Assets: This involves the systematic process of operating, maintaining, and upgrading assets to maximize their value and lifespan. It includes activities such as inspections, repairs, and replacements to ensure assets remain in good condition.

Metaheuristics: High-level problem-solving frameworks that use heuristics to find good solutions to complex optimization problems.

MOSEK: An optimization software package that provides tools for solving large-scale mathematical optimization problems.

Multi-Criteria Decision-Making (MCDM): A method that requires varying levels of knowledge and expertise to assign criteria values and combine their scores. It involves assigning values to alternatives for each criterion and multiplying them by corresponding weights to obtain a final score.

Multi-Objective Hierarchical Goal Optimization: An optimization approach that uses metaheuristics algorithms to achieve a globally optimal solution by considering multiple objectives in a hierarchical manner.

Multi-Objective Optimization: An approach that considers multiple conflicting objectives in decision-making processes. It is used extensively in the infrastructure sector for budget allocation, efficient expenditure utilization, performance enhancement, and intervention planning and scheduling.

Municipal Infrastructure: Public infrastructure managed by municipalities, including roads, water supply, sewers, and other essential services.

Nature-Based Solutions: Implementing natural defenses against climate-related hazards, such as wetland restoration and green infrastructure.

NCR (Network Coordination Ratio): Potential time savings that can be attained due to integrating common and parallel activities while undertaking the intervention.

Net Present Worth (NPW): The value of all future cash flows (positive and negative) over the entire life of an investment discounted to the present.

Network Analysis Approaches: Methods that use network-based analysis to study infrastructure interdependencies, including advanced geospatial analysis and hydraulic network modeling.

Non-Pre-Emptive Goal Optimization: An optimization approach that considers all objectives simultaneously without prioritizing one over the others.

NPI (Net Present Value): A financial metric that calculates the present value of all cash flows (both incoming and outgoing) over the life of an investment, discounted at a specific rate.

Operating Time (OT): The time between the corridor's last and next intervention.

Optimization Engine: A software tool used to find the best solution for a given problem by optimizing certain parameters.

Optimization Framework: A structured approach to solving complex problems by finding the best possible solution from a set of feasible alternatives.

Optimization: The process of making something as effective or functional as possible. In the context of asset management, it involves using various techniques to allocate resources efficiently and achieve the best possible outcomes.

Ownership: The responsibility held by an individual or entity for an indicator.

Parallel Duration (PD): The duration of intervention activities that can take place concurrently among different systems.

Pareto Optimization: A method that generates a set of Pareto near-optimal solutions in one run, allowing decision-makers to explore various scenarios with different relative weights after the optimization engine runs.

Partial Coordination: A scenario where only two out of the three networks (roads, water, sewer) are coordinated for maintenance and intervention actions.

Pavement Degradation Fees: Fees charged to agencies cutting the pavement to account for the reduction in the pavement's service life due to excavation.

Performance-Based Contracts (PBC): A type of contract designed to incentivize contractors to achieve specific performance metrics by tying payment to measurable outcomes rather than simply completing a set of predetermined tasks.

Performance Metrics: Clearly defined, measurable benchmarks tied to specific outcomes that contractors must achieve under a PBC.

Physical Interdependency: A type of interdependency where the state of one infrastructure system is dependent on the material output(s) of another infrastructure system.

Portability: The ability of an indicator to fit multi-assets with different features and attributes.

Predictive Maintenance: Maintenance approach powered by AI that anticipates maintenance needs before they become critical issues, reducing downtime and extending the lifespan of infrastructure.

Premature Replacement: The replacement of infrastructure systems before the end of their service life, which may result in economic losses.

Prequalification Phase: The phase in which the capability of proponents (general contractors or joint ventures) is checked based on prequalification criteria set earlier.

Probability of Failure (POF): The likelihood that a system or component will fail within a specified period.

Projects: These are temporary endeavors undertaken to create a unique product, service, or result. Projects have defined objectives, timelines, and resources, and they require careful planning and execution to succeed.

Public Transit: Public transportation systems that will undergo substantial upgrades, incorporating digital technology to offer real-time information and seamless payment systems.

Public-Private Partnerships (PPP): Long-term, performance-based approaches to procuring public infrastructure that engage the private sector in delivering and operating huge infrastructure projects. They transfer a significant share of risks to the private sector.

Quantum Computing: Advanced computing technology that performs rapid simulations and modeling for complex infrastructure projects and protects critical infrastructure systems from cyber threats.

Reinstating Sewer Laterals: The process of restoring sewer lateral connections after intervention activities.

Reliability (Ri): The reliability of a system at a certain point in time, considering the effects of intervention actions.

REMSOFT: A software package equipped with optimization algorithms for solving linear and mixed-integer programming problems.

Renewable Energy Investment: Investments in renewable energy sources like solar and wind, along with advanced technologies in energy storage to ensure a continuous power supply.

Residents Notification: The process of informing residents about the intervention activities and any potential impacts.

Resilience: This refers to the ability of an asset or system to withstand and recover from adverse conditions or disruptions. It involves designing and maintaining assets to ensure they can continue to function under stress.

Resilience Gap: The need for managing infrastructure in a more informed and intelligent way with respect to risks. For example, building water and sewer systems capable of meeting increasing demand due to population growth and climate change.

Resilience Preparedness: The ability to prepare for and adapt to changing conditions and withstand and recover rapidly from disruptions.

Resilient Design: Designing infrastructure to withstand natural disasters and climate change impacts, including elevated structures, flood barriers, and energy-efficient retrofits.

Resilient Infrastructure: Infrastructure designed to withstand and recover swiftly from shocks and stresses, ensuring the uninterrupted provision of essential services.

Resources: This term encompasses all the assets, personnel, and materials available to an organization to achieve its objectives. Effective resource management is essential for optimizing productivity and performance.

RIF (Risk Improvement Factor): Risk improvement due to undertaking fully coordinated intervention as opposed to several conventional interventions.

Risk Dataset: A dataset that includes the probability and consequences of failure for the assets.

Risk Exposure: This refers to the potential for loss or damage to an asset due to various risks. It involves identifying, assessing, and prioritizing risks to mitigate their impact on the asset.

Risk Index (RI): A measure used to assess the risk exposure of urban infrastructure corridors based on the probability and consequences of failure of their systems.

Risk Mitigation: Strategies to reduce potential risks associated with infrastructure projects, often involving guarantees and insurance.

Road Resurfacing: The process of applying a new surface layer to roads.

RPIF (Resilience Preparedness Improvement Factor): A metric used to evaluate the improvement in the resilience preparedness due to expanding the network capacity to meet the increasing demand within a corridor after implementing intervention actions as opposed to conventional interventions.

Scale Parameter (α): A parameter in the Weibull distribution that measures the range or spread in the distribution of data.

Service Levels: This refers to the standards or benchmarks set for the performance of a service. It includes the expected quality, availability, and responsiveness of the service provided to the users.

Sewer Backfilling and Compaction: The process of filling excavated areas and compacting the soil specifically for the sewer system.

Sewer Excavation and Shuttering: The process of excavating and installing temporary supports specifically for the sewer system.

Sewer Main Pipe Leak Repair: The process of repairing leaks in the main sewer pipes.

Sewer Pipe Bedding: The process of preparing the bed for sewer pipes.

Sewer Pipe Installation (Trenchless): The process of installing sewer pipes using trenchless technology.

Sewer Pipe Repair/Installation: The process of repairing or installing sewer pipes.

Shape Parameter (β): A parameter in the Weibull distribution that reflects the rate of failure for a system.

Simulation-Based Approaches: Computational methods used to study infrastructure interdependencies, including agent-based simulation, continuous simulation, and system dynamics.

Site Reinstatement: The process of restoring the site to its original condition after the intervention activities.

Smart Buildings: Buildings equipped with sensors that continuously collect and analyze data to enhance energy efficiency, comfort, security, and maintenance.

Smart Cities: Urban areas that use integrated digital platforms to coordinate services like transportation, healthcare, education, and public safety to improve the quality of life for residents.

Smart City Initiatives: Investments in smart technologies to improve services, reduce waste, and enhance the quality of life for residents. This includes the use of IoT sensors to monitor traffic, energy usage, and waste management in real time.

Smart Grids: Advanced energy grids that balance supply and demand while integrating diverse energy sources, including renewables.

Smart Roads: Infrastructure with sensors embedded in roadways to communicate with autonomous vehicles, helping them navigate, optimize traffic flow, and enhance safety.

Social Impact Bonds: Financial instruments that fund projects addressing social challenges, such as affordable housing and healthcare infrastructure.

Spatial Dataset: Data related to the geographic location and physical characteristics of the assets.

Spatial Interdependency: The proximity between infrastructure systems.

Spatial KPI: This KPI evaluates the space and time impacts of the intervention and computes the spatio-temporal savings arising from coordinating the interventions compared to undertaking them in silos.

Spatio-Temporal Disruption Factor (STDF): This factor integrates the spatial and temporal dimensions for conventional, partially coordinated, and fully coordinated intervention scenarios.

Stakeholder Engagement: The involvement of diverse stakeholders, including communities and organizations, in the design, implementation, and evaluation of infrastructure projects.

Standalone Duration (SD): The duration of intervention activities required only for a specific system where no other work can take place concurrently.

STIF (Spatio-Temporal Improvement Factor): Time and space savings that can be attained while undertaking a fully coordinated intervention as opposed to several conventional interventions.

Subcontracting: A method of transferring risk to a subcontractor, used in mega-scale projects when the required capabilities are too diverse for a single general contractor.

Subjectivity: The objective-oriented nature of an indicator, including a pre-defined set of rules for measuring an asset attribute.

Supersonic Travel: Rapid long-distance travel using supersonic aircraft, necessitating the development of specialized airports and flight corridors.

Surface Overlay: The process of applying a new layer of material on the surface of roads.

Sustainability: This term refers to the practice of managing resources in a way that meets current needs without compromising the ability of future generations to meet their own needs. It involves balancing economic, social, and environmental considerations in decision-making.

Sustainable Finance: Integrating sustainability criteria into investment decisions, directing capital toward projects that align with environmental and social goals.

Sustainable Fuels: Fuels such as hydrogen for fuel cell vehicles and biofuels for conventional vehicles, supported by dedicated infrastructure.

Sustainable Materials: Materials that are environmentally friendly and resource-efficient, such as recycled plastics and sustainable wood.

Sustainable Urban Planning: Urban planning that emphasizes eco-friendly designs, green spaces, efficient public transportation, and mixed-use zoning to reduce congestion and emissions.

System Level: The level at which individual systems or components are managed and optimized.

System-of-Systems Approaches: Integrated modeling approaches that consider multiple interconnected systems, such as high-level architecture (HLA) simulation.

Temporal Dataset: Data related to the timing and duration of intervention actions.

Temporal KPI: This key performance indicator (KPI) measures the intervention duration and savings resulting from coordinating interventions versus undertaking them in silos. It dynamically computes the duration of full coordinated, partially coordinated, and conventional interventions based on categorized activities and their production rates.

Tendering: A process that involves inviting contractors to bid for a project to ensure competition and obtain the best value for taxpayers' money. There are four types: single-source tendering, open tendering, selective tendering, and negotiated tendering.

Total Disruption Time (TDT): The total time to undertake all the interventions for systems.

Traffic Control Systems: Systems set up to manage and control traffic during intervention activities.

Transparent Procurement: Competitive and open procurement processes that ensure fairness, efficiency, and cost-effectiveness in infrastructure development.

Understandability: The easiness of understanding and tracing the triggers behind a sudden rise/fall throughout the assets' lifecycle.

Urbanization: The process by which cities grow and higher percentages of the population come to live in urban areas.

Water Main Pipe Leak Repair: The process of repairing leaks in the main water pipes.

Water Pipe Bedding: The process of preparing the bed for water pipes.

Water Pipe Installation (Trenchless): The process of installing water pipes using trenchless technology.

Water Pipe Repair/Installation: The process of repairing or installing water pipes.

Weibull-Based Deterioration Model: A model used to represent the deterioration pattern of water and sewer networks, considering both negative and positive impacts.

Weighted Sum Method: A method used to integrate all the KPIs into one indicator that reflects the vision of the management and develops an intervention plan that fits the municipality's priorities and preferences.

Zero-Waste Goals: Objectives aimed at eliminating waste generation in infrastructure projects through recycling and upcycling practices.

Index

Note: **Bold** page numbers refer to tables and *italic* page numbers refer to figures.

9781032355788